A Complete Guide to ENVIRONMENTAL AUDITS

Elizabeth Glass Geltman

SECTION OF NATURAL RESOURCES, ENERGY, AND ENVIRONMENTAL LAW

AMERICAN BAR ASSOCIATION

Printed in the United States of America.

Geltman, Elizabeth Glass.
Environmental audits : self policing for environmental protection / Elizabeth Glass Geltman.
p. cm.
Includes bibliographical references.
ISBN 1-57073-381-3 (pbk.)
1. Environmental law--United States. 2. Environmental audits--United States. 3. Environmental audits--United States--Forms. I. Title.
KF3775.G445 1997
344.73'046--dc21 96-50193
CIP

01 00 99 98 97 5 4 3 2 1

Dedication

For my family

Table of Contents

Preface

There is a quiet revolution occurring in corporate America. "American environmentalism" is growing. "Environmentalism" has become integrated with core American values, including American corporate values. Although American environmental values appear to be based upon a great diversity of cultural models and values, evidence shows that a new American environmental paradigm not only exists but is forcing its way into American corporate culture.

With the reorganization of EPA to make compliance and enforcement co-equals, environmental law is at a crossroads. No longer will the traditional command-and-control models meet the ever changing demands of an ever expanding, multicultural as well as multidimensional populace. EPA and equivalent state agencies must look to more creative market based controls to encourage not just mere environmental compliance but environmental leadership and other forms of environmental excellence. This is increasingly accomplished not through the traditional enforcement model but through an increasingly advanced self regulatory process.

In state after state, across the country and across the globe, the environmental audit (in one form or another), voluntary clean-up legislation, and state sponsored amnesty programs have been proposed and developed as an important tool of encouraging both mere compliance and environmental excellence. These techniques, among others, have been used as a low-budget, market based mechanism of changing corporate environmental behavior as well as wide-spread popular attitudes toward the individual's role in environmental responsibility.

This book describes the various state and federal programs defining, encouraging, and requiring environmental audits. The forms included in this book are offered with the hope that ever more companies will initiate their own efforts at improved environmental compliance.

Elizabeth Glass Geltman
Washington, D.C.
October 1996

Author's Note

The author has provided the reader with her comments concerning the current laws and regulations affecting environmental activities as of August 1996. The reader is advised that environmental law is a fast changing area of law, and others may reach different conclusions and recommendations either due to changes in laws or regulations or due to their differing views of those laws and regulations. The author does not warrant that her conclusions and recommendations will be adhered to by any or all courts and regulatory agencies that investigate or rule upon environmental matters. The reader should update his or her research (with particular emphasis on the state in question) or seek independent legal advice from his or her attorney before undertaking any activity that may be within the scope of any environmental law or regulation.

Acknowledgments

The author would like to thank Michael Greco, Winifred Hercules, Carey Mathews, John Jolley, Patrick Ryan, Stacy Wahlberg, Cynthia Webb-Manly, Mike Briggs, and Lalanthi Wijegunawardena for their research, editing, typing, and computer assistance in writing this book. In addition, the author thanks her many students and colleagues of George Washington University Law School, all of whom gave countless hours and insight, making this project possible. The author also thanks Van Carson and Geoffrey K. Barnes and the environmental group of Squire, Sanders & Dempsey for their support and encouragement while writing this book.

The author thanks Mary Sears, Donald J. Gecewicz, and the staff of the ABA SONREEL for their tireless efforts in bringing this book to press.

Finally, the author gives her greatest thanks to her children, Andy and Jeffrey, for giving her the inspiration to write this book; her husband, Edward Alan Geltman, and parents, Dr. and Mrs. Sheldon Glass, for their advice, wisdom and support through this and so many other projects; and her teachers, to whom this book is dedicated.

About the Author

Elizabeth Glass Geltman is an Associate Professor at the George Washington University Law School and Counsel to Squire, Sanders & Dempsey, Washington, D.C. Professor Geltman is the author of ten books on issues concerning environmental law, business and natural resources, including two multivolume treatises entitled *Environmental Issues in Business Transactions and Securities Disclosure of Contingent Environmental Liabilities*. She has spoken at various legal seminars and conferences and is widely published in newspapers, magazines, and law reviews on issues of environmental law and natural resources. She is a member of the Natural Resources and Environmental law sections of the District of Columbia, Maryland, and American Bar Associations. Professor Geltman is a member of the Maryland and District of Columbia bars.

Chapter 1

Why Audit?

§ 1.1 Overview

The role of the environmental audit in American business has been the focus of great attention and debate for the past few years. As environmental laws and regulations have increased in scope and complexity, American businesses have spent increasing amounts of time developing mechanisms of ensuring regulatory compliance. Similarly, as public awareness of the creeping dangers of property devaluation due to environmental contamination has grown, the environmental audit has begun to assume a more and more prominent role in daily real estate and business transactions.

Many benefits are enjoyed by those who choose to conduct audits. The following table lists some of the direct benefits as determined by EPA.[1]

[1] *See* BENEFITS TO INDUSTRY OF ENVIRONMENTAL AUDITING (EPA Report No. EPA-230-08-83-005). *See generally* JOSEPH L. BAST, ECO-SANITY: A COMMON SENSE GUIDE TO ENVIRONMENTALISM, at 185, 198-99, 238-42 (Madison Books 1994) (citing numerous real-world examples of corporate initiatives that benefit both the environmental and the corporate bottom line); LEE HARRISON, ENVIRONMENTAL, HEALTH, AND SAFETY AUDITING HANDBOOK, at 15, 31 (McGraw-Hill 2nd ed. 1994) (the shift in environmental audit focus has gone from identifying problems to verifying compliance status to confirming management system effectiveness); JOEL MAKOWER AND THE BUSINESS FOR SOCIAL RESPONSIBILITY, BEYOND THE BOTTOM LINE: PUTTING SOCIAL RESPONSIBILITY TO WORK FOR YOUR BUSINESS AND THE WORLD (Simon & Schuster 1994); JOSEPH P. CONTRELL AND BETTY J. SELDER, ENVIRONMENTAL DECISION MAKING FOR ENGINEERING AND BUSINESS MANAGERS (McGraw-Hill 1994), at 39-48; R. K. JAIN ET AL., ENVIRONMENTAL ASSESSMENT (McGraw-Hill 1993).

Generally Quantifiable Benefits	**Generally Not Quantifiable Benefits**
Improved compliance record	Improved reputation
Reduced fines	Favorable publicity
Reduced legal actions brought against the company and/or related individuals	Improved regulatory relations
Improved incident/accident statistics	
Reduced volume of environmental hazards	
Improved worker health	

The following table lists some of the indirect benefits as determined by EPA.[2]

For discussion of overseas approaches to environmental auditing *see generally* LINDA SPREDDING, ECO-MANAGEMENT AND ECO-AUDITING (Wiley Chancery Law 1993); ERNEST CALLENBACH et al., THE ELMWOOD GUIDE TO ECOLOGICAL AUDITING AND SUSTAINABLE BUSINESS (1993). *See also* LEE HARRISON, ENVIRONMENTAL, HEALTH, AND SAFETY AUDITING HANDBOOK, at 157-218 (McGraw-Hill 2nd ed. 1994). *Compare* Research Report, ENVIRONMENTAL AUDITING AND THE ROLE OF THE ACCOUNTING PROFESSION (Canadian Institute of Chartered Accountants 1992).

For an older, more historical discussion of audits, *see* ANN C. SMITH et al., ENVIRONMENTAL AUDITING QUALITY MANAGEMENT (Executive Enterprises 1989); J. LADD GREENO, et al., ENVIRONMENTAL AUDITING: FUNDAMENTALS AND TECHNIQUES (J. Wiley & Sons 1985); SIDNEY DRAGGAN et al., ENVIRONMENTAL MONITORING, ASSESSMENT AND MANAGEMENT: THE AGENDA FOR LONG-TERM RESEARCH AND DEVELOPMENT (Praeger 1987); THE NATIONAL RESEARCH COUNCIL, ENVIRONMENTAL MONITORING (National Academy of Sciences 1977); WALTER W. HELLER, ECONOMIC GROWTH AND ENVIRONMENTAL QUALITY: COLLISION OR COEXISTENCE?, in BUSINESS AND ENVIRONMENT: TOWARD COMMON GROUND, 237 (H. Jeffrey Leonard et al., eds., 1977).

[2] *See* BENEFITS TO INDUSTRY OF ENVIRONMENTAL AUDITING (EPA Report No. EPA-230-08-83-005).

Generally Quantifiable Benefits	Generally Not Quantifiable Benefits
Decrease in business interruptions by identifying problems that could affect production	Increased involvement in day-to-day environmental risks
Increased worker productivity from reduced environmental risks	Knowledge of job performance measure
Reduced insurance rates	Feeling supported by management
	Increased job satisfaction

There are many distinct purposes for businesses to either perform or cause those within their control to conduct an environmental audit. Identifying the purpose of the audit is a necessary first step in defining its scope and content.

First, businesses considering a purchase of real property or another existing business are prudent to conduct at least a Phase I environmental audit in order to meet the Comprehensive Environmental Response, Compensation and Liability Act (CERCLA)[3] "innocent purchaser defense."[4] Many lenders consider Phase I audits to be standard banking practice on any commercial loan.[5]

---

[3] 42 U.S.C. §§ 9601-9675.

[4] A purchaser may be expected to know all that is visible at time of purchase. *See, e.g.*, Smith Land & Improvement Corp. v. Rapid-American Development Corp., 851 F.2d 86 (3rd Cir. 1988) (presence of asbestos and hazardous waste disposal sites readily visible); Philadelphia Elec. Co. v. Hercules, Inc. 762 F.2d 303 (3rd Cir. 1985), *cert. denied,* 474 U.S. 980 (1985) (site contamination polluting nearby stream); Clauser v. Taylor, 44 Cal. App. 2d 453 (1941) (recision allowed for failure to disclose matter outside buyer's ability to observe). Cf., Cal. Civil Code § 20795.

[5] *See, e.g.,* Charlotte A. Biblow, *Comment: Fiduciaries Are Waiting for Relief From Exposure on Environmental Issues,* AM. BANKER, Mar. 22, 1994, at 17; Gary Hector, *A New Reason You Can't Get a Loan,* FORTUNE, Sept. 21, 1992, at 107 (explaining that the audit requirement by lenders stems from the desire to avoid PRP status and the subsequent

Second, in addition to preserving the innocent purchaser defense of CERCLA,[6] prudent investors also design an audit wide enough to uncover any and all significant potential environmental problems that could affect the value or use of the proposed acquisition. Implicit in this second purpose is gathering enough evidence to (1) establish an environmental baseline at the point of acquisition and (2) evaluate market value of the target property or business, considering all environmental problems uncovered in the audit. This goal requires evaluation of all environmental data in order to make a sound business decision concerning whether to embark on the transaction and, if so, how to structure the transaction. The types of expenses a buyer may incur from environmental contamination at a site include any or all of the following:

---

reluctance to invest in contaminated properties). Although CERCLA contains a secured creditor exemption, *see* 42 U.S.C. § 9601(20)(a) (1992), there have been a number of cases in which the courts have held lenders potentially liable for contamination on property that they held as collateral or foreclosed upon. *See, e.g.*, United States v. Fleet Factors Corp., 901 F.2d 1550 (11th Cir. 1990), *cert. denied,* 498 U.S. 1046 (1991); United States v. Maryland Bank & Trust Co., 632 F. Supp. 573, 579 (D. Md. 1986); Northeast Doran, Inc. v. Key Bank of Maine, 15 F.3d 1, 3 (1st Cir. 1994); United States v. McLamb, 5 F.3d 69, 73 (4th Cir. 1993).

The Environmental Protection Agency (EPA) has adopted modifications aimed at protecting lenders and parties that hold security interests. *See* Lender Liability Under CERCLA, 57 Fed. Reg. 18,344-85 (1992) (clarifying activities secured creditors could undertake without incurring CERCLA liability to federal government); *see also* 42 U.S.C. 9601(20)(A) (1988) (excluding from definition of "owner" or "operator" any "person, who, without participating in the management of a vessel or facility, holds indicia of ownership primarily to protect his security interest in the vessel or facility"). The EPA's proposed rule, however, was struck down as beyond the EPA's statutory rulemaking authority. Kelly v. EPA, 15 F.3d 1100, 110708 (D.C. Cir. 1994), *cert. denied,* 115 S. Ct. 900 (1995).

[6] *See* La Salle National Trust, N.A. v. Schafner, 1993 U.S. Dist. LEXIS 17026 (N.D. Ill. Nov. 23, 1993) (summary judgment denied because genuine issue of fact existed as to whether "all appropriate inquiry" was made prior to purchase of property); HRW Systems, Inc. v. Washington Gas Light Co., 823 F.Supp. 318 (D. Md. 1993) (adjoining landowner could be an innocent purchaser with respect to contamination from the neighboring site even though it contributed some contamination to its own property); Steego Corp. v. Ravenal, 830 F.Supp. 42 (D. Mass. 1993) (beneficiaries of an estate could not assert innocent landowner defense where the grandchildren beneficiaries received their interest via an instrument transferring title or possession, which is a contractual relationship, and could not acquire the interest after the disposal or placement of hazardous substances on the site).

1. financial liability for cleanup (including related investigatory and medical monitoring costs associated with cleanup) and natural resource damages;
2. lawsuits concerning liability and their related transaction costs;
3. reduction (or perceived reduction) in property value (regardless of actual level of contamination);
4. restrictions on land use;
5. adverse publicity;
6. toxic tort cases from neighboring residents.

Third, many sellers will conduct an environmental audit before selling real or business property in order to back up any documentation, such as representations and warranties concerning the absence of environmental violations or hazardous waste on the property, required in the purchase and sale agreement or other relevant documentation. Audits also assist sellers to establish an environmental baseline. Many sellers will give indemnities to purchasers for environmental conduct undertaken during the sellers' tenure on the property. Sellers who do so are well advised to be able to document the condition of the property at closing so that they do not become liable for purchasers' conduct postclosing.

Fourth, many parties conduct routine audits[7] on a periodic basis to ensure compliance with pertinent environmental laws and permits.[8] An

---

[7] For such purpose, EPA defined an environmental audit as early as 1986 as "a systematic, documented, periodic and objective review by regulated entities of facility operations and practices related to meeting environmental requirements." EPA Environmental Auditing Policy Statement, 51 Fed. Reg. 25,004, 25,006 (1986). *Compare* Department of Justice Memorandum, *Factors in Decisions on Criminal Prosecutions for Environmental Violations in the Context of Significant Voluntary Compliance or Disclosure Efforts By the Violator,* July 1, 1991 (the factors the DOJ will consider in exercising its prosecutorial discretion are voluntary disclosure, cooperation, compliance programs, pervasiveness of noncompliance, internal disciplinary measures, and subsequent compliance efforts). *See generally* Courtney M. Price and Allen J. Danzig, *Environmental Auditing: Developing a "Preventative Medicine" Approach to Environmental Compliance,* 19 LOY. L.A. L. REV. 1189, 1195 (1986).

[8] *See* Annette T. Crawley, *Environmental Auditing and the "Good Samaritan" Doctrine: Implications for Corporate Parent Corporations,* 28 GA. L. REV. 223, 229 (1993); Marianne Lavelle, *Environmental Vise: Law, Compliance,* NAT'L L.J., Aug. 30, 1993, at § 1 (over two-thirds of the 200 corporate general counsels responding to a 1993 survey conceded that, at some point during the past year, their businesses had operated in violation of federal or state environmental laws); Marianne Lavelle, *More Lawyers Expect to Urge*

environmental compliance audit is defined by EPA as "an independent assessment of the current status of a party's compliance with applicable statutory and regulatory requirements."[9] A compliance audit can focus on site activities that may require permitting by local, state, or federal agencies or extend to a thorough review of operations with respect to current regulations.[10] Compliance audits include environmental compliance record keeping, data collection, spill reports, labeling and manifesting requirements, and public notifications.[11] Compliance audits are designed to determine the level of compliance of current operations to existing environmental regulatory standards.[12]

A permutation of the general environmental compliance auditing focuses on determining compliance with a single environmental medium (such as air or water) or program (such as the Clean Air Act or the Clean Water Act). Compliance audits are often used where inspections reveal violations that raise the likelihood that environmental noncompliance might exist elsewhere within the operations.

These compliance audits may be conducted under the mandate of lenders or other financiers.[13] Often they are conducted to ensure compliance

Their Clients to Examine Compliance, NAT'L L.J., Mar. 16, 1992, at § 6 (37.4% of the 257 corporate general counsels surveyed stated that their corporations had never conducted an environmental self-audit; 16% stated that they altered their environmental auditing procedures for fear of government discovery of the audit results). *See also* Michael H. Levin et al., *Discovery and Disclosure: How to Protect Your Environmental Audit Report,* 24 ENV'T. REPT. (BNA) 1606 (Jan. 7, 1994); David T. Buente, Jr., et al., *Developing and Implementing an Environmental Corporate Compliance Program,* C868 ALI-ABA 85 (1993). *But see* Edward A. Hogan and Lisa Murtha Bromberg, *The Hidden Hazards of the Environmental Audit,* 36 PRAC. LAW. 15 (Apr. 1990).

[9] *Restatement of Policies Related to Environmental Auditing,* 59 Fed. Reg. 38,455, 38,458 (Jul. 28, 1994). To avoid confusing environmental audits with environmental assessments required by the National Environmental Policy Act, the author suggests avoiding the use of the term environmental assessment in the environmental auditing context.

[10] J. Payne, P.E., *Environmental Auditing: Beyond Due Diligence,* C831 ALI-ABA 237, 254 (Sep. 30, 1993).

[11] J. Arnold, *Compliance Auditing: Not a "Safe Harbor" but a "Course Out of Harms Way,"* 797 PLI/Corp 511 (Dec. 1992-Jan 1993).

[12] J. L. Payne, *Environmental Auditing: Beyond Due Diligence,* C945 ALI-ABA 317 (Oct. 6, 1994).

[13] *See, e.g.,* Gary Hector, *A New Reason You Can't Get a Loan,* FORTUNE, Sept. 21, 1992, at 107.

with state and federal securities laws. In such cases, the compliance results may form the basis for securities disclosures by public companies.[14] Finally, compliance audits are increasingly required in settlement agreements with environmental regulators.[15] Compliance or management audits have been the primary focus of EPA debate[16]—the question remaining the extent to which environmental audits should not be privileged from discovery.[17]

---

[14] *But see Survey Shows Most Annual Reports Will Provide Very Little Information on Environmental Compliance,* PR NEWSWIRE, Mar. 7, 1994. (Survey conducted by Annual Reports, Inc., found that 54 percent of the respondents indicated that they published an environmental policy, but they gave no indication how such information was disseminated to the shareholders or investment community. 44 percent of the companies indicated that the mention of the environmental activities of the company would be "moderately significant," while 47 percent said the mention would be "not significant." Also, 88 percent of the corporations had never produced and distributed among shareholders separate EPA reports as do some public utilities and natural resource companies. 96 percent said that they had no plans to do one in the future.) *Also see* Price Waterhouse, ACCOUNTING FOR ENVIRONMENTAL COMPLIANCE: CROSSROAD OF GAAP, ENGINEERING, AND GOVERNMENT, A SURVEY OF CORPORATE AMERICA'S ACCOUNTING FOR ENVIRONMENTAL COSTS 10-11 (1992) (Price Waterhouse survey revealed that nearly 63 percent of the publicly traded respondents admitted their financial reports did not fully reflect the known environmental problems that could have financial impact.)

[15] *See, e.g.,* Registration and Agreement for TSCA Section 8(e) Compliance Audit Program, 56 Fed. Reg. 4128 (1991) (announcing a Compliance Audit Program under section 8(e)).

Under this program, numerous companies regulated under section 8(e) signed an administrative consent agreement with the EPA under which they agreed to conduct a compliance audit and disclose the results to the EPA within 180 days. In exchange, EPA agreed to stipulate penalties for previously undiscovered violations and to cap total civil liability at $1 million.

EPA also developed a TSCA Civil Penalty Policy, under which it will adjust a proposed penalty downward by up to 15 percent if the violator made a "good faith" effort to comply with the regulations, promptly began corrective action, and assisted the EPA in minimizing the harm caused by the violation.

[16] EPA Meeting Notice, Restatement of Policies Related to Environmental Auditing, 59 Fed. Reg. 38,455, 38,459 (1994). *See generally* Patrick J. Ennis, *Environmental Audits: Protective Shields or Smoking Guns? How to Encourage the Private Sector to Perform Environmental Audits and Still Maintain Effective Enforcement,* 42 J. URB. & CONTEMP. L. 389, 395 (1992); William N. Farran III and Thomas L. Adams, Jr., *Environmental Regulatory Objective: Auditing and Compliance or Crime and Punishment,* 21 ENVTL. L. REP. (Envtl. L. Inst.) 10,239 (1991). *See generally* Lawrence B. Cahill, *Issues in Environmental Auditing,* PETROLEUM INDEPENDENT, Nov.-Dec. 1992, at A2.

Fifth, many companies undertake not only compliance audits but also conduct environmental management system audits. These are "audits" of the effectiveness of the environmental audit system itself. Such audits are systematic, documented, and objective reviews of facility operations and practices related to meeting environmental requirements. The EPA defines management audits as "an independent evaluation of a party's environmental compliance policies, practices, and controls."[18] Management system audits require ongoing effort to confirm that the company's operations are achieving and maintaining management's environmental goals. Management audits consider the need for a formal corporate environmental compliance policy, employee educational and training programs, equipment acquisitions, operation and maintenance programs, compliance officer programs, budgeting and planning systems, monitoring, record keeping and reporting systems, in-plant and community emergency plans, internal communications and control systems, and hazard identification and risk assessment.[19]

Finally, many companies will consider environmental problems in a "phased" approach, especially when the project is likely to result in environmental site remediation. Site remediation is often conducted in four phases; the first two phases essentially define the type and extent of the contamination involved.[20] The third and fourth phases define the removal and remediation process.

---

[17] *See, e.g.*, John S. Guttman, *Environmental Reviews Can Be Kept Confidential,* NAT'L L.J., June 20, 1994, at C12.

[18] *Restatement of Policies Related to Environmental Auditing,* 59 Fed. Reg. 38,455, 38,458 (Jul. 28, 1994).

[19] *Restatement of Policies Related to Environmental Auditing,* 59 Fed. Reg. 38,455, 38,458 (Jul. 28, 1994).

[20] The Phase I audit is similar in scope and substance to the transaction audit. Phase I audits generally consist of a site walk, a minimally intrusive records review, and interviews with persons having knowledge of the historical uses of the property. The Phase II audit typically requires that soil and water (both surface and subsurface) samples be taken to determine the type and vertical and lateral extent of any contamination and, where applicable, migration pathways.

Phases III and IV of the remediation process are not concerned with assessing the condition of the property. During Phase III, a remediation plan is developed. Phase IV involves the implementation of that plan and ongoing operation and maintenance activities.

Because there are many different purposes for conducting an environmental audit, the scope of what is typically called an audit can vary greatly depending on which purpose the audit is designed to satisfy.[21]

---

[21] Certain state attorneys general have urged the EPA to be even more flexible concerning environmental audits. For example, Colorado Attorney General Gale A. Norton advised EPA that a flexible approach to small business auditing processes would encourage more auditing, especially among small companies like dry cleaners. *See* EPA Should Display More Flexibility on Environmental Audits, State AG Says, 8 TOXICS L. REP. (BNA) 1256 (April 6, 1994). *See generally* Heather L. Cook and Robert R. Hearn, *Putting Together the Pieces: A Comprehensive Examination of the Legal and Policy Issues of Environmental Auditing,* 7 TUL. ENVTL. L.J. 545-593 (1994).

For further works on environmental audits, *see generally* JOHN T. WILLIG, AUDITING FOR ENVIRONMENTAL QUALITY LEADERSHIP (1995); R. K. JAIN, ENVIRONMENTAL ASSESSMENT (1993); MARTIN D. WESTERMAN, THE BUSINESS ENVIRONMENTAL HANDBOOK (1993); KATHLEEN HESS, ENVIRONMENTAL SITE ASSESSMENT, at 171 (Lewis 1993); RICHARD H. MAYS, ENVIRONMENTAL LAWS: IMPACT ON BUSINESS TRANSACTIONS (1992); STEVEN A. TASHER, ENVIRONMENTAL LAWS & REAL ESTATE HANDBOOK (1992); ALBERT R. WILSON, ENVIRONMENTAL RISK: IDENTIFICATION AND MANAGEMENT (1991); GAGE AND TUCKER, ENVIRONMENTAL DUE DILIGENCE (1991); JOEL S. MOSKOWITZ, ENVIRONMENTAL LIABILITY AND REAL PROPERTY TRANSACTIONS (1989); FRANK D. GOSS, ENVIRONMENTAL EVALUATIONS FOR REAL ESTATE TRANSACTIONS (1989); MORTON GORDON, ENVIRONMENTAL MANAGEMENT (1972).

For discussions of industry success stories in setting up productive environmental audit protocols *see generally* TIMOTHY C. JACOBSON, WASTE MANAGEMENT: AN AMERICAN SUCCESS STORY (1993); ANN RAPPAPORT, DEVELOPMENT AND TRANSFER OF POLLUTION PREVENTION TECHNOLOGY (1993); MICHAEL S. BARRAM, ET AL., MANAGING CHEMICAL RISKS: CORPORATE RESPONSES TO SARA TITLE III (1992). *See also* M. JIMMEE KILLINGSWORTH, ECOSPEAK (1992); W. MICHAEL HOFFMAN, ET AL., BUSINESS, ETHICS AND THE ENVIRONMENT (1990); R. E. HESTER, INDUSTRY AND THE ENVIRONMENT IN PERSPECTIVE (1983); DONALD M. MCALLISTER, EVALUATION IN ENVIRONMENTAL PLANNING: ASSESSING ENVIRONMENTAL, SOCIAL, ECONOMIC AND POLITICAL TRADE-OFFS (1980).

For information concerning a home audit or the audit of a small business *see generally* DONALD W. LOTTER, EARTHSCORE: YOUR PERSON ENVIRONMENTAL AUDIT & GUIDE (1993); STEPHEN J. MARTIN AND INGRID RITCHIE, THE HEALTHY HOME KIT (1995); LYNN LAWSON, STAYING WELL IN A TOXIC WORLD (1993); G. B. LESLIE AND F. W. LUNAU, INDOOR AIR POLLUTION: PROBLEMS AND PRIORITIES, AT 117-31 (Cambridge 1992).

§ 1.2 Transaction Triggered Audits

Transaction triggered environmental audits are typically conducted in phases. There is no uniformity as to what is examined or what terms are used to delineate each audit phase. The following chart describes the purpose of each audit phase and certain commonly used terms to describe each such phase.

Phase	Common Terminology	Objective
Preliminary Screening	Preliminary evaluation Screening-level evaluation Initial assessment Property screening Transaction screen Paper audit Environmental risk assessment	To determine if a full environmental audit is necessary
Phase I	Preacquisition assessment Phase I assessment Environmental (site) assessment Property transfer assessment Due diligence evaluation Innocent purchaser assessment Establishing baseline	To establish the CERCLA innocent purchaser defense, to begin to quantify potential liability, to verify property cost valuation, and to begin to establish the likelihood of environmental degradation of the property
Phase II	Site investigation Intrusive investigation Intrusive sampling Phase II assessment Subsurface investigation Baseline environmental assessment	To identify the vertical and lateral extent of area(s) of property affected by environmental contamination and degradation. This phase may also include removal actions to prevent further release, leaching, or other forms of spread of contamination.

Phase	Common Terminology	Objective
Phase III	Remedial investigation (and design) Site characterization Cleanup Corrective action Remediation	Full site characterization to formulate complete cleanup plan and eventual cleanup of site

§ 1.3 The "Paper Audit" or "Transaction Screen"

It is important to note that all these enumerated factors need not be considered by an environmental consultant in every audit. To do so would make most transactions too expensive to contemplate.

All the factors should, however, be evaluated by an environmental lawyer in a paper "transaction screen" before directing the consultant or engineer in what to pay particular attention to when conducting the Phase I audit. Such a screen will necessarily involve detailed discussions with the client concerning perceived site history, intended site use, and, if at all possible, a site walk on the targeted property by the attorney conducting the transaction screen. Consultants who are not provided guidance may merely conduct an audit designed to allay the innocent purchaser defense and not consider other critical factors in conducting environmental due diligence.

Below is a checklist that should be used as the basis for the transaction screen.

✓ **Consider the type of land and facilities involved.** Is the facility likely to generate hazardous waste? What is the hydrogeology of the land? Is the land conducive to storage of hazardous waste? How likely is it that contamination existing on the land will spread? How far? At what rate? Is it likely to affect residential or agricultural land?

✓ **Consider the age and ownership history of the land.** For what purposes is the land currently used? What have been past uses? Is there a trash heap or a waste site? Does the site contain any drums or stagnant pools? Does the land have any unusual bumps that do not follow the general terrain? Are there or have there been any abandoned roads? In other words, might there be any buried or concealed drums on the site?

✓ **Consider the security of the site.** How large is the developed portion of the land compared with the vacant portion? Is there public access to the undeveloped portions? Is there much traffic by the site? How often has the site been checked, by foot or by air, for changes in the terrain? Is the topography of the land such that an unauthorized dumping would be readily discovered? Could this land be, or might it have been, the subject of a midnight dumping?

✓ **Consider the neighboring facilities.** Are there any facilities nearby that may have generated, transported, or stored hazardous waste? Does any nearby facility currently generate, transport, or store hazardous waste? Are there any homes or farms nearby that may be the source of personal injury suits if the release should spread?

✓ **Consider the physical composition of the site.** Are there chemicals or materials used on the site that may have spilled accidentally? Are there any chemicals used in the plant, or transported to or from the plant, that might leak into the environment? What type of wastes does the facility generate? Do the neighboring facilities use state-of-the-art technology? Are neighboring facilities likely to contaminate ground water?

✓ **Consider machines used on the land.** Do they contain substances that might leak as they get older or are abandoned?

✓ **Consider substances used in the building.** Many old buildings are insulated with asbestos and many transformers contain PCBs.

✓ **Consider the types of permits and licenses the past and current owners and operators held concerning the land and the facility.** Do any permits or licenses indicate a source of environmental liability? Was the owner conscientious in maintaining the correct permits? Bad records may indicate a general lack of maintenance of the property.

✓ **Consider the insurance history of the site.** Who insured the land? For what time period? What did the policy say? Is the policy available? Is the policy likely to cover liability under CERCLA (in light of the most recent insurance coverage cases)? If the policy covers CERCLA liability, retain copies of the policy and consider using it in bargaining for an indemnity agreement.

✓ **Consider the future plans for the site.** Is the site the subject of any short-term or long-range county plans? What is the development trend of the area? Is it possible that future development might damage existing waste sites? Will future activities in the area increase the likelihood of a release, add to existing waste sites, or increase the risk of a midnight dumping? Will future development make the cleanup of the site more difficult (and, hence, more expensive)?

✓ Consider the likelihood of suit. Is the site listed on the National Contingency Plan? Is the property the subject of state or local concern? Are there any other potentially responsible parties who are solvent and might share in the liability for response costs?

✓ Consider the contamination itself. How bad is it? What is the contamination's depth and width? How easily and at what expense can the problem be corrected? Would further use of the property require environmental permits (for instance under RCRA)?

Although this checklist concerns primarily purchasers of real property, the considerations raised are equally applicable to lenders as well as to transferees in mergers, consolidations, and acquisitions where the acquired corporation owns any real property. Transferees, however, should contemplate certain additional considerations if they acquire real estate by a transaction other than a simple purchase of property. For example, if a merger or consolidation (or any other transaction in which stock is exchanged) is contemplated, consider the other assets and operations of the company to be acquired. Does the company to be acquired engage or has it ever engaged in any activities that generate hazardous waste? If so, how and where were these wastes disposed? Who handled the disposal? Is that company still solvent? What method of disposal was used? Does the company to be acquired have reason to believe that the disposal company may not have complied with the disposal agreement or used a methodology other than what it claimed to have used? What did the disposal agreement say?

§ 1.4 Sources of Environmental Data

The sources used to gather information can be as revealing to the buyer as the information itself. The smart seller will cooperate in giving the buyer all the necessary information. The seller who cooperates is less likely to be saddled with the entire liability, because the buyer will have acted as a prudent businessperson and his or her decision to buy the property will be deemed a calculated business risk. The seller who discloses the requested information will be more likely to be shielded from liability on the basis of fraud.

Although the smart seller cooperates in disclosing information, many sellers are not cooperative. Thus, the buyer must be able to obtain the necessary information from other sources. The following is a list of potential sources a prospective buyer may use to obtain information about the site despite an uncooperative seller and to ascertain whether the information received is truthful and accurate.

✓ **Check state and federal environmental records** to see if the site has ever been listed as having been in need of cleanup. The National Priority list maintained by the EPA is printed in the Federal Register. The Maryland Register or equivalents in other states list local sites. The prospective buyer should make Freedom of Information Act requests to the EPA to determine if past owners made CERCLA section 103(c) notifications or if there is a record of report of a release of hazardous substances filed under CERCLA section 103(a). The National Response Center also may have such records. In addition, EPA maintains a computer list of PRPs at CERCLA NPL sites; a Superfund Comprehensive Action Plan (SCAP) that gives the status of NPL sites; and summary reports on Clean Water Act violators, Clean Air Act violators, and RCRA violators. The investigation could include a review of remedial investigations and feasibility study reports (and associated records of decision) at NPL sites, submissions of Part A applications under RCRA, and RCRA "corrective action" facility investigations and feasibility assessments. Finally, the RCRA section 3010 notification list and the Remedial Response Information System listing of hazardous waste sites also may provide additional sources of information about a site.

✓ **Interview parties** who might be familiar with the site. Former employers and employees who have worked with the company for a long time frequently have a better knowledge than management of how waste is actually handled. They often know the location of spills and what action, if any, was taken to clean or cover up these spills. Management often lacks knowledge of daily mishaps that might result in liability, but employees frequently are aware of such occurrences. Employees no longer employed with the company often are willing to talk about possible problems at the site.

✓ **Neighbors** often know a great deal about the activities of the site's landowners—especially if they are concerned about the activities which occur in the area. Inquire about any organized, or unorganized, grievances by neighboring landowners concerning activities on the site. Neighbors are also good sources of information about past activities on the land.

✓ **Discussions with management** can be useful in obtaining information regarding site history, environmental compliance, processes used on the premises, raw materials used, the types of wastes and by-products produced, and how and where the company disposed of these wastes and by-products, as well as how the company handled complaints from neighbors.

✓ **Discussions with citizens or environmental groups** may yield information about the site or its surrounding area. Contact with such groups should be handled with caution, however, as inquiry may lead to activity by the group.

✓ **Discussions with local real estate brokers and agents** may be helpful in determining information concerning the site and its relative value over the past few years. Real estate agents are often familiar with who bought and sold land, how they did so, and for what purpose. If the property values in the area have changed for reasons attributable to the site or its surrounding area, an experienced real estate agent in the locality is likely to know about that change and why it occurred.

✓ **County records** can be very useful in discovering past uses of the premises. Building permit records, tax assessment work sheets, and the planning and zoning boards are all fertile sources of information regarding the past, present, and potential uses of the property and adjacent properties.

✓ **Check the SEC records** of the target. Review the company's SEC filings, such as the 10-K and 10-Q reports, and information relating to environmental policies and liabilities. Such legal computer services as LEXIS contain full-text 10-Ks, 10-Qs, and 8-Ks, without exhibits, for more than 11,000 public companies. These files are updated regularly. There are also specialty companies that provide access to all SEC public filings.

✓ **Environmental documents** of interest would include permits issued; records of product spills and losses; reports to, by, or for a state, local, or federal environmental agency; enforcement actions by environmental agencies; records of waste analysis; foundation boring and well drilling records and tests; drawings and specifications for landfills; and environmental audits.

✓ **Consider speaking with an environmental engineer** who has had occasion to check groundwater characteristics and surrounding land uses in the area, specifically to answer the following questions: Do any nearby waterways show any evidence of contamination? Does the topography indicate that the site in question might be the source of the contamination? If a spill occurred on a neighboring site, is the waste likely to migrate to the facility?

✓ **Check an atlas, maps, and newspapers** regarding past usage and owners of the land. If the site is a point of economic importance to the community, check the records in the local library for sources of historical information (many libraries have "vertical" files about important people, places, and industry in the area).

✓ **Request and review documents** relating to the targeted property. To the extent possible, buyers should obtain from the seller and review the following documents relating to the property being transferred: environmental audits and inspection reports; consultant's reports and correspondence; remedial investigations, feasibility studies, corrective action programs, and cost estimates for ongoing studies and remedial alternatives; groundwater assessments; permits and permit applications (and related memoranda and correspondence); monitoring reports (groundwater, surface water discharges, sewer discharges, air emissions); notices (including notices of violation) to and from environmental regulatory authorities; correspondence with environmental regulatory authorities; spill reports; manifests and other documents relating to hazardous waste management; notifications and tests of underground and aboveground storage tanks; and documents relating to any environmental litigation involving the property (including administrative proceedings). If hazardous substances have been present on the site in the past, and if it is determined that the property is still worth buying, verify that such substances are no longer present on-site and that they have been disposed of off-site in accordance with applicable law.

✓ **Review the land records and liens** concerning the targeted property to determine whether any restrictions were placed on the deed regarding waste disposal or whether the deed includes a notice of prior waste disposal at the site. Also, determine whether any liens have been placed on the site pursuant to federal law or a state "Superlien Statute." This is important even if the state in which the site is located does not have a "Superlien Statute," as federal law may nevertheless cause a lien to attach in that state.

✓ Finally, **on-site investigation** provides one of the best sources of information. Besides walking the site and looking for overt evidence of environmental damage (via use of the senses), certain scientific tests may be used to check for the presence of hazardous waste. These tests include hydrogeological studies, material analysis, aerial photography, and metal detection (to look for buried drums or storage tanks).

These items can be used in varying degrees by the attorney as part of the transaction screen or by the consultant during the environmental audit actually conducted.

§ 1.5 Market Valuation Based on Audit Results

There will be situations, of course, where a party will make a business decision to acquire a property or use property as collateral for a loan even if it exhibits some degree—maybe even a significant degree—of environmental impairment. The purpose of the well-structured environmental audit is not to eliminate the sale of environmentally impaired property or prohibit its use as collateral. Rather, the purpose is to qualify and quantify the nature and magnitude of the associated risks to allow the auditing party to make informed decisions regarding the value of the property in question, the extent of contingent liability, and the type and scope of work necessary to come into compliance with law, protect human health and the environment, and use the targeted property for its intended purpose.

§ 1.6 Tables

§ 1.6.1 Environmental Hotlines

Below is a table of hotline numbers for obtaining further environmental information.

Hotline	Telephone	Description
ASHAA Coordination Center	(800) 462-6706 (301) 530-5603	Provides information and support concerning the Asbestos School Hazard Abatement Act (ASHAA) loan and grant program. Managed by the Office of Pollution Prevention and Toxics (OPPT).
Air RISC Hotline	(919) 541-0888	Provides technical assistance and information to state and local air pollution control agencies in areas of health, risk, and exposure for both toxic and criteria pollutants.
Asbestos Ombudsman Clearinghouse/ Hotline; Small Business Ombudsman Clearinghouse/ Hotline	(800) 368-5888 (703) 305-5938 (703) 305-6824	Provides information to the public on the handling and abatement of asbestos. Assists small business in complying with EPA regulations.
Automotive Imports Helpline	(202) 233-9660	Provides information to the public about the federal emission requirements for vehicles imported into the U.S. Contractor operated.

Hotline	Telephone	Description
CERCLIS Helpline	(703) 538-7234	Provides technical support and referrals to users of the CERCLIS (Comprehensive Environmental Response, Compensation and Liability Information System) database, CleanLAN and WasteLAN. Contractor operated.
CleanLAN/ WasteLAN Hotline	(703) 538-7234	Provides information about the CleanLAN and WasteLAN programs. Contractor operated.
Control Technology Center	(919) 541-0800	Provides technical support and guidance concerning air pollution emissions and control technology for all air pollutants, including air toxins emitted by stationary sources.
Customer Technical Support National Computer Center (NCC)	(800) 334-2405 (919) 541-7862	Provides information about the Customer Technical Support National Computer Center (NCC) program and provides NCC customers with technical assistance and tracking. Contractor operated.
Data Processing Support Services National Computer Center (NCC)	(800) 334-9700 (919) 541-2385	
EMF (Electromagnetic Field) Infoline	(800) 363-2383	Answers inquiries concerning EMFs.
Emergency Planning and Community Right to Know (Title III of SARA) and Superfund	(800) 535-0202 (703) 412-9877	Provides regulatory, policy, and technical assistance to federal agencies, states and local governments, the public and the regulatory community. Contractor operated.
Emergency Response Section Notification Line	(415) 744-2000	Attendant screens calls and transfers appropriate callers to EPA's technical assistance team.

Hotline	Telephone	Description
EPA Action Line	(800) 223-0425 (913) 551-7122	Referral service to appropriate program offices.Provides information about the CleanLAN and WasteLAN programs. Contractor operated.
Ethics Program Advice	(202) 260-4550	Provides information and advice concerning the conflict of interest law and standards of conduct that apply to all EPA employees. Attorneys supply advice.
Facility Index System (FINDS)	(800) 980-2493	Technical support for FINDs users.
Green Lights Energy Star Program	(202) 775-6650	Information about the Green Lights Program, a voluntary program that provides extensive information and technical support to encourage businesses and government to install energy-efficient lighting. Considerable cost savings and improved lighting quality are among the many benefits generated by the program.
Hazardous Waste Ombudsman	(800) 262-7937 (202) 260-9361	Handles complaints from citizens and the regulatory community concerning administration of the Hazardous Waste Management Program, obtains facts, sorts information, and substantiates policy.
Indoor Air Quality Information Clearinghouse	(800) 438-4318	Provides information on indoor air pollutants and their sources, health effects, testing, measuring, and controlling. Information about constructing and maintaining homes and buildings to minimize indoor air quality problems also available.

Hotline	Telephone	Description
Mobile Sources—Region 1	(800) 821-1237 (617) 565-9145	Handles complaints relating to automobile emission violations, as well as warranties, fuel issues, CSC recycling, and automobile air conditioning.
National Lead Information Center Hotline	(800) 532-3394 (800) 526-5456	Provides information to help parents protect their children from lead poisoning. Lists state and local contacts. Managed by the National Safety Council and sponsored by OPPT.
National Pesticides Telecommunication Network (800) 858-PEST	(800) 858-7378 (806) 743-3091	Provides information about the National Pesticides Telecommunications Network program. Provides the general public, medical, veterinary, and professional communities with information on pesticide product information, recognition and management of pesticide poisonings, toxicology and symptomatic reviews, safety information, health and environmental effects, and clean-up and disposal procedures. Contractor managed.
National Radon Hotline	(800) 767-7236 (800) SOS-RADON	Provides information about the National Radon program. Contractor operated.
National Response Center—U.S. Coast Guard Oil and Hazardous Materials Spills	(800) 424-8802 (202) 267-2675 (202) 267-2185	Used to report oil and hazardous materials spills.
Office of Environmental Justice	(800) 962-6215	Coordinates public communication, outreach, and education on equity issues. Provides technical and financial assistance on equity concerns. Serves as a repository of environmental equity information.

Hotline	Telephone	Description
Office of Inspector General Whistle-Blower Hotline	(800) 424-4000 (202) 260-4977	Handles allegations of fraud, abuse, or mismanagement of EPA funds.
Pollution Prevention Information Clearinghouse (PPIC)	(202) 260-1023	Established as part of EPA's Pollution Prevention Act of 1990, which directed the EPA to make information on source reduction and analysis available to the public. Provides references and referrals in response to pollution prevention questions from the public. Contractor operated.
Records of Decision System (RODS)	(703) 538-7227	Provides information about Records of Decision System (RODS). Contractor operated.
RCRA Information Hotline (RIL)	(415) 744-2074	Provides information on hazardous waste identification, generators, transporters, treatment, storage and disposal facilities, recycling sites, and export and import.
RCRA/Superfund/UST and EPCRA Hotline	(800) 424-9346 (703) 412-9810 (800) 553-7672	Provides answers to the regulated community, other interested parties, and the public about EPA's RCRA regulations and practices. Provides referrals for obtaining documents concerning RCRA, underground storage tanks (USTs), Superfund/CERCLA, and Pollution/Waste Minimization. Contractor operated.
Region 1 General Information	(617) 565-3420	Referral service.
Region 2 Potential Responsible Party Hotline	(212) 264-2738 (800) 245-2738	Provides Superfund civil investigators with information relevant to Superfund site enforcement investigations.

Hotline	Telephone	Description
Safe DrinkingWater Hotline	(800) 426-4791 (202) 260-7908	Provides information about the Safe Drinking Water Act program to the regulated community (public water systems) and the public.
Solid Waste Assistance Program (SWAP)	(800) 677-9424 (301) 585-2898 (301) 585-0297	Provides information on all aspects of solid waste management, including source reduction, public participation, legislation, regulation, waste combustion, collection, transfer and disposal, landfill gas, and special wastes.
STORET	(800) 424-9067 (202) 260-8161	Provides information and technical support for STORET users.
Storm WaterHotline	(703) 821-4823	Discontinued April 15, 1994. Recording directs callers to alternative sources of information.
Stratospheric Ozone Information Hotline	(800) 296-1996	Provides in-depth information about ozone protection regulations and requirements under title VI of the Clean Air Act Amendments of 1990. Serves as a distribution center and referral source concerning stratospheric ozone protection and depletion.
Region 9 Superfund/RCRA Community Relations Office	(415) 744-2423 (800) 231-3075	Provides information concerning Superfund abandoned waste sites.
Toxic Substances Control Act (TSCA) Assistance Information Service	(202) 554-1404 (202) 554-0551	Provides technical and general information concerning TSCA regulation information to the chemical industry, labor and trade organizations, environmental groups, and the public.

Hotline	Telephone	Description
Wetlands Protection	(800) 832-7828 (703) 525-0985	Provides information about the value and function of wetlands, options for wetlands protection, and other data concerning the Wetlands Protection program. Contractor operated.
Waste Isolation Pilot Plant (WIPP) Information Line	(202) 233-9716	Provides recorded information to the public about upcoming EPA public meetings and hearings involving EPA oversight of the Waste Isolation Pilot Plant.
WASTEWI$E Hotline	(800) EPA-WISE (202) 260-4180	Provides information about Waste Wi$e, EPA's voluntary program encouraging businesses to reduce solid waste. The program offers recognition and technical assistance to U.S. companies that commit to undertake new waste reduction efforts.

§ 1.6.2 EPA Regions

Below is a table of information concerning EPA regions and how to get in touch with them.

Region	Phone	Address
1	(617) 565-3420	EPA John F. Kennedy Federal Bldg. Boston, MA 02203
2	(212) 264-2657	EPA Jacob K. Javits Federal Bldg. 26 Federal Plaza New York, NY 10278
3	(215) 597-9800	EPA 841 Chestnut Bldg. Philadelphia, PA 19107

Region	**Phone**	**Address**
4	(404) 347-4727	EPA 345 Courtland St., NE Atlanta, GA 30365
5	(312) 353-2000	EPA 77 West Jackson Blvd. Chicago, IL 60604-3507
6	(214) 665-6444	EPA First Interstate Bank Tower at Fountain Place 1445 Ross Ave, 12th Floor Suite 1200 Dallas, TX 75202-2733
7	(913) 551-7000	EPA 726 Minnesota Ave. Kansas City, KS 66101
8	(303) 293-1647	EPA 999 18th St, Suite 500 Denver, CO 80202-2405
9	(415) 744-1305	EPA 75 Hawthorne St. San Francisco, CA 94105
10	(206) 553-4973	EPA 1200 Sixth Ave. Seattle, WA 98101

Chapter 2

Transaction Triggered Audits

§ 2.1 Overview of the Transaction Triggered Audit

Environmental audits may be conducted for many reasons, but in the transaction context audits are conducted in order to identify and quantify potential environmental liabilities. Transaction triggered audits may be conducted by would-be sellers in order to identify and correct problems before putting the property up for sale. They are almost always conducted by potential commercial buyers as part of the "due diligence" process—to quantify the value of the target project. Most lenders will require an environmental audit (and often environmental cleanup) as a condition to entering into a loan transaction. Potential landlords and tenants may want to conduct an environmental audit in order to define the environmental baseline of the property being rented—so that each party is sure when (if any) contamination occurs. These baseline audits are particularly attractive when the lease (or sale) occurs in a commercial or industrial park.

Transaction triggered audits are typically conducted in phases. While the CERCLA innocent purchaser defense has been a driving force in helping to define Phase I transaction triggered audits, CERCLA is by no means the sole source for environmental concern. Other typical concerns include the presence of asbestos,[1] wetlands,[2] and PCBs. These concerns

[1] For issues concerning asbestos in buildings, *see generally* Courtney M. Price, Asbestos in Building, Facility & Industry (1990).

[2] For issues concerning wetlands, *see generally* Robert E. Steinberg, Wetlands and Real Estate Development Handbook (1991).

implicate the Clean Air Act, the Clean Water Act, and the TSCA respectively. Accordingly, the minimum components for a Phase I environmental site assessment should include at least the following:[3]

Step	**Description**
Environmental Setting	Topography Geology Hydrogeology Hydrology
Historic Usage of the Property	Title Search Aerial Photographs Fire Insurance Maps Local Street Directories
Regulatory Agency Listings	NPL CERCLIS RCRA Lists ERNS State Hazardous Waste Site List Solid Waste Disposal Sites Underground Storage Tank (UST) List Leaking UST (LUST) List
Site Reconnaissance	Current Use Prior Use Hazardous Substances Present Signs of Property Misuse Effluence and Air Emissions Waste Disposal Techniques Surface Water Drainage Transformers Surrounding Area Use
Interviews	Owner Occupants Operations Management Local Governments

[3] *See* KATHLEEN HESS, ENVIRONMENTAL SITE ASSESSMENT—PHASE I: A BASIC GUIDE, at 42-3 (Lewis 1993).

§ 2.2 Preserving the Innocent Purchaser Defense

Notwithstanding the CERCLA third-party defense, landowners have been held liable for environmental problems even where they were caused by third parties.[4] Congress enacted CERCLA in 1980 in response to the environmental and public health hazards posed by improper disposal of hazardous wastes. In general terms, the statute established the "Superfund," which is financed primarily through excise taxes on the oil and chemical industries.[5] The federal government is authorized to use the Superfund to finance governmental responses to hazardous waste problems, to pay claims arising from the response activities of private parties, and to compensate federal or state governmental entities for damages to natural resources.[6] The government may then recover Superfund expenditures from those responsible for the generation, transportation, or disposal of the hazardous substances.[7] The federal government also has the option of pursuing a direct action under section 106 to require the "responsible party" to conduct the cleanup.

Section 107(a) of CERCLA[8] provides, in part, as follows:

> Notwithstanding any other provision or rule of law, and subject only to the defenses set forth in subsection (b) of this section—
>
> 1. the owner and operator of a vessel or a facility,
> 2. any person who at the time of disposal of any hazardous substance owned or operated any facility at which such hazardous substances were disposed of,

---

[4] *See, e.g.*, United States v. Messer Oil Corp., 391 F. Supp. 557 (W.D. Pa. 1975) (spill caused by pranks of trespassing kids); Lansco v. Dept of Envtl. Protection, 350 A.2d 520 (N.J. 1975) (release of oil caused when someone opened tank valves; storage of oil considered an ultrahazardous activity); United States v. Pacific Hide & Fur Depot, Inc., 768 F.2d 1096 (9th Cir. 1985) (TSCA liability for improper handling of PCBs).

[5] Section 517 of the Superfund Amendments and Reauthorization Act of 1986 (SARA), Pub. L. 99-499, § 517, 100 Stat. 1772 (1986), established the Hazardous Substances Superfund, which is in effect a continuation of the Hazardous Substance Response Trust Fund established by Section 221 of CERCLA, 42 U.S.C. § 9631 (repealed).

[6] 42 U.S.C. § 9611(a).

[7] *Id.* § 9607(a).

[8] *Id.*

3. any person who by contract, agreement, or otherwise arranged for disposal or treatment, or arranged with a transporter for transport for disposal or treatment, of hazardous substances owned or possessed by such person, by any other party or entity, at any facility or incineration vessel owned or operated by another party or entity and containing such hazardous substances, and
4. any person who accepts or accepted any hazardous substances for transport to disposal or treatment facilities, incineration vessels or sites selected by such person, from which there is a release, or a threatened release which causes the incurrence of response costs, of a hazardous substance, shall be liable....

In construing the terms of section 107(a), the courts have uniformly imposed strict liability, subject only to the defenses set forth in section 107(b).[9] Accordingly, in this case, a *prima facie* case under section 107(a) is established if the United States shows that

1. the site is a "facility";[10]
2. a "release" or threatened release of a "hazardous substance" from the site has occurred;[11]

---

[9] Artesian Water Co. v. Gov't. of New Castle County, 659 F. Supp. 1269 (D. Del. 1987), *aff'd,* 851 F.2d 643 (3d Cir. 1988).

[10] 42 U.S.C. § 9601(9) defines "facility" as follows:

> (A) any building, structure, installation, equipment, pipe or pipeline (including any pipe into a sewer or publicly owned treatment works), well, pit, pond, lagoon, impoundment, ditch, landfill, storage container, motor vehicle, rolling stock, or aircraft, or
>
> (B) any site or area where a hazardous substance has been deposited, stored, disposed of, or placed, or otherwise come to be located; but does not include any consumer product in consumer use or any vessel.

[11] 42 U.S.C. § 9601(22) defines "release" as follows:

> [A]ny spilling, leaking, pumping, pouring, emitting, emptying, discharging, injecting, escaping, leaching, dumping or disposing into the environment (including the abandonment or discarding of barrels, containers, and other closed receptacles containing any hazardous substance or pollutant or contaminant)....

42 U.S.C. § 9602(a) provides that the EPA shall promulgate regulations designating materials as hazardous substances. The regulations at 40 C.F.R. § 302.4 designate substances found at the Taylor site, in United States v. Serafini, 706 F. Supp. 346 (M.D. Pa. 1988) as hazardous substances.

3. the release or threatened release has caused the United States to incur "response costs";[12] and
4. the defendants are "owners" of a facility.[13]

CERCLA section 107(b) sets out what has come to be known as the "third party defense."[14] The statute states that there shall be no liability for

---

[12] "Response" means "remove, removal, remedy and remedial action," including "enforcement activities related thereto." 42 U.S.C. § 9601(25).

[13] "Owner" means "in the case of an onshore facility..., any person owning or operating such facility...." *Id.* § 9601(20). The term "person" includes an individual and a partnership. *Id.* § 9601(21). *See also* United States v. Maryland Bank & Trust Co., 632 F. Supp. 573 (D. Md. 1986).

[14] 42 U.S.C. § 9607(b). *See also* H.R. 2787, 101st Cong., 1st Sess. (introduced by Congressman Curt Weldon); H.R. 1217, 102d Cong., 1st Sess. (reintroduced by Congressman Weldon, Feb. 28, 1991); H.R. 1643, 102d Cong., 1st Sess. (reintroduced by Congressmen Owens, Weldon, Meyers, March 22, 1991).

Note that there is also a third-party defense under Oil Spill Prevention Statute FWPCA § 311(f), 33 U.S.C. § 1321(f) (no liability where owner of facility can prove the discharge was caused *solely* by "an act or omission of a third party without regard to whether such act or omission was or was not negligent"). For case law interpreting the FWPCA third-party defense, *see* Burgess v. M/V Tamano, 564 F.2d 964, 981-82 (1st Cir. 1977), *cert. denied,* 435 U.S. 941 (1978); United States v. LeBeouf Bros. Towing Co., 621 F.2d 787 (5th Cir. 1980), *cert. denied,* 452 U.S. 906 (1981); Union Petroleum Corp. v. United States, 651 F.2d 734 (Ct. Cl. 1981).

Notwithstanding the CERCLA third-party defense, landowners have been held liable for environmental problems even where they were caused by third parties. *See, e.g.,* United States v. Messer Oil Corp., 391 F. Supp. 557 (W.D. Pa. 1975) (spill caused by pranks of trespassing kids); Lansco v. Dept of Envtl. Protection, 350 A.2d 520 (N.J. 1975) (release of oil caused when someone opened tank valves; storage of oil considered an ultrahazardous activity); United States v. Pacific Hide & Fur Depot, Inc., 768 F.2d 1096 (9th Cir. 1985) (TSCA liability for improper handling of PCBs). *But see* City of New York v. Chemical Waste Disposal Corp., 1993 WL 463174 (E.D.N.Y. Sept. 28, 1993) (court rejected third-party defense by a solvent-reclamation site owner who attempted to blame the release on the cleanup contractor); Hatco Corp. v. W.R. Grace & Co.-Connecticut, 1993 WL 394391 (D.N.J. Oct. 4, 1993) (court rejected a prior landowner's assertion of the third-party defense because the prior landowner had not shown that the release or threatened release were caused solely by the acts or omissions of a third party).

a person otherwise liable under CERCLA, if that person can establish[15] "that the release or threat of release of a hazardous substance, and the damages resulting therefrom, were caused solely by an act or omission of a third party."[16] The defendant must also establish by a preponderance of the evidence that

(a) he exercised due care with respect to the hazardous substance concerned, taking into consideration the characteristics of such hazardous substance, in light of all relevant facts and circumstances; and

(b) he took precautions against foreseeable acts or omissions of any third party and the consequences that could foreseeably result from such acts or omissions.[17]

The "innocent purchaser defense" is one form of "third-party defense." The scope of the CERCLA "innocent purchaser" defense was clarified in the Superfund Amendment and Reauthorization Act of 1986 (SARA) by adding a definition of "contractual relationship."[18] Under

---

[15] The proof must be established by a preponderance of the evidence. 42 U.S.C. § 9607(b). Note that there is also a third-party defense under Oil Spill Prevention Statute FWPCA § 311(f), 33 U.S.C. § 1321(f) (1982) (no liability where owner of facility can prove the discharge was caused *solely* by "an act or omission of a third party without regard to whether such act or omission was or was not negligent"). For case law interpreting the FWPCA third-party defense, *see* Burgess v. M/V Tamano, 564 F.2d 964, 981-88 (1st Cir. 1977), *cert. denied,* 435 U.S. 941 (1978); United States v. LeBoeuf Bros. Towing Co., 621 F.2d 787 (5th Cir. 1980); Union Petroleum Corp. v. United States, 651 F.2d 734 (Ct. Cl. 1981).

[16] 42 U.S.C. 9607(b)(3). A third party is a person "other than an employee or agent of the defendant, or than one whose act or omission occurs in connection with a contractual relationship, existing directly or indirectly, with the defendant (except where the sole contractual arrangement arises from a published tariff and acceptance for carriage by a common carrier by rail)." *Id.* No case law prior to enacting SARA successfully raised the innocent purchaser defense under CERCLA. *See, e.g.,* New York v. Shore Realty, 759 F.2d 1032 (2d Cir. 1985); United States v. Argent, 21 Env't Rep. Cas. (BNA) 1354 (D.N.M. 1984); United States v. South Carolina Recycling & Disposal, Inc. (SCRDI), 653 F. Supp. 984 (D.S.C. 1984), *cert. denied,* 490 U.S. 1106 (1989).

[17] 42 U.S.C. § 9607(b)(3) (emphasis added).

[18] *Id.* § 9601(35). For a discussion of the impact of SARA on the innocent purchaser defense, see, J. B. Ruhl, *The Third Party Defense: Narrowing the Contractual Relationship Exception,* 29 S. TEX. L. REV. 291 (1988). For post-SARA cases concerning the innocent purchaser defense, *see, e.g.,* United States v. Serafini, 711 F. Supp. 197, 198 (M.D. Pa. 1988); United States v. Serafini, 706 F. Supp. 346, 348 (M.D. Pa. 1988) (a genuine issue of

SARA, the definition of contractual relationship includes, but is not limited to, land contracts, deeds, or other instruments transferring title or possession, provided at the time of acquisition the purchaser "did not know and had no reason to know that any hazardous substance . . . was disposed of on, in, or at the facility."[19] To establish that a potentially responsible party (PRP) had no "reason to know" of the existence of any hazardous substances on the property, the PRP "must have undertaken, at the time of acquisition, all appropriate inquiry into the previous ownership and uses of the property consistent with good commercial or customary practice in an effort to minimize liability."[20] For purposes of determining what is "appropriate inquiry,"[21] Congress dictated that the court shall take into account

---

material fact existed as to whether it was the "customary or good commercial practice" in 1969 to visually inspect the property prior to a purchase); Pacific Hide v. Fur Dept, Inc., 716 F. Supp. 1341, 1348 (D. Idaho 1989) (children of corporation owners who received interest in contaminated land by gift, had no reason to suspect that the land was contaminated with PCBs and were entitled to innocent landowner defense); Jersey City Redevelopment Auth. v. PPG Indus., 655 F. Supp. 1257, 1261 (D.N.J. 1987) (issue of fact existed as to whether purchaser of site previously contaminated with waste mud had requisite knowledge to preclude assertion of the innocent landowner defense); United States v. Tyson, 25 Env't. Rep. Cas. (BNA) 1897, 1905 (E.D. Pa. 1986).

[19] 42 U.S.C. § 9601(35)(A)(i).

[20] *Id.* § 9601(35)(B).

[21] The legislative history of SARA explains Congress's intent in revising the innocent landowner defense:

> The duty to inquire under this provision shall be judged as of the time of acquisition. Defendants shall be held to a higher standard as public awareness of the hazards associated with hazardous substance releases has grown, as reflected by this Act, the 1980 Act and other Federal and State statutes.
>
> Moreover, good commercial or customary practice with respect to inquiry in an effort to minimize liability shall mean that a reasonable inquiry must have been made in all circumstances, in light of best business and land transfer principles.
>
> Those engaged in commercial transactions should, however, be held to a higher standard than those who are engaged in private residential transactions. Similarly, those who acquire property through inheritance or bequest without actual knowledge may rely upon this section if they engage in a reasonable inquiry, but they need not be held

1. any specialized knowledge or experience on the part of the defendant;
2. the relationship of the purchase price to the value of the property if uncontaminated;
3. commonly known or reasonably ascertainable information about the property;
4. the obviousness of the presence or likely presence of contamination at the property; and
5. the ability to detect such contamination by appropriate inspection.[22]

By so defining "contractual relationship," Congress created an obligation on the part of a purchaser to undertake diligent inquiry before purchasing property. There is no statutory or regulatory pronouncement as to what constitutes diligent inspection.[23] EPA has stated that what constitutes "all appropriate" inquiry must be determined on a case-by-case basis.[24]

> to the same standard as part of a commercial or private transaction and those who acquire property by inheritance without knowing of the inheritance shall not be liable, if they satisfy the remaining requirements of section 107(b)(3).

Superfund Amendments and Reauthorization Act of 1986, Pub. L. No. 99-499, 1986 U.S.C.C.A.N. 3280-81.

[22] 42 U.S.C. § 9601(35)(B). Moreover, the statute stated that "Nothing in this paragraph shall affect the liability under this chapter of a defendant who, by any act or omission, caused or contributed to the release or threatened release of a hazardous substance which is the subject of the action relating to the facility." *Id.* § 9601(35)(D).

[23] The language of some courts suggests that the combination of a contract selling property "as is" coupled with the purchasers contractual right to inspect the property prior to sale may shift environmental responsibility to the purchaser. *See, e.g.*, Smith Land and Improvement Corp. v. Celotex Corp., 851 F.2d 86 (3rd Cir. 1988); Shapiro v. Hu, 233 Cal. Rptr. 470 (1987). *Compare* United States v. Norris, 937 F.2d 286 (6th Cir. 1991) (foreclosing lender held to standard of care where title search and document review were insufficient due diligence; actual sight inspection would have disclosed dikes and levees that violated Clean Water Act requirement). *Cf.* Aliff v. Joy Technologies, Inc., 873 F.2d 1437 (4th Cir. 1989) (buyer prevailed against seller on theory of misrepresentation for failure to disclose PCBs; court upheld valuation of property).

[24] EPA explained:

> a determination as to what constitutes "all appropriate inquiry" under all the circumstances is to be made on a case-by-case basis. Generally,

Current industry standards, however, seem to require that at least a "Phase I Environmental Audit" be conducted.[25]

Little case law has developed concerning the innocent purchaser defense, although it is a driving force in commercial practice. Before the enactment of SARA, no defendant had successfully raised the innocent purchaser defense under CERCLA.[26] Courts acknowledging the defense tend to construe it narrowly.[27] The innocent purchaser defense has,

> when determining whether a landowner has conducted "all appropriate inquiry," the Agency will require a more comprehensive inquiry for those involved in commercial transactions than for those involved in residential transactions for personal use. For example, an investigation along the lines of a survey for contamination may be recommended in some commercial transactions, whereas this type of inquiry would not typically be recommended for the purchaser of personal residential property. In sum, the determination will be made on the basis of what is reasonable under all of the circumstances.

Public Notice, 54 Fed. Reg. 34,235, 34,238 (Aug. 18, 1989). For an analysis of EPA's guidance document on landowner liability, *see,* O'Brien, *EPA's Landowner Liability Guidance,* TOXICS L. REP. 184 (Jul. 19, 1989).

[25] For a discussion of the impact of SARA on the Innocent Purchaser Defense, *see* Ruhl, *The Third Party Defense: Narrowing the Contractual Relationship Exception,* 29 S. TEX. L. REV. 291 (1988). *See generally* Wolf, *Energy Contours of the CERCLA "Innocent Purchaser" Defense,* 20 ENVTL. L. REP. (Envtl. L. Inst.) 10485 (1990); Note, *The Innocent Too Shall Pay: EPA's Settlement Policy Under CERCLA for De Minimis Landowner Liability,* 51 U. PITT. L. REV. 727 (1990); Note, *Liabilities of Innocent Current Owner of Toxic Property Under CERCLA,* 23 U. RICH. L. REV. 403 (1989); Mays, *The Blessed State of Innocence: The Innocent Landowner Defense Under Superfund,* 18 CHEM. WASTE LITIG. REP. 864 (1989); Marcus, *The Price of Innocence: Under CERCLA & SARA,* 6 TEMP. ENVTL. L. & TECH. J. 117 (1987).

[26] *See, e.g.,* New York v. Shore Realty, 759 F.2d 1032 (2d Cir. 1985); United States v. Argent, 21 Env't Rep. Cas. (BNA) 1354 (D.N.M. 1984); United States v. South Carolina Recycling & Disposal, Inc. (SCRDI), 20 Env't Rep. Cas. (BNA) 1753 (D.S.C. 1984).

[27] *See* Westwood Pharmaceuticals v. National Fuel Gas Distrib., Inc., 964 F.2d 85, 91 (2d Cir. 1992) (stating that landowner must "take precautions against the foreseeable acts or omissions of third parties" for innocent purchaser defense to apply after sale of property); Acme Printing Ink Co. v. Menard, Inc., 870 F. Supp. 1465, 1480 (E.D. Wis. 1994) (holding that landowner who accidentally ruptured waste barrels dumped by unidentified third party could not assert innocent purchaser defense); United States v. Broderick Inv. Co., 862 F. Supp. :72, 27576 (D. Colo. 1994) (holding that purchaser could not assert innocent purchaser defense because predecessor had knowledge of tenant's hazardous waste disposal activities); *see also* Pacific Hide & Fur Depot, 716 F. Supp. at 134849 (applying innocent

therefore, been described by many as merely a "safe harbor with no water."[28]

§ 2.3 Elements of a Phase I Audit Designed to Meet the Innocent Purchaser Defense

The customary procedures that are widely followed for a Phase I environmental audit designed to meet the innocent purchaser defense vary somewhat, but have been set out by Congressman Weldon in a bill before Congress to further define the innocent purchaser defense.[29] This proposal both codified what was perceived to be appropriate practice and (by its very proposal) served as a driving force to standardize the conduct. The customary procedures are as follows:

✓ **Review of recorded chain of title documents** regarding the real property, including all deeds, easements, leases, restrictions, and covenants for a period of 50 years.

---

purchaser defense in situation where landowners obtained property through inheritance when they were "barely out of their teenage years").

[28] *See* Albert R. Wilson, *The Environmental Opinion: Basis for an Impaired Value Opinion,* 62 APPRAISAL J. 410, 411 (1994). *See also* Debra L. Baker and Theodore G. Baroody, *What Price Innocence? A Realistic View of the Innocent Landowner Defense Under CERCLA,* 22 ST. MARY'S L.J. 115, 125 (1990); Michael J. Gergen, *The Failed Promise of the "Polluter Pays" Principle: An Economic Analysis of Landowner Liability for Hazardous Waste,* 69 N.Y.U. L. REV. 625, 652 (1994); L. Jager Smith, Jr., *CERCLA's Innocent Landowner Defense: Oasis or Mirage?,* 18 Colum. J. ENVTL. L. 155, 166 (1993); Richard H. Mays, *The Blessed State of Innocence: The Innocent Landowner Defense Under the Superfund,* 20 ENV'T REP. (BNA) No. 19, at 809, 811 (Sept. 8, 1989)

[29] See, e.g., H.R. 1285, 104th Cong., 1st Sess. (reintroduced by Congressman Weldon 3/21/95). The stated purpose of the bill was:

> To amend the Comprehensive Environmental Response, Compensation, and Liability Act of 1980 to provide a specific definition of the requirement that a purchaser of real property make all appropriate inquiry into the previous ownership and uses of the real property in order to qualify for the "innocent landowner" defense.

See also H.R. 2787, 101st Cong., 1st Sess. (introduced by Congressman Curt Weldon, 1990); H.R. 1217, 102d Cong., 1st Sess. (reintroduced by Congressman Weldon, Feb. 28, 1991).

✓ **Review of aerial photographs** that may reflect prior uses of the real property and that are reasonably obtainable through state or local government agencies.

✓ **Determination of the existence of recorded environmental cleanup liens** against the real property that have arisen pursuant to federal, state, and local statutes.

✓ **Review of reasonably obtainable federal, state, and local government records** of sites or facilities where there has been a release of hazardous substances and that are likely to cause or contribute to a release or threatened release of hazardous substances on the real property, including investigation reports for such sites or facilities; reasonably obtainable federal, state, and local government environmental records of activities likely to cause or contribute to a release or a threatened release of hazardous substances on the real property, including landfill and other disposal location records, underground storage tank records, hazardous waste handler and generator records, and spill reporting records; and such other reasonably obtainable federal, state, and local government environmental records that report incidents or activities that are likely to cause or contribute to a release or threatened release of hazardous substances on the real property.

✓ **A visual site inspection** of the real property and all facilities and improvements on the real property, and a visual inspection of immediately adjacent properties from the real property, including an investigation of any chemical use, storage, treatment, and disposal practices on the property.

In order to be deemed "reasonably obtainable," a copy or reasonable facsimile of the aerial photography or other records must be obtainable from the government agency by request.[30]

Numerous types of maps or aerial photographs may be used to provide the environmental auditor with appropriate environmental information.[31]

---

[30] *Id.*

[31] *See generally* KATHLEEN HESS, ENVIRONMENTAL SITE ASSESSMENT—PHASE I: A BASIC GUIDE, at 78-90 (Lewis 1993).

Source	**Address**	**Description**
United States Geological Survey (USGS) seven-and-one-half-minute quadrangle (quad) maps	United States Geological Survey (USGS) Earth Sciences Information Office 507 National Center Reston, Virginia 22092	These topographical maps, with an approximate scale of one inch to 2,000 feet, show many physical features.
United States Department of Agriculture, Agricultural Stabilization and Conservation Services photographs	United States Department of Agriculture, Agricultural Stabilization and Conservation Services Salt Lake City, Utah	Aerial photographs of various areas throughout the country dated back to 1950 are available for a small fee. Exact site location should be provided. Expect 4 to 6 weeks for a response.
National Archives and Records Service; Cartographical Archives Division	National Archives and Records Service; Cartographical Archives Division General Services AdministrationWashington, DC 20408	Aerial photographs are maintained of various areas in the country dating back to 1935.

Source	Address	Description
Sanborn Mapping and Geographical Information Service	Sanborn Mapping and Geographical Information Service 629 Fifth Avenue Dept. OSC Pelham, NY 10803	Produced detailed Fire Insurance Maps dating back to 1867. These are available from Sanborn for a small fee. The Library of Congress presently holds over 700,000 maps, most of which are submitted through Sanborn's copyrighting procedures.[32]
As-Built Plan		Normally obtained from the property owner, these drawings cover property boundaries, streets bordering the target site, and building layouts and provide accurate scale and a north arrow. They tend to be most useful in evaluating developed property.

[32] *See* RISTOW, WALTER W., FIRE INSURANCE MAPS IN THE LIBRARY OF CONGRESS—PLANS OF NORTH AMERICAN CITIES AND TOWNS PRODUCED BY THE SANBORN MAP COMPANY. Library of Congress, Washington, D.C. (1981).

Source	Address	Description
Assessor's Map		Often used when more formal maps are not available, these maps provide property boundaries and give the location of properties bordering the subject site.
Zoning Records and Maps		Often used to assess how property use has evolved as well as to evaluate permissible future uses.[33]

§ 2.4 The Inheritance Defense

The "inheritance defense"[34] is a subset of the innocent landowner defense. In order for this defense to be utilized, the PRP must prove by the preponderance of the evidence that

✓ The PRP acquired the property after hazardous substances came to be located on the property;

✓ The PRP received the property by "inheritance or bequest"; and

✓ The PRP exercised due care with respect to the hazardous substances concerned, taking into consideration the characteristics of such substances, in light of all relevant facts and circumstances.

The inheritance defense differs from the innocent landowner defense in that the former eliminates two elements required in the latter. First, there is no

---

[33] These documents are generally available through the city or county planning department, as appropriate. Other documents that may be of use that are also available from the local planning departments include

✓ Building Permits and

✓ Certificates of Occupancy.

See Kathleen Hess, Environmental Site Assessment—Phase I: A Basic Guide, at 88-90 (Lewis 1993).

[34] 42 U.S.C. § 9601 (35)(A)(iii).

express statutory requirement that the PRP be without knowledge of the contamination at the time of inheritance. Second, there is no express statutory requirement that the PRP have undertaken any investigation at the time of the inheritance.[35]

A question has been raised concerning whether the inheritance defense can be used effectively to defend a testamentary trustee. Since the testamentary trustee is not the person benefited by the will, it is not clear whether the defense should apply to the trustee as well as the beneficiary under the trust.[36]

§ 2.5 ASTM: Standard Practice Guides

Since CERCLA does not define the "all appropriate inquiry" necessary to prove the "due diligence" required to prevail as an "innocent purchaser," several industry associations have established guidelines to do so. The American Society for Testing and Materials (ASTM) was among the first to develop guidelines. The Environmental Protection Agency (EPA) has neither sanctioned these guidelines[37] nor issued its own. Accordingly, ties following industry guidelines are still subject to case by case interpretation by courts in hind site as to whether their methods constitute "all appropriate inquiry."[38]

The ASTM developed two standard practice guides for the purpose of developing a customary practice for environmental site assessments for parcels of commercial real estate with respect to the range of contaminants within the scope of CERCLA and petroleum products (collectively known

---

[35] 42 U.S.C. § 9601 (35)(A)(i) and (iii).

[36] Steven L. Leifer and Nancy E. Allin, *Trustee Executor Liability under Superfund* [20 Current Developments], ENV'T REP. at 1787 (Feb. 16, 1990).

[37] *See Superfund: Standard for Innocent Landowner Will Not Be Clarified, EPA Says in Memo,* DAILY ENV'T REP. (BNA) 7, Apr. 21, 1994.

[38] *See* Hemingway Transp., Inc. v. Juniper Dev. Group (In re Hemingway Transp., Inc.), 174 B.R. 148, 166 (Bankr. D. Mass. 1994) (legislative history of CERCLA indicates that Congress intended standard for all appropriate inquiry to be flexible and evolve); United States v. Pacific Hide & Fur Depot, Inc., 716 F. Supp. 1341, 134849 (D. Idaho 1989) (Congress's failure to define all appropriate inquiry indicates its desire for courts to decide scope of inquiry on case-by-case basis).

as "recognized environmental conditions").[39] The practice guides, "Standard Practicc for Environmental Site Assessments: Phase I Environmental Site Process (E 1527)" and "Standard Practice for Environmental Site Assessments: Transaction Screen Process (E 1528)" (the "Practices"), are intended to permit an entity to satisfy one of the requirements to qualify for the innocent landowner defense to CERCLA liability. The Practices are designed to constitute "all appropriate inquiry into the previous ownership and uses of the property consistent with good commercial or customary practice" as defined in 42 U.S.C. § 9601(35)(B).[40]

In defining a standard of good commercial and customary practice for conducting an environmental site assessment of a parcel of property, the goal of the processes established by the Practices is to identify recognized environmental conditions. ASTM has defined the objectives of the Practices as follows:

1. to synthesize and put in writing good commercial and customary practice for environmental site assessments for commercial real estate;
2. to facilitate high-quality, standardized environmental site assessments;
3. to ensure that the standard of appropriate inquiry is practical and reasonable; and
4. to clarify an industry standard for "appropriate inquiry" in an effort to guide legal interpretation of CERCLA's innocent landowner defense.

ASTM notes that the Practices have broad utility for a wide range of persons who are trying to develop information about the environmental condition of a property, beyond those who may claim the CERCLA defense. The Practices advise that not every property will warrant the same level of assessment. Consistent with good commercial or customary practice, the appropriate level of environmental site assessment will be

---

[39] *See* Janet S. Kole, *Superfund and Risk Management for Owners of Real Estate,* RISK MANAGEMENT, Nov. 1993, at 30, 40.

[40] The Practices warn that they do not address requirements of any state or local laws or of any federal laws other than the appropriate inquiry provisions of CERCLA's innocent landowner defense. Users are cautioned that federal, state, and local laws may impose environmental assessment obligations that are beyond the scope of the Practices.

dependent upon the type of property subject to assessment, the expertise and risk tolerance of the party seeking to perform "appropriate inquiry" with respect to the property (the "user") (including an agent or employee of the user), and the information developed in the course of the inquiry.

The user may commence inquiry to identify recognized environmental conditions in connection with a property by performing either the Transaction Screen Process or the more comprehensive Phase I Environmental Site Assessment (a "Phase I"). A Phase I also may be commenced based on the findings of a completed Transaction Screen Process.

§ 2.6 The ASTM's Transaction Screen Process

The Transaction Screen Process consists of (1) asking questions contained within the transaction screen questionnaire of owners and occupants of the property, (2) observing the site conditions of the property with direction provided by the transaction screen questionnaire, and (3) to the extent reasonably ascertainable, conducting limited research regarding particular government records and certain standard historical sources. The Transaction Screen Process may be conducted either by the user or wholly or partially by an environmental professional.[41]

The transaction screen questionnaire is followed by a guide designed to assist the person completing the transaction screen questionnaire. The guide is divided into three sections: (1) Guide for Owner/Occupant Inquiry, (2) Guide to Site Visit, and (3) Guide to Government Records/Historical Sources Inquiry. The guide repeats each of the questions set out in the questionnaire to assist the user in preparing a report.

If any of the questions set forth in the transaction screen questionnaire are answered in the affirmative, the user must document the reason for the affirmative answer. If any of the questions are not answered or the answer is unknown, the user is instructed to document and evaluate each, in light of the other information obtained in the Transaction Screen Process, including, in particular, the site visit and the government records/historical

---

[41] ASTM defines an environmental professional as "a person possessing sufficient training and experience necessary to conduct a site reconnaissance, interviews, and other activities . . . and having the ability to develop conclusions regarding recognized environmental conditions in connection with the property in question."

sources inquiry. If the user decides no further inquiry is warranted after receiving no response, an answer of unknown, or an affirmative answer, the user must document the reasons for such conclusion.

If either the guide to the question or other information obtained during the Transaction Screen Process does not permit a user to conclude that no further inquiry is appropriate with respect to a particular question, then the user must determine whether further inquiry may be limited to those specific issues identified as of concern or whether to proceed with a full Phase I. This analysis requires exercise of the user's reasonable business judgment and is based upon the totality of unresolved affirmative answers or answers of unknown received during the Transaction Screen Process. Upon completing the transaction screen questionnaire, if the user concludes that a Phase I is needed, the user is instructed to proceed with such inquiry with the advice and guidance of an environmental professional.

§ 2.7 The ASTM's Phase I Environmental Site Assessment Process

A companion practice to the Transaction Screen Process is the Phase I. The four components of a Phase I are as follows: (1) records review; (2) site reconnaissance; (3) interviews with current owners and occupants of the property and interviews with local government officials and (4) a written report of the findings and conclusions. The Phase I Practice provides comprehensive guidance for undertaking each of these components.

While the Transaction Screen Process may be conducted either by the user (including an agent or employee of the user) or wholly or partially by an environmental professional, a Phase I must be conducted by an environmental professional. The interviews and site reconnaissance, the review and interpretation of information upon which the report is based, and overseeing the writing of the report are all portions of a Phase I that must be performed by an environmental professional(s). The user participates in the Phase I by checking title records for environmental liens and communicating any specialized knowledge or experience to the environmental professional.

A records review is conducted to obtain and review records that will help identify recognized environmental conditions in connection with the property. Such a review must encompass records relating to adjoining properties within an approximate minimum search distance of the property under assessment. In addition, a review of historical sources is necessary to identify those uses or occupancies that are likely to have led to recognized environmental conditions.

A site reconnaissance involves an environmental professional visually and physically inspecting the property and any structure(s) located on the property to the extent not obstructed by bodies of water, adjacent buildings, or other obstacles. In addition, the environmental professional must observe and identify current and past uses of both the property and adjoining properties and provide a general description of the site and structures on the property.

The environmental professional also must interview owners and a reasonable number of occupants of the property concerning the uses and conditions of the property. While the person conducting the interview has an obligation to ask questions, in many instances the persons to whom the questions are addressed will have no obligation to answer them. However, if the person to be interviewed is the user (the person on whose behalf the Phase I is being conducted) the user has an obligation to answer all questions posed by the person conducting the interview, in good faith, to the extent of his or her actual knowledge or to designate a key site manager to do so.

The environmental professional also must interview local government officials. The content of such questions is left to the discretion of the environmental professional, provided that the questions should be directed toward identifying recognized environmental conditions in connection with the property.

The information from site reconnaissance and the interviews can provide data regarding hazardous substances in connection with identified uses, hazardous substance containers and unidentified substance containers, storage tanks, indications of PCBs, and evidence of solid waste disposal. The Phase I Practice requires that a report of findings be completed in accordance with the report format set forth in the appendix of the Phase I Practice. Such a report must have documentation to support the analysis, opinions, and conclusions in the report. In addition, the report must include an opinion of the impact of recognized environmental conditions in connection with the property.

§ 2.8 Other Elements of a Phase I Audit

An audit designed to satisfy the innocent purchaser defense is the minimum audit that a prospective purchaser will want to conduct in order to determine the potential environmental problems on a property or business being purchased. An audit designed to (1) allow a prospective seller to establish an environmental baseline prior to sale or (2) ensure that the seller's representations and warranties to the buyer are in fact truthful will probably go well beyond the scope of the innocent purchaser defense. The scope of these additional audit procedures will vary, depending upon the present and past usages of the property and the practices of the business or property that is the subject of the transaction. Thus, additional state and federal environmental statutes and common law should be considered when designing an environmental audit. Although these laws would not affect the availability of the "innocent purchaser defense," depending on the circumstances, one or more of these statutes may be important in establishing baseline or determining market value.

The following chart describes the environmental attributes that should be considered when designing an audit.

Environmental Attributes[42]	
Air	Diffusion factor Particulates Sulfur oxides Hydrocarbons Nitrogen oxide Carbon monoxide Photochemical oxidants Hazardous toxicants Odors

[42] *See* R. K. JAIN ET AL., ENVIRONMENTAL ASSESSMENT, at 85 (McGraw-Hill 1993).

Environmental Attributes	
Water	Aquifer safe yield Flow variations Oil Radioactivity Suspended solids Thermal pollution Acid and alkali Biochemical oxygen demand (BOD) Dissolved oxygen (DO) Dissolved solids Nutrients Toxic compounds Aquatic life Fecal coliforms
Land	Soil stability Natural hazard Land-use patterns
Ecology	Large animals Predatory birds Small game Fish, shellfish, and waterfowl Field crops Threatened species Natural land vegetation Aquatic plants
Sound	Physical effects Psychological effects Communication effects Performance effects Social behavior effects
Sociology	Lifestyles Psychological needs Physiological systems Community needs
Economics	Regional economic stability Public sector review Per capita consumption
Resources	Fuel resources Nonfuel resources Aesthetics

These can also be referred to as "environmental parameters" or "environmental elements." The importance of each factor will vary from case to case.

Although Superfund is perhaps the most publicized of all environmental laws threatening persons with liability, many other environmental laws—statutory and common law—could also affect the price and structure of a deal and should be considered in designing a Phase I environmental audit. A very brief description of the more salient environmental laws and the "red flags" to look for when conducting an audit are listed below. Red flags to consider include the following:

1. the site was once used for commercial or industrial activity;
2. oil and gas wells were once drilled on the site;
3. there is an unexplained vegetative stress on the property;
4. there are USTs on the site;
5. there are landfills or land disposal sites on the property or premises;
6. the site was once used for waste treatment or storage;
7. there is evidence of site clearing in the past;
8. there are chemical odors on the site;
9. there are visible stains or evidence of seepage on the site.

§ 2.9 Resource Conservation and Recovery Act (RCRA)[43]

Like Superfund, RCRA imposes obligations on any person who manages solid and hazardous wastes (i.e., generators, transporters, and owners or operators of a RCRA facility). For example, every person who generates or transports hazardous wastes or who owns or operates a RCRA treatment, storage, or disposal facility (TSD facility) must notify EPA (and the state agency if the state has implemented a state program) of its hazardous waste activities, obtain an EPA identification number,[44] and maintain records of the waste generated.[45] Regulated persons must also

[43] 42 U.S.C. §§ 6901-6992k.
[44] RCRA § 3010, 42 U.S.C. § 6930.
[45] 40 C.F.R. § 262.40.

abide by labeling and storage requirements and report each off-site shipment of hazardous waste to the government via a nationwide manifest system involving a series of documents in which they must specify the generator, the transporter, and the licensed TSD facility to which the waste is to be sent.[46] RCRA also mandates reporting hazardous waste spills.[47]

Permits are required for all TSD facilities.[48] A generator who accumulates wastes listed under RCRA or defined as hazardous in excess of 90 days will be deemed an operator of a facility for purposes of requiring a TSD permit.[49] EPA also has authority under RCRA to require cleanup ("corrective action") at TSD facilities, and also may bring suit against any person who has contributed, or is contributing, to the past or present handling, storage, treatment, transportation, or disposal of solid or hazardous waste that may present an "imminent and substantial endangerment to health or the environment."[50]

---

[46] *Id.* §§ 262.20-262.23. Under RCRA, generators are permitted to use only transporters and TSD facilities that are regulated under RCRA and have obtained an EPA identification number. *Id.* § 262.12(c).

[47] *See id.* §§ 264.196, 265.196.

[48] *See* RCRA § 3005, 42 U.S.C. § 6925; 40 C.F.R. Parts 264 and 265.

[49] 40 C.F.R. § 262.34.

[50] RCRA § 7003, 42 U.S.C. § 6973.The "imminent hazard" provision of RCRA authorizes EPA to bring suit for injunctive relief against any person who has contributed, or is contributing, to the past or present handling, storage, treatment, transportation, or disposal of solid or hazardous waste that may present an "imminent and substantial endangerment to health or the environment." Such persons may either be restrained from the activities causing the imminent and substantial endangerment or be ordered to take other action that may be necessary to deal with the problems, including, among other things, reimbursement of funds expended by EPA in connection with the site cleanup and restoration of groundwater and other contaminated resources. This can result in millions of dollars of liability, since the cost of cleaning up a hazardous waste site is considerable. Liability of subsequent landowners is limited, however, to situations where they "could not reasonably be expected to have actual knowledge of the presence of hazardous waste at such facility or site and of its potential for release." RCRA § 3013(b), 42 U.S.C. § 6934(b). *See also* United States v. Price, 523 F. Supp. 1055, 1069-74 (D.N.J. 1981), *aff'd.*, 688 F.2d 204 (3d Cir. 1982) (holding both the past and present owners of a dormant landfill liable under RCRA; sellers held liable due to dumping on the site; buyers held liable because they "contribut[ed] to the disposal (*i.e.*, leaking) of wastes merely by virtue of their studied indifference to the hazardous condition" that existed on the site).

Citizens may also bring suit against persons who violate any RCRA permit, regulation, or provision, or whose past or present handling, storage, treatment,

EPA uses the term "facility" loosely and has included real property and improvements owned or operated by other parties. Certain commercial uses (*e.g.*, dry cleaners, printers, copy centers, opticians); industrial uses; and sites with lead paint or any chemicals, solvents, metals, plastics, or other man-made materials stored on the premises are likely to be subject to RCRA requirements. Landfills on the property are also likely to be subject to RCRA, including the requirement to obtain a TSD permit and undertake corrective actions for all past releases of hazardous wastes.[51]

The following are a list of items that should be included in the Phase I environmental audit if RCRA regulations are likely to be of concern:

✓ **Investigate** whether a landfill is on site; and if so, whether it is lined, closed properly and contains appropriate leachate control systems.

✓ **Investigate** whether the facility stores, transports, or handles hazardous substances or hazardous wastes; and if so, whether they are being transported, stored, used, handled, generated, and disposed of in compliance with applicable legal requirements.

✓ **Investigate** whether the facility is (1) a conditionally exempt small quantity generator (*i.e.*, does the facility generate less than 100 kg of hazardous waste per month that is stored for less than 90 days), (2) generates 100 to 1,000 kg of hazardous waste per month (*i.e.*, a small

---

transportation, or disposal of hazardous waste may present an "imminent and substantial endangerment to health or the environment." RCRA § 7002 (B), 42 U.S.C. § 6972 (B). Citizens are required, however, to provide notice (60 days for suits alleging violations and 90 days for suits alleging imminent and substantial endangerment) to EPA, the state involved, and the offender before filing suit. RCRA § 7002 (b)(1)(A)-(2)(A), 42 U.S.C. § 6972 (b)(1)(A)-(2)(A). Citizen suits may also be brought against EPA for failure to perform its duties.

[51] The 1984 amendments expanded EPA's authority to require cleanup of environmental problems caused by hazardous waste facilities. Corrective action is required for *releases* of hazardous waste or constituents, including those that have migrated "beyond the facility boundary," from any *solid waste management unit* at a TSD *facility* that is seeking or otherwise subject to a RCRA permit. The amendments also authorize EPA to require corrective action at *interim status facilities* whenever there is, or has been, a release of hazardous waste or constituents. RCRA permits issued after enactment of the HSWA amendments must require an operator to take corrective action for all releases of hazardous constituents at the TSD facility regardless of when the waste was placed in a unit or whether the unit is currently active.

quantity generator), or (3) requires RCRA generator permits (*i.e.*, does it generate more than 1,000 kg of hazardous waste per month?).[52]

✓ **Investigate** whether the facility has an EPA Identification number; if so, what is it?

✓ **Investigate** the "cradle-to-grave" history of each hazardous substance and hazardous waste associated with the facility; including how the facility handles spent solvents and waste oils, *e.g.*, Safety Kleen.

✓ **Investigate** whether there is any reason to know that the facility's solid waste management and disposal practices have caused a release of hazardous Substances or hazardous Wastes into the environment.

✓ **Check** for other factors that may indicate the existence of a RCRA problem, including

- ✓ discolored soil;
- ✓ dead or stressed vegetation;
- ✓ spills around loading docks, fuel areas or surface impoundments;
- ✓ stained or discolored sinks, drains or sumps;
- ✓ discolored or turbid surface water on property;
- ✓ noxious odors and/or leaks from pipes, electrical transformers, barrels or containers; and
- ✓ inadequate "housekeeping" in and around buildings.

§ 2.10 Underground Storage Tanks (USTs) and Aboveground Storage Tanks (ASTs) Regulations

In 1984, Congress amended RCRA to establish new requirements for dealing with environmental problems associated with USTs containing petroleum or hazardous substances.[53] In 1986, the passage of the Superfund Amendment Reauthorization Act[54] established an Underground Storage Tank Trust Fund that provided for a cleanup liability scheme similar to that imposed under CERCLA.

---

[52] 40 C.F.R. § 262.34(b)-262.34(d).

[53] 42 U.S.C. §§ 6991-6991i.

[54] Pub. L. No. 99-499, 100 Stat. 1613 (1986).

A UST is a tank, including underground piping connected to the tank, that has at least 10 percent of its volume underground.[55] Several types of tanks are excluded by law or regulation from UST requirements. These include tanks used to store heating oil consumed on the premises where stored; emergency spill and overflow tanks; and tanks holding 110 gallons or less.[56]

The following items should be included in the audit if USTs or ASTs are known (or likely) to be on the property—especially if there are USTs that are over 10 years old or that do not have secondary containment, leak detection, spill/overflow prevention mechanisms, and cathodic protection.

✓ **Investigate** the size, location, number, age, construction materials, contents, and integrity of all USTs, ASTs, and associated piping.

✓ **Investigate** whether there are any known or reported spills, leaks, or other releases from any USTs or ASTs. If so, determine what response measures have been taken.

✓ **Evaluate** compliance with legal requirements including, but not limited to, preparation and filing of all applicable registrations.

✓ **Evaluate** spill/overflow protection, cathodic protection, secondary containment, inventory controls, provision of financial assurances, filling practices, and integrity testing.

✓ **Determine** whether any mandatory tank upgrades are necessary.

✓ **Determine** whether the ASTs require a spill prevention, control, and countermeasures (SPCC) plan.

§ 2.11 Clean Water Act[57]

The Clean Water Act (CWA or the Act), as amended,[58] establishes a comprehensive scheme for regulating all sources of "pollution" and discharges of "pollutants" to "navigable waters."[59] Any addition of pollution to navigable waters from "any discernible, confined, and discrete conveyance, including but not limited to any pipe, ditch, channel, tunnel,

---

55 42 U.S.C. § 6991(1).

56 *Id.; See also* 40 C.F.R. Part 280.10.

57 33 U.S.C. §§ 1251-1387.

58 *Id.*

59 *Id.* § 1362 (12).

conduit, well, discrete fissure, or container"[60] (*i.e.*, a "point source"), is prohibited, except as provided in the conditions of a National Pollutant Discharge Elimination System permit issued by EPA or an authorized state agency. The CWA also regulates storm water point source discharges, the discharge of dredged or fill material, and oil spills, among other things.

§ 2.12 Wetlands (Section 404)[61]

Section 404 of the CWA gives the Army Corps of Engineers (Corps) jurisdiction to issue permits for the discharge of any "dredged" or "fill material" into "waters of the United States,"[62] including wetlands. In addition, section 10 of the Rivers and Harbors Act of 1899 requires a Corps permit for any work in waters that are "navigable" in fact.[63]

Under section 404 of the federal Clean Water Act, before "dredged" or "fill" material may be discharged in any wetland, an individual permit must be obtained from the United States Army Corps.[64] "Wetlands" are defined as those areas that are "inundated or saturated" with water frequently enough so that they support "vegetation . . . adapted for life in saturated soil conditions."[65] Wetlands usually are associated with swamps and marshes, but they often are located in areas where there is little visible surface water. Wetlands permits come in different varieties: nationwide, areawide, general, or individual.

States may assume the section 404 permit program, but EPA retains extensive oversight rights. To date, only Michigan has adopted the section 404 program.

The U.S. EPA has broad authority to "veto" any individual permits issued by the Corps where the proposed projects will entail "unacceptable adverse effects."[66] In general, the Corps will issue no permits if there is a

[60] *Id.* § 1362 (14).
[61] 33 U.S.C. § 1344.
[62] *Id.* § 1344(a).
[63] 33 U.S.C. § 403.
[64] 33 U.S.C. § 1344(a); 33 C.F.R. § 323.3.
[65] *See* 33 C.F.R. § 328.3(b).
[66] 33 U.S.C. 1344(c).

"practicable alternative" to the proposed project that would have a less adverse impact on the aquatic ecosystem. For projects that do not require access to the water, there is a presumption that "practicable alternatives" to the proposed wetlands site exist, and that these alternatives will entail a less adverse impact. The penalties for ignoring or violating wetlands regulations can be harsh, including large civil penalties and restoration of damaged wetlands to their original appearance and condition. Consequently, it is important that any investigation of property to be acquired in the United States that could conceivably involve wetlands include an assessment by a qualified expert of whether the property qualifies as wetlands. If so, the permitting process should be initiated well in advance (6 to 10 months) of any closing.

Since a person cannot always spot a wetland by visual inspection, environmental consultants are often needed to determine if an area may be considered a wetland. Section 404 may affect fills for parking lots, buildings, roads, pipelines, bridges, etc. To shed light on whether the property (especially farms, raw land, streams, swamps, marshes, and stormwater management ponds) contains wetlands:

✓ **Review national inventories** of wetlands (incomplete for some areas), satellite data, and other appropriate information.

✓ **Determine** whether wetlands vegetation (*e.g.*, cattails and willows), hydric soils, periodic inundations of water, or other indicators of wetlands are found on the property.

✓ **Determine** whether there has been or the borrower is planning to engage in any "dredging or filling" activity on the property. If so, determine whether the owner/operator of the property has filed (or may file) a Section 404 permit to engage in such activities.

§ 2.13 Stormwater and Chiller Water Discharges

EPA has struggled for many years to develop a regulatory program for point source discharges of stormwater runoff. Each attempt to issue NPDES regulations for stormwater discharges has met with legal challenges from industry and environmental groups. In 1987, Congress amended the CWA to place a moratorium until October 1992 on NPDES permit requirements for stormwater, except for five classes of stormwater point sources that EPA is specifically directed to regulate.[67] Included among these regulated discharges are discharges of stormwater "associated with industrial activity"[68] and discharges from large and medium-sized "municipal separate storm sewer systems."[69]

Some facilities discharge into a municipal or private sewer system rather than directly into a waterway. Users of publicly owned treatment works (POTWs) are not subject to NPDES permitting requirements, but may be subject to pretreatment standards under section 307 of the Clean Water Act.[70] To determine whether the business or property will be affected by EPA's NPDES regulations, the following should be covered in the environmental audit:

✓ **Examine** the stormwater and chiller waters flows and how they are discharged.

✓ **Determine** whether a permit is necessary.

§ 2.14 Effluent (a/k/a/ Waste Stream) Discharges to Waters, Sewers, and Treatment Works

The CWA provides for development and application of several kinds of effluent limitations and standards. Most of these are incorporated in a source's NPDES permit as numerical restrictions on the concentration or mass of pollutants discharged.

Discharges from new or significantly modified sources directly into navigable waters must comply with "new source performance standards" (NSPS). These standards reflect the greatest degree of effluent reduction achievable through application of the "best available demonstrated control technology."[71]

---

[67] *See* 33 U.S.C. § 1342(p).

[68] *Id.* § 1342(p)(2)(B).

[69] *Id.* § 1342(p)(2)(C)-(D).

[70] For a general discussion of policy under the Clean Water Act, *see* Robert W. Adler et al., The Clean Water Act 20 Years Later (Island Press 1993).

[71] A "new source" is one that commenced construction after publication of proposed NSPS that will apply to it, if the standards are in fact thereafter adopted in accordance with the Act, 33 U.S.C. § 1316(a)(2).

The CWA also contains a separate water quality standards program. The states may designate appropriate uses for waters within their borders, and adopt criteria (in effect, ambient pollutant levels) that will allow such uses. These use designations and criteria constitute "water-quality standards" that are subject to EPA review. If a state fails to adopt these standards, EPA may adopt them for the state.[72]

Sources discharging wastes to municipal sewers for treatment by POTWs must comply with general and categorical "pretreatment standards," published by EPA at 40 C.F.R. Parts 400-471. Under EPA's regulations, POTWs are required to develop "pretreatment programs" designed to ensure industrial users' compliance with pretreatment standards, and the terms of such programs are incorporated in the POTWs' NPDES permits.[73]

To identify potential effluent-based problems, the auditor should:

✓ **Examine** where the sewers and drains flow.

✓ **Determine** whether the facility has all applicable permits, including, but not limited to, NPDES permits.

✓ **Determine** whether effluent discharges are in compliance with all applicable laws, rules, and regulations.

✓ **Industries discharging into a POTW should ascertain** whether the municipality operating the POTW has adopted an industrial waste ordinance limiting the quality of the wastewater they may send to the POTW.

§ 2.15 Oil Pollution Prevention (OPA)[74]

The OPA was signed into law on August 18, 1990, in response to public awareness and concern regarding oil spills in the wake of the *Exxon Valdez* grounding. The OPA contains significant modifications to many of the provisions of section 311 of the CWA. Included among these provisions

---

[72] 33 U.S.C. § 1313(c).

[73] *See* 40 C.F.R. Part 403.

[74] 33 U.S.C. §§ 2701-2761; 40 C.F.R. § 112. The SPCC regulations under the CWA were revised on October 22, 1991, to include the regulations implementing the Oil Pollution Act of 1990. *See generally* RONALD B. MITCHELL, INTERNATIONAL OIL POLLUTION AT SEA: ENVIRONMENTAL POLICY AND TREATY COMPLIANCE (MIT 1994).

are limited liability provisions for owners and operators of vessels and facilities that spill oil into water of the United States. These provisions allow the imposition of penalties of up to $350 million for onshore facilities and deepwater ports; $75 million plus removal cost for offshore facilities; and $1,200 per gross ton or up to $10 million for tank vessels, whichever amount is greater. The following matters should be considered when evaluating potential OPA problems:

✓ **Determine whether the facility has**

1. a potential to discharge oil into the waters of the united States;
2. underground storage capacity of more than 42,000 gallons of petroleum products;
3. above-ground storage capacity of more than 1,320 gallons of petroleum products;
4. a single above-ground storage capacity of more than 660 gallons.

If so, determine whether the facility has an SPCC plan in place. If not, determine whether the facility fulfills the requirements of an exemption under the statute and its implementing regulations.

✓ **Examine the SPCC plan:**

1. Is a copy of the SPCC plan readily available at the facility or nearest field office?
2. Is the SPCC plan signed, dated, and stamped by a Registered Professional Engineer?
3. Is the SPCC plan current (does it meet the three-year requirement; are the facility contacts current)?
4. If the facility contact list or SPCC plan was amended, determine whether a Registered Engineer reviewed and approved the update.
5. Determine whether the facility has environmental incident and notification procedures established.
6. Determine whether storage areas with potential oil spill potentials are properly identified in the SPCC plan (are the spill pathways themselves identified, along with the volumes?).
7. Determine whether operators are properly briefed on spill prevention and the facilities SPCC plan.

8. Determine whether the facility generally conforms to the requirements and commitment described in the SPCC plan.

✓ **Concerning spill reporting, consider the following:**

1. Determine whether a reportable release occurred at the facility. If so, determine when it occurred, whether it was properly reported and whether audit procedures were changed in response to the incident.
2. Determine whether any release from the facility exceeded 1,000 gallons or whether the facility has had two reportable releases within any 12-month period. If so, determine whether the required written report was submitted to the Regional Administrator within 60 days and determine whether the report contains the required information.
3. Determine whether the facility maintains a current spill history.
4. Examine whether the loading and unloading areas have sufficient spill containment to contain or minimize oil spills.
5. Determine whether the loading and unloading procedures are available to personnel and whether they are followed.
6. Examine whether storm drains that service a potential spill area flow into catch basins or treatment facilities that would prevent release of oil to waters of the United States.
7. Examine whether the facility has adequate security measures (*e.g.*, fences, gates, lock valves, lights, guards, etc).
8. Determine whether the facility has a site drawing that shows the size, volume, and locations of tanks.
9. Determine whether contained rainwater around bermed tank areas is inspected prior to discharge and whether records of such discharges are kept.
10. Determine whether records are properly maintained concerning tank and pipe inspection and testing.
11. Examine loading and unloading areas to determine whether they are kept clean or not. If there is indication of recent or chronic spills or leaks around the loading and unloading zones or around loading equipment, determine the source and procedures used to handle these problems.

12. Determine whether oil traps, trap sumps, containment areas are monitored and maintained. If so, do they appear oil free? If not, can the oil source be identified?

§ 2.16 Safe Drinking Water Act[75]

States may assume responsibility for the safe drinking water program and many have elected to do so. The Safe Drinking Water Act (SDWA)[76] provides substantive and procedural requirements for "public water supply systems," which are defined as "a system for the provision of piped water for human consumption, if such system has at least fifteen service connections or regularly serves at least twenty-five individuals."[77] Regulations promulgated under the SDWA authorize EPA to set standards for maximum contaminant levels for specific chemicals as well as turbidity, coliform bacteria, and certain types of radiation. The following should be considered as an initial assessment of the water quality on the property:

✓ Check to see if any part of the water distribution system contains lead.

✓ Check to see if the property is located in an area where radon[78] is a concern (or better yet check to be sure there is no radon problem on the target property).

✓ Determine whether each fixture on the property has a backflow prevention device.

✓ Consider the overall taste, color, odor, or other significant characteristics of the water on the premises (especially if it has changed over time).[79]

---

[75] 42 U.S.C. §§ 300f to 300j-26.

[76] *Id.*

[77] *Id.* § 300f(4).

[78] *See generally* Note, *Clearing the Air on Radon Testing: The Duty of Real Estate Brokers to Protect Prospective Homebuyers,* 15 FORDHAM URB. L.J. 767 (1987). *See also* LEONARD A. COLE, ELEMENT OF RISK: THE POLITICS OF RADON (Oxford 1993); G. B. LESLIE AND F. W. LUNAU, INDOOR AIR POLLUTION: PROBLEMS AND PRIORITIES, at 99–116 (Cambridge 1992).

[79] For a checklist and discussion of these environmental issues in the context of the home, see INGRID RITCHIE AND STEPHEN J. MARTIN, THE HEALTHY HOME KIT (1995).

The definition of "public water supply system" includes a broad range of facilities and systems that are not traditionally considered to be "water systems," such as schools, hotels, restaurants, gas stations, factories, private housing developments, and other facilities that have their own wells or surface water supplies and have the requisite number of service connections or persons. Thus, careful consideration must be given to the use of a site when evaluating whether this law will be applicable.

If drinking water is obtained from a groundwater well, cistern, or privately owned treatment works, then the SDWA could be invoked if there is a problem with the groundwater on the property.[80] Because problems with drinking water could seriously impair the value of the property, an audit should include the following if the property contains a well, cistern, or private treatment works:

✓ Evaluate the source and quality of drinking water.

✓ Test the drinking water for lead or other contaminants.

✓ Determine whether the building was built before 1930.

✓ Check the pipes in the building to see if they are made of a soft, dull gray metal that can be easily scratched with a key.

✓ Check to see if there are, or there was a history of, frequent leaks, rust-colored water, stained dishes, or stained laundry.

✓ Check to see if the plumbing is made of a nonplastic and is less than five years old.

✓ Evaluate the softness of the water.

✓ Check to see whether the electrical system is grounded to the water pipes.

✓ Check to see if the building has a submersible well pump.

✓ Check to see if the building is located near a busy road or in the area of a lead smelter, lead battery plant, or other industry that releases lead into the environment.

✓ Check to see if the interior or exterior surfaces of a pre-1980 building have recently been dry-sanded, heat-treated or otherwise treated to remove paint without lead testing.[81]

---

[80] *See, e.g.*, O'Neil v. Q.L.C.R.I., Inc., 750 F. Supp. 551 (D.R.I. 1990).

[81] For a checklist and discussion of these environmental issues in the context of the home, see INGRID RITCHIE AND STEPHEN J. MARTIN, THE HEALTHY HOME KIT (1995).

Local water suppliers are required to test water for copper and lead under recent regulations. If excessive levels are detected, the water supplies must be treated to reduce the likelihood that metals from pipes will leach into and contaminate drinking water.

Some buildings, however, have been discovered to contain water pipes that, through corrosion or otherwise, cause violations of the "end of pipe" drinking water quality standards. In other words, a building may supply drinking water exceeding lead standards even though the drinking water supplied to the building meets the applicable standards.

✓ Conduct end-of-the-pipe checks to be sure drinking water meets the guidelines set by the SDWA.

✓ Check to be sure any wells are not located close to any septic systems, underground storage tanks, or other sources of water contamination (including pesticides, fertilizers, and livestock).

✓ Check the hardness of the well water.

✓ Determine whether the property development project will require installation of a well.

✓ Check to see that the water test results *exceed* the EPA standards and guidelines.[82]

There are special requirements under the SDWA for control of development and land use in areas that recharge aquifers that are the principal drinking water supplies (also known as "sole source aquifers"):

✓ Check to see whether the drinking water comes from a sole source aquifer.

The SDWA establishes a regulatory scheme for underground injection control. The act establishes a permitting scheme based on the type of materials being injected into the ground. The program may be administered by the state.

✓ Check to see if there are or have been in the past any underground injection wells on or near the property.

✓ If the property contains a septic system, check how old it is (is it more than ten years old?) and whether there is a "glugging" sound when the fixtures are used.

---

[82] For a checklist and discussion of these environmental issues in the context of the home, see INGRID RITCHIE AND STEPHEN J. MARTIN, THE HEALTHY HOME KIT (1995).

✓ Determine whether there are any fixtures with S-shaped traps or no traps?
✓ Determine whether there are any fixtures without air gaps.
✓ Determine whether there has been a problem on the property with a cross connection. Also determine whether the cross connection exists.
✓ Determine whether flooding is a problem on the lot.
✓ Evaluate whether the lot has proper surface drainage.
✓ Consider whether the ground slopes away from buildings on the lot.
✓ Evaluate whether the drainfield is outlined by green grass, the lot has an odor typical of waste water, or the ground is spongy even when there has been no recent rain.
✓ Determine when the septic system was last pumped (has it been more than five years?).[83]

§ 2.17 Clean Air Act (CAA)[84]

Although both asbestos in its natural form and asbestos-containing materials (ACM) are exempt from regulation under CERCLA, asbestos is subject to EPA's Clean Air Act and Occupational Safety and Health Act (OSHA)[85] regulations.

EPA promulgated a National Emission Standard for Hazardous Air Pollutants (NESHAP) for Asbestos[86] pursuant to Clean Air Act §§ 112 and 301(a).[87] The NESHAP applies to manufacturing, demolition, renovation, asbestos mills, roadways, spraying, fabricating, insulating materials, and waste disposal. For demolition and renovation operations, the NESHAP

---

[83] For a checklist and discussion of these environmental issues in the context of the home, see INGRID RITCHIE AND STEPHEN J. MARTIN, THE HEALTHY HOME KIT (1995).

[84] 42 U.S.C. §§ 7401-7671q.

[85] 29 U.S.C. § 651-78. According to EPA "asbestos-containing materials" are any materials that have greater than 1 percent asbestos by volume. *See* KATHLEEN HESS, ENVIRONMENTAL SITE ASSESSMENT: PHASE I : A BASIC GUIDE, at 173 (Lewis 1993).

[86] 40 C.F.R. §§ 61.140-156.

[87] 42 U.S.C.A. §§ 7412, 7601(a).

imposes notification requirements, prescribes work procedures for asbestos emission control, and regulates disposal of ACM. The 1990 Amendments to the Clean Air Act completely revised the system for controlling hazardous air pollutants.[88]

In addition, some states require inspection for and abatement of asbestos in buildings used for particular purposes (in addition to the federal and state programs regarding schools).[89] The requirement for asbestos abatement may be triggered by the renovation or demolition of buildings. Many states require inspection for asbestos and, if necessary, the formulation of an abatement plan as a condition to obtaining a building permit to renovate or demolish a building built before a certain date (usually around 1978).[90]

If the party seeking the audit is intending to renovate or demolish a structure on the property used as collateral, substantial costs may be incurred due to regulations governing asbestos removal. Thus, the following items should be included in an audit of a building:

✓ Determine whether the property contains natural asbestos or ACM.

✓ Determine whether the asbestos is friable—*i.e.*, can it be crumbled under hand pressure?

✓ Determine whether the concentration of airborne fibers exceeds .01 fiber/cc.

✓ Determine whether the owner/operator of the property anticipates renovation, alteration, repair, modification, or demolition activity that may disturb more than 260 linear feet or 160 square feet of ACM.

---

[88] P.L. 101-549, Title III, § 301, 104 Stat. 2531 (1990).

[89] In New York, for example, the prospectus for a proposed conversion of rental property to a condominium or cooperative must disclose the presence of any ACM in the building and recommend a protocol for asbestos abatement. The New York City Administrative Code also requires some form of asbestos abatement as a condition precedent to obtaining a building permit in certain situations. New York City is currently attempting to pass regulations imposing inspection and abatement requirements on virtually all building owners.

[90] Many states also operate licensing programs requiring training and licensing for members of asbestos-related occupations such as asbestos contractors or workers, asbestos inspectors, and asbestos management planners. Some states are beginning to require a permit for each specific asbestos abatement project.

✓ Determine the date the building was built. Was it before 1973? Between 1973 and 1985?
✓ Check the insulation on the heating ducts and/or hot water pipes in the building.
✓ Check to see if the building contains any other blown-in insulation, vinyl floor tiles, or linoleum sheeting.
✓ Check to see if the building has asbestos siding, shingles, or roofing that is deteriorating or will need to be replaced due to renovation or demolition.
✓ Check to see whether the building has plaster that is sprayed on or troweled on walls or ceilings.
✓ Check to see if the building has caulking, putty, cement, mastic, joint compound, or similar material.[91]

The following table[92] summarizes the classification of asbestos and ACM:

Types of ACM	Description
Friable ACM	Crumbled, pulverized, or reduced to powder by hand pressure and > 1% asbestos.
Nonfriable ACM	Cannot be crumbled, pulverized, or reduced to powder by hand pressure when dry and > 1% asbestos.
Category I	Asbestos-containing packings, gaskets, resilient floor coverings, and asphalt roofing products
Category II	Any nonfriable ACM not designated as Category I

[91] For a checklist and discussion of these environmental issues in the context of the home, see INGRID RITCHIE AND STEPHEN J. MARTIN, THE HEALTHY HOME KIT (1995).

[92] *See generally* US EPA: Asbestos/NESHAP Regulated Asbestos Containing Materials Guidance. Office of Toxic Substances, Washington, D.C. EPA 340/1-90-018, at 12; KATHLEEEN HESS, ENVIRONMENTAL SITE ASSESSMENT: PHASE I : A BASIC GUIDE, at 173 (Lewis 1993).

Companies with greater than 100 employees that are located in nonattainment areas must comply with the Employer Trip Reduction (ETR) Program of the Clean Air Act Amendments of 1990. The goal of the ETR program is to reduce the number of cars traveling to work sites by 25% by 1996. This would be done by encouraging people to travel to work in car pools or by using public transportation. Company plans for implementing the ETR program are already past due, with many employers not in compliance.

✓ Check to see how many employees are located at each facility.

✓ Check to see whether the facility is located in a nonattainment area.

✓ Check to see whether the facility has adopted a plan or has the means to encourage employee trip reduction programs.

The CAA also requires permits for certain types of emissions. Thus, if any emissions are seen emanating from the subject collateral, the following should be examined:

✓ Determine whether the facility generates any air emissions subject to EPA National Ambient Air Quality Standards, National Emission Standards for Hazardous Air Pollutants, acid deposition control measures, or other applicable emission restrictions or permit requirements; if so, determine whether the owner/operator of the facility complied with all applicable emission restrictions or permit requirements.

§ 2.18 Toxic Substances Control Act (TSCA) (*e.g.*, regulation of PCBs)[93]

Under the Toxic Substances Control Act,[94] EPA is authorized to prohibit, limit, or otherwise regulate the manufacture, processing, distribution, use, or disposal of a chemical that presents an unreasonable risk to health or the environment. All but small manufacturers who manufacture or process chemical substances must keep certain records and submit reports to EPA. Unless exempted under § 2604(h), all producers of a new chemical, or of an existing chemical for a significant new use, must notify EPA 90 days before producing a new chemical and may be required to submit test data before production begins. In addition, if anyone producing or distributing a chemical obtains information reasonably supporting the conclusion that the chemical presents a substantial risk of injury to health or the environment, they must notify the EPA Administrator immediately, unless they have actual knowledge that the Administrator has already been adequately informed. To examine the possible impact of TSCA liability, the auditor should:

✓ Examine records kept pursuant to TSCA.

✓ Request a list of all chemicals stored on site along with the use, disposal, and average volume kept on site.

✓ Evaluate whether any chemicals pose a potential threat to health.

If the property contains any transformers,[95] capacitors, hydraulic equipment, heat transfer systems, or other equipment containing dielectric fluids that may contain PCBs, then TSCA could become an issue. Accordingly, suspected equipment should be tested for PCBs.

---

93 15 U.S.C. §§ 2601-2692.

94 *Id.*

95 If a transformer is labeled "non-PCB," it may still contain up to 50 ppm of PCBs and therefore may present a potential for a release that could be costly to remediate.

§ 2.19 Occupational Safety and Health Act (OSHA)[96]

A variety of environmental issues are indirectly regulated under OSHA. To date, OSHA is the only federal statute addressing the issue of indoor air pollution, and it does so only if employees are injured. OSHA does not protect the public at large from indoor air pollution problems.

§ 2.20 Asbestos

OSHA has promulgated regulations and standards applicable to all occupational exposures to ACM.[97] In particular, OSHA has established permissible exposure levels (PELs), monitoring and recordkeeping procedures, medical surveillance protocols, and warning requirements.[98]

§ 2.21 OSHA Hazard Communication Program

Violations of the OSHA Hazard Communication Program account for approximately one-third of all OSHA violations, fines, and penalties. Few auditors will look for such violations unless asked to do so, even though they are relatively easy to detect. Accordingly, the following investigation might be added when appropriate:

- ✓ Determine whether the facility receives Material Safety Data Sheets (MSDSs) when it receives materials that are used on-site; if so, determine whether the facility has a formal written Hazard Communication Program that conforms to OSHA requirements.
- ✓ Determine whether the MSDSs are available and stored in a central location.
- ✓ Determine whether the facility has a formal training program to teach employees how to handle hazardous substances and hazardous wastes.

---

[96] 29 U.S.C. § 651, *et seq.*

[97] *See* 29 C.F.R. §§ 1910.1001, 1926.58 (general industry and construction industry requirements, respectively).

[98] Twenty-three states and territories have set up their own OSHA plans, which are either as rigorous as or more so than the federal standards.

✓ Determine whether incompatible materials are stored together—*e.g.*, acids and bases, or chlorine and ammonia.
✓ Determine whether hazardous materials are stored near floor drains.
✓ Determine whether the fire protection system is in compliance with all applicable requirements.
✓ Examine how halon is handled.
✓ Determine whether the facility has appropriate personal protective gear, eye washes, confined spaces *(e.g.*, vaults), and means of ingress and egress.

§ 2.22 Emergency Planning and Community Right-to-Know Act (SARA Title III)[99]

The Emergency Planning and Community Right-to-Know Act of 1986 (EPCRA) was enacted as Title III of the 1986 Superfund Amendments.[100] EPCRA imposes detailed chemical release and inventory reporting requirements on businesses that handle "toxic chemicals," "hazardous chemicals," and "extremely hazardous chemicals."

EPCRA establishes four principal reporting or notification requirements for owners or operators of facilities at which certain substances are produced, processed, used, or stored. Under specified circumstances, facility owners or operators must provide

1. notification to the state emergency response commission of the presence of a substance or substances at the facility in excess of EPA-established threshold planning quantities (TPQs) for extremely hazardous substances and emergency notification of releases of extremely hazardous chemicals and hazardous substances;
2. information from MSDSs for hazardous chemicals;
3. emergency and hazardous chemical inventory forms; and
4. toxic chemical release forms.

---

[99] 42 U.S.C.A. §§ 11001-11050.
[100] *Id.*

EPA's regulations implementing the EPCRA require the owner or operator of a facility at which extremely hazardous substances are present in amounts equal to or in excess of the TPQs to provide notice to the State Emergency Response Commission (SERC) by May 17, 1987, or within sixty days after the facility first becomes subject to the requirements of this section, whichever is later.[101] The facility owner or operator also must designate a facility representative who will participate in the local emergency planning process as a facility emergency response coordinator.

EPA's regulations implementing MSDS and chemical inventory forms apply to the owner or operator of any facility at which a hazardous chemical for which an MSDS is required for amounts equal to or greater than 10,000 lbs., or at which an extremely hazardous substance is present in amounts greater than 500 lbs. (or 55 gallons) or the TPQ, whichever is less.[102] The owner or operator must submit either (1) an MSDS for each chemical or (2) a list of hazardous chemicals with other appropriate information.[103] The MSDSs or chemical list must be submitted to the SERC, the LEPC, and the fire department with jurisdiction over the facility.

Owners or operators of facilities at which hazardous chemicals are present in amounts over 10,000 lbs. during the preceding calendar year, or at which extremely hazardous substances are present in amounts greater than or equal to 500 lbs. (or 55 gallons) or the TPQ during the previous calendar year, must have filed an emergency and hazardous chemical inventory form[104] by March 1, 1988 (or March 1 of the first year after the facility first becomes subject to the subpart). Owners and operators of facilities at which hazardous chemicals or extremely hazardous substances are present in amounts less than these levels, but greater than 0, must also file these reports at a later date.[105]

---

[101] 40 C.F.R. Part 355.

[102] 40 C.F.R. Part 370.

[103] *See* 40 C.F.R. § 370.20-.21. That information must have been submitted by October 17, 1987 (or three months after the facility first becomes subject to the rule).

[104] There are two types of inventory forms. Tier I reports require information on the maximum amount, daily average amount, number of days on site, and general location on site of all hazardous chemicals and extremely hazardous substances. Tier II reports are much more detailed.

[105] MSDSs or chemical lists for such chemicals or substances were due by October 17, 1989. Inventory forms were due by March 1, 1989.

By March 1990, or March 1 of the third year after the facility first becomes subject to the subpart as stated in the regulations, and annually thereafter, facilities must file inventory forms for all hazardous chemicals present at the facility during the preceding calendar year in amounts equal to or greater than 10,000 lbs., or for extremely hazardous substances present at the facility in an amount equal to or greater than 500 lbs. (or 55 gallons) or the TPQ, whichever is less.

Owners or operators of facilities at which hazardous chemicals or extremely hazardous substances are located also must be prepared to submit a Tier II form providing more detailed information to the SERC, LEPC, or the fire department, or any other person upon written request.[106]

SARA Title III imposes significant legal requirements on facilities that handle chemicals and other hazardous substances. There are four programs under SARA: (1) emergency planning notification, (2) emergency release notification, (3) community right-to-know reporting, (4) and toxic chemical release reporting.

✓ Check to see whether the facility contains any "extremely hazardous substances," as that term is defined under SARA, or has filed any reports with the state emergency response commission indicating the presence of such materials.

✓ Check to see whether the facility contains any "hazardous chemical," as that term has been defined under SARA, or has notified the state emergency response commission of the release of such materials into the environment.

✓ If the facility is subject to OSHA regulations, check to see whether the facility owner has submitted to the local emergency planning committee, the state emergency response commission, and the fire department with jurisdiction over the facility, a list of all substances present at the facility for which an MSDS is required.

---

[106] Persons who fail to comply with the emergency planning notification and release reporting requirements are subject to civil penalties of up to $25,000 for each violation or $25,000 for each day during which the violation continues. 42 U.S.C. § 11045(a). Subsequent violations are subject to civil penalties up to $75,000 for each day of violation. *Id.* § 11045(b)(2). Knowing or willful failure to provide the required notices is punishable by fines of up to $25,000 or imprisonment for not more than two years, or both. *Id.* § 11045(b)(4) Repeat offenses are subject to fines of up to $50,000 or imprisonment for not more than five years or both. *Id.*

✓ Check to see if the facility manufactures, imports, processes, or otherwise uses any chemical listed in SARA.

✓ Check to see if the facility has notified its employees of hazardous chemicals that may be present in the workplace and consequences of exposure.

✓ If the facility has ten or more "full-time" employees, check to see if a toxic chemical release form was filed.

Facilities that have made reports under SARA Title III may run into problems with community perception. The value of collateral could be greatly impaired if a property becomes unsalable due to community activism. Even worse, a facility that is obligated to report under SARA Title III but fails to do so could run into even greater public perception problems if the violations are later discovered. Accordingly, the following inquiry should be considered:

✓ Determine whether the facility is in compliance with the Emergency Planning and Community Right-to-Know provisions set forth in SARA Title III and other applicable laws and regulations.

✓ Determine whether the facility sends notices to the local emergency response planning commission, fire department, and other applicable authorities to inform them of the presence and location of hazardous substances.[107]

[107] If you are representing an owner or operator, you may want to consider reporting at a "zero" threshold to avoid having to ascertain local threshold quantities, to avoid having to calculate the volumes of the hazardous constituents and, most importantly, to reduce potential tort liabilities.

§ 2.23 Indoor Air Pollution

Americans spend up to 90 percent of each day indoors, and as a result there is great potential for exposure to indoor contaminants (whether in the workplace, schools, public buildings, residences, or transportation vehicles).[108] For this reason, Congress proposed to develop and coordinate, through the EPA and other appropriate federal departments and agencies, a comprehensive program of research and development concerning the seriousness and extent of indoor air contamination. Under the proposed legislation, the agencies would establish a process using the existing federal statutes to reduce the health effects of indoor air contaminants.

Indoor air pollution, or "Tight Building Syndrome," is usually a function of inadequate heating, ventilation, and air-conditioning (HVAC) systems. A building that is perceived as "sick" can be very hard to sell, even if it is thoroughly cleaned up. In our experience, few auditors will examine indoor air problems unless specifically asked. The following are items to include in an audit:

✓ Determine whether the building has an adequate number of air changes per hour (more than two, if possible).

✓ Determine whether there have been any complaints of dizziness, headaches, or other physical problems.

✓ Examine whether formaldehyde glue is used on the carpets or if the building has urea formaldehyde insulation.

✓ Examine whether the facility has a "smoking" policy in effect and how the designated smoking areas are ventilated.

---

[108] S. 455 § 2(1)-(2).

Indoor air pollution has become a significant issue in building devaluation and a rising cause of action in tort. Below is a table[109] of certain hazardous building materials that were widely used, the year they generally became commercially available, and the year their usage was banned.

Building Material	Year Introduced Commercially	Year Banned
Friable asbestos	1897	Partial ban: 1975-1978
Nonfriable asbestos	1890	Phase-out ban
Lead-containing paint	Early 1900s	Partial ban: 1978
Leaded gasoline	Early 1900s	1996
Lead-soldered water pipes	Early 1900s	1986
Polychlorinated biphenyls (PCBs)	1929	Manufacturing: 1979

§ 2.24 Formaldehyde

Formaldehyde glue is often used in construction of buildings. The fumes from this glue can make people ill. To check for indoor air problems due to formaldehyde:

✓ Determine whether the building was constructed within the last five years.

✓ Determine whether the building was renovated or remodeled within the last 6 to 12 months. If so, consider whether new carpeting, furniture, drapes, paint, vinyl flooring, or cabinets were installed.

✓ Determine whether the building has urea-formaldehyde foam insulation (UFFI) installation by examining, inter alia, whether there are plugged holes at uniform distances on the exterior of the building

[109] *See* Kathleen Hess, Environmental Site Assessment, at 171 (Lewis 1993).

or whether there is a spongy material behind the electrical outlets. If so, determine whether the UFFI insulation is ten or more years old.

✓ Determine whether mobile homes, trailers, or prefabricated building are presently on or will be placed on the property.[110]

§ 2.25 Volatile Chemicals in Construction Materials, Furnishings, and Consumer Products

There are many volatile chemicals[111] in construction materials, furnishings, and consumer products. The following are factors to consider as sources of indoor air pollution:[112]

✓ Determine whether the building was constructed within the last five years.

✓ Determine whether the building was renovated or remodeled within the last 6 to 12 months. If so, consider whether new carpeting, furniture, drapes, paint, vinyl flooring, or cabinets were installed.

✓ Determine whether pesticides, solvents, or cleaning materials were used or stored in the building. If so, examine the storage procedures.[113]

---

[110] *See generally* G. B. LESLIE AND F.W. LUNAU, INDOOR AIR POLLUTION: PROBLEMS AND PRIORITIES at 117-31 (Cambridge 1992). For a checklist and discussion of these environmental issues in the context of the home, see INGRID RITCHIE AND STEPHEN J. MARTIN, THE HEALTHY HOME KIT (1995). *See also* LYNN LAWSON, STAYING WELL IN A TOXIC WORLD (1993).

[111] *See generally* G. B. LESLIE AND F.W. LUNAU, INDOOR AIR POLLUTION: PROBLEMS AND PRIORITIES (Cambridge 1992). For a checklist and discussion of these environmental issues in the context of the home, *see* INGRID RITCHIE AND STEPHEN J. MARTIN, THE HEALTHY HOME KIT (1995). *See also* LYNN LAWSON, STAYING WELL IN A TOXIC WORLD (1993).

[112] *See generally* STEVE HAYES ET AL., INDOOR AIR QUALITY: SOLUTIONS AND STRATEGIES (McGraw-Hill 1995); ANTHONY L. HINES, ET AL., INDOOR AIR QUALITY: QUANTITY & CONTROL (Prentice-Hall 1993); JACK G. KAY, ET AL., INDOOR AIR POLLUTION: RADON, BIOAEROSOLS & VOCs (Lewis 1991).

[113] For a critical discussion of pesticide contamination see ROBERT VAN DEN BOSCH, PESTICIDE CONSPIRACY (1978).

§ 2.26 Moisture[114]

Moisture is a significant problem leading to environmental health hazards as well as damage to the building. In order to protect against indoor air pollution problems due to moisture:

✓ Examine the building for indications of the existence of a moisture problem using the following chart as an aid:

Yes	No	Indicator
		Lingering odors Dampness Humidity greater than 50% Mold and mildew Rotting and decaying wood Discolored, warped, or cracked wood Water-carrying fungus Condensation on windows Sweating pipes Corrosion of metal surfaces Water leaks

✓ Check to be sure the ground surrounding the foundation does not slope toward the foundation of the building and that no water collects around the foundation.

✓ Check to be sure no trees or shrubs crowd the foundation of the building.

✓ Examine whether the lot has low spots on it where water collects or is likely to collect.

✓ Examine the building for the following items, using the following chart as a guide:

---

[114] *See generally* G. B. LESLIE AND F. W. LUNAU, INDOOR AIR POLLUTION: PROBLEMS AND PRIORITIES (Cambridge 1992). For a checklist and discussion of these environmental issues in the context of the home, see INGRID RITCHIE AND STEPHEN J. MARTIN, THE HEALTHY HOME KIT (1995). *See also* LYNN LAWSON, STAYING WELL IN A TOXIC WORLD (1993).

Area of Building	**Concerns**
Indications of moisture	Lingering odors Dampness Humidity greater than 50% Mold and mildew Rotting and decaying wood Discolored, warped, or cracked wood Water-carrying fungus Condensation on windows Sweating pipes Corrosion of metal surfaces Water leaks
Exterior of the building	Examine the exterior walls to determine whether they are warped, peeling, or blistering; whether there is rotted wood on the surface; whether there are dissolved areas of wood or mold on the surface. Examine the concrete or foundation to see if it is crumbling or any efflorescence is present; examine the foundation for evidence of water stains or rotted wood.
Roof	Determine whether the roof is peeling or flaking, warped, or has ventilation vents. Determine whether the flashing around the fireplace or other openings is missing or deteriorated. Determine if ice accumulated during the winter in any spots on the roof. Determine whether downspouts are missing. If they are present, determine whether they extend less than six inches from the building.
Interior space	Determine whether the relative humidity is greater than 50%. Determine whether the building is damp or smells musty. For example, do cooking odors linger in various units? Determine whether there is visible mold or rot on walls, doors, or other areas. Determine whether there is visible evidence of water damage or dampness in the building. Determine whether there is evidence of moisture dripping around windows, doors, pipes, or other surfaces. Determine whether there is carpeting located in any rooms in the building where water is used or may collect.

Area of Building	Concerns
Basement and crawl spaces	Determine whether there is any standing water, dampness, or signs of mold in the basement or crawl space of the building. Examine whether the basement or crawl space has vapor retarders present. Examine the vents in the crawl space or basement. Determine whether the sumps discharge close to the building. Check the crawl space and basement for debris. Determine whether fiberboard was used to make part of the supply or return ductwork in the crawl space. Determine whether the sumps and drains in the basement and crawl space back up or remove water slowly.[115]

§ 2.27 Mold and Other Biological Contaminants[116]

Mold and other biological contaminants are a great source of indoor air pollution. They can contaminate HVAC systems, are often expensive to fix, and may be a significant problem leading to environmental health hazards as well as damage to the building. In order to protect against indoor air pollution problems due to mold and other biological contaminants:

✓ Examine for evidence of dust mites, dander, excretions, or feather contaminations in the building:

1. Determine whether the people living or working in the building have any unexplained allergic symptoms that seem to be related to being in the building.
2. Examine whether bird droppings accumulate on or near the building.

[115] For a checklist and discussion of these environmental issues in the context of the home, see INGRID RITCHIE AND STEPHEN J. MARTIN, THE HEALTHY HOME KIT (1995).

[116] *See generally* G. B. LESLIE AND F. W. LUNAU, INDOOR AIR POLLUTION: PROBLEMS AND PRIORITIES (Cambridge 1992); FRANK E. JONES, TOXIC ORGANIC VAPORS IN THE WORKPLACE (Lewis 1993); JACK G. KAY, ET AL., INDOOR AIR POLLUTION: RADON, BIOAEROSOLS & VOCs (Lewis 1991). For a checklist and discussion of these environmental issues in the context of the home, see INGRID RITCHIE AND STEPHEN J. MARTIN, THE HEALTHY HOME KIT (1995). *See also* LYNN LAWSON, STAYING WELL IN A TOXIC WORLD (1993).

✓ Examine for evidence of infestation by rodents or hantavirus:
1. Examine whether there is evidence of mice or rats in the building or within 100 feet of the building.
2. Examine cabinets and closets for droppings.
3. Examine for potential external nesting cites in shrubs, wood piles, and other external debris piles.

✓ Examine for mold contamination:
1. Check to see whether there are visible signs of mold, rot, or discolored areas on wood and other surfaces.
2. Check all attics, basements, and crawl spaces for signs of mold.
3. Examine areas of covered insulation for mold.
4. Even if there is no evidence of mold, check for potential moisture problems that may give rise to future conditions favorable for growth of mold.[117]

§ 2.28 Combustion Contaminants from Furnaces and Other Fuel-Burning Appliances[118]

Combustion contaminants from furnaces and other fuel-burning appliances are also a source of indoor air pollution. These problems often result from improper venting in buildings. These may lead to significant environmental health hazards, including death. There are five principal sources of combustion contamination. These include

1. vented fuel-fired heating appliances;
2. vents, chimneys and flues;
3. vented woodstoves, fireplaces, gas heaters, and kerosene heaters;

[117] For a checklist and discussion of these environmental issues in the context of the home, see INGRID RITCHIE AND STEPHEN J. MARTIN, THE HEALTHY HOME KIT (1995).

[118] *See generally* G. B. LESLIE AND F. W. LUNAU, INDOOR AIR POLLUTION: PROBLEMS AND PRIORITIES (Cambridge 1992); JAMES W. PATTERSON, ET AL., INDUSTRIAL WASTE MANAGEMENT: PROCESS MODIFICATIONS FOR INDUSTRIAL SOURCE REDUCTION (Lewis 1985). For a checklist and discussion of these environmental issues in the context of the home, see INGRID RITCHIE AND STEPHEN J. MARTIN, THE HEALTHY HOME KIT (1995). *See also* LYNN LAWSON, STAYING WELL IN A TOXIC WORLD (1993).

4. industrial and household appliances (such as gas-fired cooking stoves, water heaters, and clothes dryers);
5. unvented gas or kerosene heaters.[119]

Obviously, the single most important thing that can be done to prevent damage from these type of devices is the existence of adequate fire alarms, fire extinguishers, and carbon monoxide detectors in the building.

Source	**Factors to Consider**
Vented fuel-fired heating appliances	Determine whether the central-heating appliance is located in a small, confined space, such as under a stairway or in a closet. Examine whether there are signs of any of the following present in the building: —condensation on windows or other surfaces; —yellow flames that are not uniform in shape; —black soot in or around the appliance or burner assembly; —corrosion in the vent flue or burner assembly. Examine the draft hood and conduct draft hood tests to evaluate combustion air. Examine the return ducts for deterioration.
Vents, chimneys, and flues	Determine whether the crawl space is used as part of the ducting system. Determine whether fiberboard or cardboard was used as part of the ductwork in the crawl space or basement.
Vented woodstoves, fireplaces, gas heaters, and kerosene heaters	Determine whether the use of a fireplace, woodstove, or kerosene heater produces indoor smoke or places fine ash deposits in the building. Determine whether the appliance is less than three feet away from any combustible materials. If there is a wood-burning stove, determine whether it has a single walled metal pipe that acts as a chimney. If there is a gas heater, determine whether it was made before June 1, 1984.

[119] For a checklist and discussion of these environmental issues in the context of the home, see INGRID RITCHIE AND STEPHEN J. MARTIN, THE HEALTHY HOME KIT (1995).

Source	Factors to Consider
Industrial and household appliances (such as gas-fired cooking stoves, water heaters, and clothes dryers)	Determine whether the flames of any pilot lights and appliance burners located in the building look yellow or poorly defined. Determine whether a gas-fired cooking stove is used for heat or if there are gas fired dryers in the building. If so, determine whether the cooking stove or clothes dryer vents end in any area inside the building. Determine the location of any water heaters in the building to be sure they are not located in any close spaces. Check to be sure the air heater is located at least 18 inches above the floor if it is located in a garage, basement, or crawl space.[120]

§ 2.29 Radon

Like indoor air pollution, the presence of radon on a site can greatly impair the value of collateral. Radon is a colorless, odorless gas naturally occurring in the environment.[121] Radon contains alpha particle (short wavelength) radiation. Radon gas is often found to leak into and accumulate in buildings. EPA estimates that 5,000 to 30,000 cases of lung cancer a year are produced at least partially due to radon.

Currently, there are no laws or regulations governing radon. EPA has published a recommended policy that action should be taken to reduce radon levels to 4 picocuries per liter (Pci/l). These recommendations include information on EPA's preferred radon remediation techniques.

There is no way to predict which buildings will contain high radon levels without testing. Certain geographic regions, however, have been found to have high levels of radon on a random, but frequent, basis. These exist throughout the United States, but areas of particular concern include Washington, D.C., portions of Pennsylvania, and Delaware.

[120] For a checklist and discussion of these environmental issues in the context of the home, see INGRID RITCHIE AND STEPHEN J. MARTIN, THE HEALTHY HOME KIT (1995).

[121] *See generally* Note, *Clearing the Air on Radon Testing: The Duty of Real Estate Brokers to Protect Prospective Homebuyers*, 15 FORDHAM URB. L.J. 767 (1987). *See also* LEONARD A. COLE, ELEMENT OF RISK: THE POLITICS OF RADON (Oxford 1993); G. B. LESLIE AND F.W. LUNAU, INDOOR AIR POLLUTION: PROBLEMS AND PRIORITIES, at 99–116 (Cambridge 1992).

The law of liability for selling real estate containing high radon levels is not yet developed.[122] Buyers and sellers are both wise to conduct testing, since remediation efforts range from very inexpensive to extraordinarily expensive (often without accomplishing any increase in environmental efforts). Purchasers wishing to conduct radon testing should be sure that the purchase contract is drafted broadly enough to allow such testing. Some commercial residential contracts may not be broad enough to allow radon testing and the contract may need to be revised before contracting to allow radon tests.

The following facts should be considered when evaluating radon:

✓ Check to see whether the property has been tested for radon.

✓ Consider whether the basement, floor, or any crawl spaces contain exposed soil or gravel, since these can be sources for radon entrance into the building. If so, check to see if the floor or crawl space (1) has a plastic covering over the dirt and (2) is secured around the perimeter of the floor. Such plastic liners may be covered with several inches of gravel to secure the liner.

✓ Check to see whether the crawl space or basement has return air heating and cooling ducts in it. If so, evaluate whether they contain any openings and determined if the openings are for utilities or for ducts that go through the building.

✓ Evaluate whether the crawl space or basement contains a dry well or a sump that is not sealed.

✓ Evaluate whether there are any utilities or ducts that go through the crawl space into the building.

✓ Check to see if the basement walls and floors have cracks or other unsealed openings.

✓ Check to see if the basement contains standard drains without water traps.

✓ Check to see if the basement has a chimney chase.

---

[122] About 30 percent of all radon tests are conducted in conjunction with real estate transactions. In residential transactions, many lenders now require radon tests. *See* Ingrid Ritchie and Stephen J. Martin, The Healthy Home Kit at 70 (1995).

✓ Evaluate whether there are openings for utility ducts between the basement and the building.

✓ Evaluate whether the basement is damp or there are any signs of recent water damage.[123]

Accordingly, the following investigation should be considered:

✓ Determine whether the geographic region where the property is located is in a high, medium, or low radon concentration region.

✓ Conduct a radon test, where appropriate.

§ 2.30 Lead

Lead is also an important source of indoor air pollution. The following should be considered when checking for lead:

✓ Determine when the building was built (before 1950? 1950-80? after 1980?).

✓ If the building was built before 1980, determine whether the interior or exterior paint is peeling, cracking, or lifting away or if any renovations are planned that would require paint stripping.

✓ If the paint is deteriorating, check to see if the cracking is in a pattern of fairly uniform squares.[124]

§ 2.31 Electromagnetic Fields

Electromagnetic fields (EMF), a form of nonionizing radiation created by the electric and magnetic fields surrounding power transmission lines and electronic appliances, have been investigated for potential to cause adverse health effects. In 1990, EPA published an inconclusive report on the possible link between low-level EMF and cancer. Congress has ordered another study that has yet to be concluded.

---

[123] For a checklist and discussion of these environmental issues in the context of the home, see INGRID RITCHIE AND STEPHEN J. MARTIN, THE HEALTHY HOME KIT (1995).

[124] *Id.*

Certain states have enacted legislation governing EMF. Following is a table of the guidelines set by certain states for EMF:

State	Electric Fields	Magnetic Fields
California	Conducting research	
Florida	2 kV/M at edge of right of way	230 kV lines: 150 mG at edge of right of way 500 kV lines: 200 mG at edge of right of way Double circuit 500 kV lines: 250 mG at edge of right of way
Georgia	Conducting research	
Minnesota	8 kV/M maximum within right of way	
Montana	1 kV/M at edge of right of way in residential area	
New Jersey	3 kV/M at edge of right of way	
New York	1.6 kV/M at edge of right of way (informal limit)	
North Dakota	9 kV/M at edge of right of way	
Oregon	9 kV/M maximum within right of way	

Regardless of health effects, siting of power transmission lines and other sources of EMF may lead to property devaluation even if there is no property damage or adverse health effects. The perceived fear due to EMF has been successfully raised by landowners seeking compensation for devaluation of their property beyond the physically condemned land in eminent domain and other condemnation cases. Accordingly, the following should be examined:

✓ Check to see whether any high-intensity electric utility poles, substations, or overhead wires are located on, near, or adjacent to the property (this is especially important if the property will be used as a school, playground, or other place where children can be expected to play frequently.

✓ If the electric wires entering the building are buried, check to see where the metal electrical box was placed (*i.e.*, its proximity to the building and its occupants).

✓ Check the building for appliances or gadgets that operate on small motors, including, but not limited to, computers and color TVs.

✓ Encourage building residents not to place beds or cribs next to walls that contain electrical wiring or next to places where the wires enter the building.[125]

§ 2.32 Federal Insecticide, Fungicide and Rodenticide Act (FIFRA)[126]

Under the Federal Insecticide, Fungicide, and Rodenticide Act (FIFRA),[127] any distributor or seller of an insecticide, fungicide, or rodenticide must register and submit certain data to EPA. EPA then determines whether use of the pesticide will be allowed and, if so, under what conditions. Distributors or sellers must comply with certain labeling requirements. EPA also has broad authority to prohibit or limit the sale or use of a registered pesticide and may seize products posing imminent hazards to the public.

Golf courses and facilities with substantial landscaped grounds may be unwitting subjects of FIFRA liabilities. Accordingly, the following should be examined in the audit when such facilities are involved:

✓ Determine whether the facility uses, stores, or handles substantial quantities of herbicides, pesticides, growth inhibitors, etc.

✓ Determine whether the facility has recently been treated with a termicide or other pesticide inside the facility. If so, is there a lingering odor.

✓ Determine whether any pesticides other than disinfectants are stored inside facilities located on the property. If so, what are storage conditions? What quantities are present?

---

[125] For a checklist and discussion of these environmental issues in the context of the home, see INGRID RITCHIE AND STEPHEN J. MARTIN, THE HEALTHY HOME KIT (1995).

[126] 7 U.S.C. § 136-136y.

[127] *Id.*

✓ Examine whether pest strips (especially those containing dichlorvos) are used indoors or if insect sprays or lotions that contain DEET are used on the premises. If so, examine frequency and quantity of use.
✓ Determine whether pets are sprayed with pesticides indoors or pets are allowed to use flea collars indoors. Again, if so, in what quantity and with what frequency?
✓ Determine whether the facility has a history of serious pest infestation and if so, how it has been treated in the past.[128]
✓ Inquire whether insect sprays or lotions that contained DEET or pesticide insect strips with dichlorvos were used on the property.

§ 2.33 Endangered Species Act

The Department of Interior has primary responsibility for listing endangered species. Federal agencies must use their authority so as to not jeopardize the continue existence of endangered or threatened species. This includes ensuring that federal projects—no matter how large or small—do not destroy or modify the habitat of such species. Where a proposed project may jeopardize an endangered or threatened species, the U.S. Fish and Wildlife Service is required to identify alternatives that would allow the project to proceed. Therefore, final federal approval may be conditioned on compliance with various recommendations to protect affected wildlife or plants.[129]

---

[128] For a checklist and discussion of these environmental issues in the context of the home, see INGRID RITCHIE AND STEPHEN J. MARTIN, THE HEALTHY HOME KIT (1995).

[129] For more information concerning the endangered species act and its potential impact on development *see generally* CHARLES C. MANN AND MARK L. PLUMMER, NOAH'S CHOICE: THE FUTURE OF ENDANGERED SPECIES (1995); TIM W. CLARK ET AL., ENDANGERED SPECIES RECOVERY: FINDING THE LESSONS, IMPROVING THE PROCESS (1994); KE CHUNG KIM AND ROBERT D. WEAVER, BIODIVERSITY AND LANDSCAPES (1994); R. EDWARD GRUMBINE, ENVIRONMENTAL POLICY AND BIODIVERSITY (1994); TIMOTHY BEATLEY, HABITAT CONSERVATION PLANNING (1994); LOWELL W. ADAMS, URBAN WILDLIFE HABITATS: A LANDSCAPE PERSPECTIVE (1994); RUTHERFORD H. PLATT ET AL., THE ECOLOGICAL CITY: PRESERVING AND RESTORING URBAN BIODIVERSITY (1994); ROCKY BARKER, SAVING ALL THE PARTS: RECONCILING ECONOMICS AND THE ENDANGERED SPECIES ACT (1993); LISA MIGHETTO, WILD ANIMALS AND AMERICAN ENVIRONMENTAL ETHICS (1991); KATHRYN A. KOHM, BALANCING ON THE BRINK OF EXTINCTION: THE ENDANGERED SPECIES ACT AND LESSONS FOR THE FUTURE (1991).

✓ Check to see if the project is planned in an area that endangered or threatened species are known to inhabit.

✓ Since a disproportionate number of endangered species live in wetlands, check to see if the property contains or is located nearby a wetland.

§ 2.34 Fish and Wildlife Coordination Act

This statute dictates that whenever waters of any stream or other body of water will be modified or controlled for any purpose, Fish and Wildlife must be consulted. This statute applies to private parties proposing the modifications as well as to the federal government. The purpose of the consultation is to conserve wildlife resources by preventing loss of and damage to such resources. The "consultation" process with Fish & Wildlife can be expensive and protracted.

✓ Consider whether the proposed project will modify or control any stream or body of water.

§ 2.35 The National Historic Preservation Act[130]

The National Historic Preservation Act imposes land use restrictions on property listed on the National Register of Historic Places. This law may cause considerable delay in a proposed project. The law requires only consultation and comment, but surveys may be required at applicant's expense to determine if a site is eligible for inclusion in the Register.

✓ Check the age of the building.

✓ Check to see if the target property is known in the community to be of historical significance.

✓ Check to see if the property is listed on the National Register of Historic Places.

For a more scientific assessment of this tension *see generally* NATIONAL RESOURCE COUNCIL, SETTING PRIORITIES FOR LAND CONSERVATION (1993); IAN F. SPELLERBERG AND STEVEN R. HARDES, BIOLOGICAL CONSERVATION (1992); WENDY E. HUDSON, LANDSCAPE LINKAGES AND BIODIVERSITY (1991); ANNE E. MAGURRAN, ECOLOGICAL DIVERSITY AND ITS MEASUREMENT (1988); R. V. O'NEILL ET AL., A HIERARCHICAL CONCEPT OF ECOSYSTEMS (1986).

[130] 16 U.S.C. § 470 *et seq.* (1985).

§ 2.36 National Flood Insurance Act[131]

The availability of flood insurance is premised on local regulation of all land use activities in the floodplains, including building prohibitions in regulatory floodways. By executive order, federal agencies are precluded unless it is the only practicable alternative.

✓ Check to see if the target property is located on a flood plain. If so, is it a 100-year floodplain?

✓ Check to see whether the local community has enacted floodplain development ordinances.

§ 2.37 Coastal Barriers Resources System Act[132]

This statute precludes federal financial assistance for roads, bridges, and sewers on designated undeveloped coastal barrier islands. In addition, federally guaranteed loans and flood insurance would also be unavailable on sites designated under the act.

✓ Is the target a barrier island?

✓ If so, has it been identified under the Coastal Barriers Resources System Act?

§ 2.38 Coastal Zone Management Act[133]

In a state with an approved coastal zone management plan, with certain limited exceptions, each federally funded agency conducting or supporting activities "directly affecting" the coastal zone must gain certification from the state that the action is consistent with the Coastal Zone Plan before project approval is granted.

States may regulate a variety of land uses in the coastal zone pursuant to this law. Some states use the coastal zone management plans to set up regulatory barriers for offshore oil and gas development.[134]

---

[131] 42 U.S.C. § 4016.

[132] 42 U.S.C. § 4028.

[133] 16 U.S.C. § 1451 *et seq*. (1985).

[134] *See generally* TIMOTHY BEATLEY ET AL., AN INTRODUCTION TO COASTAL ZONE MANAGEMENT (Island Press 1994).

✓ Check to see whether the target property is located in a state with an approved coastal zone management plan.

§ 2.39 Transaction Triggered Environmental Statutes/Environmental Policy Acts[135]

One of the most important trends in recent years has been the emergence of state legislation that seeks to deal with potential environmental concerns by imposing preconditions on the consummation of common forms of business transactions involving the transfer of real property. Such legislation regulates the transfer or abandonment of contaminated property, including transfer of control in mergers and acquisitions and foreclosure on contaminated real estate.

This important development was initiated by the enactment, in late 1983, of the New Jersey Environmental Cleanup Responsibility Act (ECRA).[136] This statute has been modified over time and is now known as the Industrial Site Recovery Act (ISRA). Under that statute, owners of certain types of industrial and commercial property must notify the New Jersey Department of Environmental Protection (NJDEP) before the sale, transfer, or closure of the site. They also must submit either a statement that the site is clean or a cleanup plan. New Jersey takes the position that this law encompasses certain out-of-state business transactions if they affect a covered site in New Jersey. If the seller fails to comply with any ECRA provision, the sale may be voided by either the NJDEP or the buyer, and the buyer can recover damages against the seller. Violators also can be fined up to $25,000 for each offense.

Although only a minority of states have enacted such legislation, several states have provisions requiring notice and disclosure before transfer of an interest in land that has been used as a hazardous waste site.[137] Although Environmental Policy Acts and transaction triggered statutes will

---

[135] *See* chapter 5 for a more detailed discussion of these state statutes.

[136] N.J. STAT. ANN. § 13:1K-6 *et seq.*

[137] *See* GA. CODE ANN. § 12-8-96.1 (Michie); KY. REV. STAT. ANN. §224.46-580 (Michie); MASS. ANN. LAWS ch. 21C § 7 (Law Co-op); MINN. STAT. § 115 B.16; OHIO REV. CODE ANN. § 5302.30 (Baldwin).

not necessarily give rise to direct environmental liabilities depending on the party involved, it can take a considerable period of time to process the required documentation under these statutes. Therefore, any party transferring title—legal or equitable—should be aware a borrower is subject to such provisions, since they may significantly delay closing. More importantly, there are substantial penalties for failure to comply with these acts. In some states (New Jersey, for example), failure to comply with such statutes will render the transaction voidable. Thus, the following should be added to audit procedures involving the transfer of businesses:

- ✓ Determine whether the proposed transaction triggers any environmental statutes.
- ✓ Evaluate whether the property would benefit from participation in any state voluntary cleanup program or in one of the EPA pilot programs for brownfields cleanup and redevelopment.
- ✓ Check to be sure the audit and the report will be sufficient to satisfy the state or local transaction triggered environmental statute.
- ✓ Determine whether it is necessary to file or record a disclosure statement.
- ✓ Determine whether the owner/operator/borrower needs to prepare an environmental impact statement to obtain any permits or approvals.

§ 2.40 Permit or Other "Advance Approval" Requirements

Taken together the state and federal environmental laws and regulations contain a comprehensive system precluding certain types of emissions. "Permits" or "prior approval" requirements grant parties a limited license to undertake certain specified emissions. Most federal and state environmental laws contain "permit" or other "prior approval" requirements. Permit or approval requirements can be triggered by certain facilities or land uses involving, among other things:

Permit Trigger	**Indicators**
Pollutant discharges	(a) Water pollution discharges (b) Air pollution discharges (c) Solid or hazardous waste disposal

Permit Trigger	Indicators
Participation by or in conjunction with	(a) Federal government (b) State or local governments
Certain locations	(a) Use of wetlands, swamps, bogs, critical areas, or prairie potholes (b) Work in navigable (or state) waters (c) Use of state-owned bottomlands of state waters (subaqueous beds) (d) Use of flood plains (e) Withdrawal of water in "surface or groundwater management areas" (f) Construction or activities in aquifer recharge areas (g) Use of "critical environmental areas" (h) Building within a "historic district" (i) Activity within the state's coastal zone management area (j) Activities on federal lands (k) Activities in marine sanctuaries (l) Construction on a "wild and scenic" river
Certain activities	(a) Construction or operation of water supply systems of a certain size (b) Construction or operation of sewerage systems and treatment works (c) Construction of dams or other impounding structures (d) Open burning (e) Land disturbing activity likely to lead to land erosion or stream sedimentation (f) Strip mining of certain materials (g) Discharge of dredged or fill material (h) Construction of wells (i) Construction or operation of nuclear power plants (j) Construction or operation of hydroelectric power plants (k) Construction or operation of tall buildings or towers
Certain materials	(a) Manufacture or use of toxic or hazardous substances (b) Manufacture or use of pesticides (c) Manufacture or use of radioactive substances
Certain protected items	(a) Destruction of rare or endangered species (b) Destruction of historical buildings (c) Alteration of terrain involving significant historical or archeological data

Permits normally must be obtained before construction or operation of facilities requiring them can be commenced. The definition of "commencing" construction or operation is not uniform among environmental laws, and this determination is frequently one of the key determinations that must be made.

Any one project may require a multitude of permits from separate local, state, and federal agencies. Failure to obtain any one of these permits may be problematic. Obtaining a permit from any one agency generally will not relieve responsibility for obtaining a needed permit from another agency. Obviously, the regulated party bears the burden of determining if a permit is needed, and ignorance of the law is not a defense for failure to obtain a permit. Permit requirements are usually multitudinous and uncoordinated. Bureaucratic delays are frequent and can be significant. Public intervention in the process is also common and may contribute further to delays, so it is critical to plan ahead by applying early for permits.

§ 2.41 The National Environmental Policy Act (NEPA)

Another critical environmental law that is in the nature of a permit is the National Environmental Policy Act of 1969 (NEPA),[138] which set forth a requirement that federal agencies prepare a "detailed statement" (known as an "Environmental Impact Statement" or "EIS") analyzing the environmental effects and potential alternatives of all "major Federal actions significantly affecting the quality of the human environment."[139] An action is "federal" under NEPA if it is (1) undertaken directly by a federal entity *or* (2) undertaken by a private party that is effectively "in partnership" with the federal government. This can result when the government provides significant funding for the project or issues a lease, permit, license, insurance, contract extension or modification, conveyance, assistance authorization, or other entitlement that is a legal precondition to the activity. Thus, any activity or commercial transaction that involves government licensing, approval, or participation can trigger environmental assessment responsibilities.

---

[138] 42 U.S.C. § 4321-4370d.

[139] 42 U.S.C. § 4332(2)(C). Most states have their own version of NEPA.

When environmental assessment responsibilities are triggered and a federal agency is required to prepare an EIS, the agency normally requires any private party involved to prepare a broad-ranging and expensive back-up study that will serve as a basis for the agency's preparation of its independent EIS. Agency rules generally provide for public notice, review, and comment on an agency's written statement of environmental impacts and alternatives.

An EIS must include an adequate discussion of

1. the environmental impact of the proposed action;
2. any adverse environmental effects that cannot be avoided should the proposal be implemented;
3. alternatives to the proposed action;
4. the relationship between local short-term uses of man's environment and the maintenance and enhancement of long-term productivity; and
5. any irreversible and irretrievable commitments of resources that would be involved in the proposed action should it be implemented.[140]

Any failure by the agency to comply with these requirements can lead to court challenges and to delay or derailment of the private transaction in question. Counsel for a private participant must, therefore, be concerned that the environmental studies of both the client and the federal agency conform to NEPA requirements[141] or that the transaction is structured so that NEPA requirements are not triggered.

Although many permitting processes are exempt from NEPA review, issuances of a section 404 wetlands permit, an NPDES permit, or an air permit under the New Source Review program may, however, trigger NEPA review.

---

[140] 42 U.S.C. § 4332(2)(c)(i-v).

[141] The Council on Environmental Quality (CEQ) is responsible for supervising the preparation of EISs by agencies and has promulgated regulations governing federal agency compliance with NEPA. The regulations establish uniform standards applicable throughout the federal government for conducting environmental reviews. *See* 40 C.F.R. Parts 1500-1508. In addition, most federal agencies have regulations governing their own NEPA procedures, which should be consulted in appropriate cases. EPA's NEPA regulations, for example, are set forth at 40 C.F.R. Part 6, and those of the Corps of Engineers are at 33 C.F.R. Part 230.

Transaction Triggered Audits

Form 2.1
Purchaser's Right of Inspection, Including Environmental Audit

INSPECTION. During the pendency of this Contract, Purchaser shall have access to the Real Property for the purpose of conducting such inspections, investigations, soil tests, environmental audits, engineering and feasibility studies of the Real Property as Purchaser deems necessary or advisable in connection with the purchase of the Property. Further, Seller agrees to make available to Purchaser or to its duly authorized agents or representatives, the originals of all Tenant Leases and all applicable books and records relating to the Real Property and the operation and maintenance thereof. Such items may be examined at all reasonable times during normal business hours. In this regard, Seller agrees that during the pendency of this Contract, Purchaser and its agents and representatives shall be entitled to enter upon the Real Property for inspection and examination; provided, however, that: (i) neither Purchaser nor its agents shall interfere with Seller's operation of the Real Property or any tenant's occupancy of same; (ii) Purchaser shall be responsible for any damages to the property or injuries to persons caused by Purchaser or Purchaser's agents and representatives; (iii) Purchaser shall, at Purchaser's sole cost and expense, restore the Real Property to its condition prior to Purchaser's inspection and testing of same; (iv) if this Contract does not close for reasons other than Seller's default, Purchaser shall furnish to Seller copies of all reports, test results, etc. obtained by Purchaser in the conduct of Purchaser's inspection and testing (without representation or warranty of any kind, express or implied); and (v) Purchaser shall indemnify and hold Seller harmless from and against any loss, cost, damage or expense, including reasonable attorneys' fees and court costs, suffered or incurred by Seller as a result of any activities conducted by Purchaser, its agents, employees, contractors or representatives on or about the Real Property. Purchaser's obligations contained in this paragraph shall survive the Closing or other termination of this Contract.

Form 2.2
Phase I Audit as Condition Precedent to Closing Land Acquisition—Alternate 1

Environmental Audit. Purchaser shall have received, at its cost and expense, an Environmental Audit (as defined below), or other evidence satisfactory to Purchaser, in its sole discretion, which confirms to Purchaser's satisfaction that the Company, its business and assets and the Real Estate are in compliance with all Environmental Laws. For purposes hereof, "Environmental Audit" shall mean a review for purposes of determining whether the Company, its business and assets and the Real Estate are in compliance with Environmental Laws and whether there exists any condition or circumstance which requires or will require a cleanup, removal or other remedial action under Environmental Laws on the part of the Company, including, but not limited to, the subsurface investigation described in the supplemental letter agreement with ______________________ dated _______________.

Form 2.3
Phase I Audit as Condition Precedent to Closing Land Acquisition—Alternate 2

ENVIRONMENTAL INSPECTION: To determine the environmental status of this Property, Purchaser wishes to undertake an environmental audit of the Property. Purchaser shall select an engineering firm specializing in environmental surveys. The fees for the environmental audit shall be paid by Purchaser. All preliminary environmental audits or inspections shall be completed during the feasibility period stated in Article 11.

The environmental audit of the Property shall, at a minimum, consist of:

A) Land use inspection, to include investigating all prior owners and their use of the Property, for as far back in time as records exist; and

B) Site inspection, which shall consist of a personal surface inspection of the Property for any apparent or obvious hazard or contamination of any kind.

Upon receipt of the results of the environmental audit, Purchaser shall immediately notify Seller of its decision to:

A) Terminate this Contract, in which case the Deposit and interest thereon shall be returned to Purchaser, and all parties shall be released from any further obligation, liabilities, and performances under this Contract, except for the surviving provisions set forth in this Contract; or

B) Perform, at Purchaser's sole cost and expense, all inspections, soil tests, or remedial activities recommended by the environmental audit or required by all applicable federal and state laws in order to remove any environmental hazard.

Form 2.4
Phase I and II Audits as Condition Precedent to Closing Land Acquisition with Cost Sharing Provisions

ENVIRONMENTAL AUDIT. ____ shall deliver to ____ a Phase I Environmental Assessment Report with respect to the property listed on Schedule 4.9 (the "Phase I Reports"), in a form reasonably satisfactory to ____. The Parties agree to share equally any and all costs incurred in obtaining the Phase I Reports requested after the date hereof in connection with this Section _.__. Based on the Parties' review of the Phase I Reports, the Parties shall determine, in good faith based on a reasonable assessment of the results of the Phase I Reports, whether to obtain Phase II Site Assessments ("Phase II Assessments") for any ____ Real Property. The Parties shall mutually agree on the environmental consultant to prepare any such Phase II Assessment and shall share equally any and all costs incurred in obtaining such Phase II Assessments. If the amount of remediation expenses related to the ____ Real Property, as set forth in the Phase II Assessments, is estimated to exceed $1,000,000, then ____ shall have an additional ninety (90) days from the date of receipt of such Phase II Assessments within which to cure the excessive remediation costs and to reduce the estimated costs of remediation to an amount below $1,000,000.

Form 2.5
Environmental Audit and Remediation as Condition Precedent to Closing

Section _.__. Environmental Matters.

(a) For purposes of this Section _.__, the following definitions shall be applicable:

(i) "Applicable Environmental Laws" shall mean any and all laws, statutes, regulations, and judicial interpretations thereof of the United States, of any state in which the Assets, or any portion thereof, or the Station's Business, are located, and of any other government or quasi-government authority having jurisdiction, that relate to the prevention, abatement and elimination of pollution and/or protection of the environment, including, but not limited to, the federal Comprehensive Environmental Response, Compensation, and Liability Act, the Resource Conservation and Recovery Act, the Clean

Water Act, the Clean Air Act, the Safe Drinking Water Act, the Toxic Substances Control Act, and the Hazardous Materials Transportation Act, together with all state statutes serving any similar or related purposes, as in effect on or before the Closing Date.

(ii) "Hazardous Substance" means any substance now or hereafter designated pursuant to Sections 307(a) and 311(b)(2)(A) of the federal Clean Water Act, 33 U.S.C. §§ 1317(a), 1321(b)(2)(A), Section 112 of the federal Clean Air Act, 42 U.S.C. § 3412, Section 3001 of the federal Resource Conservation and Recovery Act, 42 U.S.C. § 6921, Section 7 of the federal Toxic Substances Control Act, 15 U.S.C. § 2606, or Section 101(14) and Section 102 of the Comprehensive Environmental Response, Compensation, and Liability Act, 42 U.S.C. §§ 9601(14), 9602, as amended by the Superfund Amendments and Reorganization Act of 1986.

(b) Seller has supplied to Buyer a copy of the report for each environmental inspection or audit that it has caused to be conducted with respect to the Assets as listed on Schedule _.__.

(c) (i) Since the date of their acquisition by Seller, the Assets have been used by Seller for its current business operations and for no other purposes and at no time during that period have the Assets been used by the Seller for the treatment, storage or disposal of Hazardous Substances in violation of Applicable Environmental Laws or as a landfill or other solid or hazardous waste disposal site. Seller has no knowledge that prior to its acquisition of the Assets the Assets were used for the treatment, storage or disposal of Hazardous Substances in violation of Applicable Environmental Laws or as a landfill or other solid or hazardous waste disposal site.

(ii) To the Seller's knowledge, except as set forth in Schedule _.__, there are no underground storage tanks on the Real Property.

(iii) Except as set forth on Schedule _.__, Seller has no knowledge that any of the improvements owned or used by Seller on the Real Property contain any asbestos, or that any equipment owned or used by Seller on the Real Property contains any polychlorinated biphenyls.

(iv) Seller has no knowledge that any material contamination exists on or under the Real Property and Assets, or affecting any natural resources therein that would constitute a material violation of or noncompliance with any Applicable Environmental Law.

(v) Seller has no knowledge that Seller, the Assets, or the Station's Business are not in material compliance with all Applicable Environmental Laws.

(vi) Except as set forth on Schedule _.__, Seller has no knowledge of any agreements, consent orders, decrees, judgments, license or permit conditions, or other directives of Government Authorities directed to Seller, that are based on or arise out of Applicable Environmental Laws and relate to the future use of the Assets or the Station's Business or that require any material change in the present conditions of the Assets or the Station's Business.

(vii) Except as set forth on Schedule _.__, Seller has no knowledge of any Actions, suits, claims or proceedings seeking money damages, injunctive relief, remedial action or other remedy pending or threatened relating to an alleged violation of or noncompliance with any Applicable Environmental Laws; or the disposal, discharge or release of Hazardous Substances by Seller, the Assets or the Station's Business or Seller's ownership or use of the Assets or the Station's Business.

(viii) Seller has given to pertinent Government Authorities all notices required pursuant to Applicable Environmental Laws in connection with the Station's Business. Except as listed on Schedule _.__, Seller has not prior to the date hereof received any order or notice of violation or noncompliance from, or been the subject of any regulatory audit or investigation (other than any periodic investigation or inspection of a routine nature) by, any Government Authority in connection with ownership or operation of the Station's Business.

(ix) To the Seller's knowledge, no consent or approval is needed from any Government Authority under any Applicable Environmental Laws for the transfer of the Assets from Seller to Buyer; and neither the execution of this Agreement nor the closing of the transactions contemplated hereby will materially violate any Applicable Environmental Laws.

(x) All necessary plans for development, applications, inspection reports, certificates and other instruments required under Applicable Environmental Laws to be filed by Seller in connection with the conduct of the Station's Business on the Real Property have been filed with the appropriate Government Authority and all permits, licenses, or other authorizations necessary for the lawful conduct of the Station's Business in compliance with all Applicable Environmental Laws have been obtained.

(d) Between the time of Seller's execution of this Agreement and the time of Closing, Seller shall not take or permit any entity under its control to take any actions that would result in the Assets being in violation of any

Applicable Environmental Laws, nor will Seller omit to take any action necessary to prevent the Assets from becoming in violation of Applicable Environmental Laws.

(e) (i) Subsequent to Seller's execution of this Agreement, Seller shall give Buyer, and its agents and consultants, reasonable access during Seller's normal business hours to: (y) the Assets and (z) Seller's personnel to enable Buyer to conduct, at Buyer's sole expense, an environmental investigation and/or audit as Buyer reasonably deems appropriate with regard to compliance with Applicable Environmental Laws. Buyer shall give Seller reasonable advance notice of its need for access for such investigation and/or audit and Buyer and Seller shall cooperate to schedule and conduct such investigations so as to minimize interference with the Station's Business and to allow Buyer to complete such investigations promptly.

(ii) If Buyer determines through its environmental investigation and/or audit or other due diligence activities that a condition with respect to the Assets exists that constitutes a violation of Applicable Environmental Laws, Buyer shall, by no later than________________, provide Seller with a written notification identifying with specificity any such violation that the Buyer believes exists and the basis for that belief, including copies of any written reports prepared by the consultant who performed any environmental investigation and/or audit for Buyer, together with a good faith estimate by that consultant or an independent consultant retained by Buyer, of the cumulative total costs of bringing the Assets into compliance with Applicable Environmental Laws (including, without limitation, attorneys' and consultants' fees, plan/design preparation costs, and removal, treatment, storage, transportation, and disposal expenses). If Buyer fails to give timely notice to Seller as required by this Section _.________: (A) Seller shall have no further obligations under this Agreement with respect to violations of Applicable Environmental Laws other than its indemnity obligations under Article XI; (B) all other provisions of this Agreement shall remain in full force and effect without change; and (C) except pursuant to Article XI, Buyer agrees to and does hereby release Seller from any obligations with respect to, and waives any claims against Seller by reason of discovery of, any violations of Applicable Environmental Laws.

(iii) Within twenty (20) days after Seller's receipt of the notification and information provided by Buyer pursuant to Section _.________ above, Seller may obtain and deliver to Buyer a good faith estimate by an independent consultant retained by Seller of the cumulative total costs to bring the Assets into compliance with Applicable Environmental Laws (including, without limitation, attorneys' and consultants' fees, plan/design preparation costs, and removal, treatment, storage, transportation, and disposal expenses). Buyer shall have the right to review the good faith cost estimate.

(iv) If the cumulative total costs to bring the Assets into compliance with Applicable Environmental Laws, as estimated by the independent consultant retained by Seller (or as estimated by a consultant retained by Buyer if Seller does not timely deliver to Buyer an estimate by an independent consultant retained by Seller), is less than or equal to $750,000, Seller shall take such remedial measures as it reasonably deems necessary to bring the Assets into compliance with Applicable Environmental Laws within a reasonable time under the circumstances (taking into consideration the type, nature, location, and extent of the conditions causing the noncompliance, government approvals, contractor schedules, and the like), and Closing shall proceed as otherwise provided in this Agreement (with the Seller diligently pursuing and completing the remedial work either before or after the Closing Date). Seller's expenses in discharging its liability under this Agreement for such remedial measures shall be considered to be an indemnification obligation that is counted toward and subject to the $12 million overall limitation on Seller's indemnification obligations as set forth in Article XI; provided, however, Seller hereby agrees that its obligations under this Section _.________ shall not be subject to the $250,000 deductible otherwise applicable to Seller's indemnification obligations as set forth in Article XI.

(v) If the cumulative total costs to bring the Assets into compliance with Applicable Environmental Laws, as estimated by the independent consultant selected by Seller (or as estimated by a consultant retained by Buyer if Seller does not timely deliver to Buyer an estimate by an independent consultant retained by Seller), exceeds $750,000:

(A) Seller may, by written notice to Buyer, elect to bring the Assets into compliance with Applicable Environmental Laws. In this case, (1) Seller shall have liability under this Agreement for the cost of such remedial measures as Seller reasonably deems necessary to bring the Assets into compliance with Applicable Environmental Laws up to $750,000, (2) Buyer shall have liability under this Agreement for the next $250,000 of additional costs above $750,000 for such remedial measures, and (3) Seller shall have liability under this Agreement for all additional costs for such remedial measures above $1 million; provided, however, that Seller's expenses in discharging its liability under this Agreement for such remedial measures shall be considered to be an indemnification obligation that is counted toward and subject to the $12 million limitation on Seller's indemnification obligations as set forth in Article XI; or

(B) Seller may elect to give Buyer notice of termination in accordance with Section 12.1(c), which notice shall become effective unless Buyer delivers to Seller a written election (within twenty (20) business days after the date of Seller's notice of termination) that Buyer will reimburse Seller for the first $750,000 of costs for remedial measures the Seller reasonably deems necessary for the Assets to achieve substantial compliance with Applicable Environmental Laws. In this case, Seller shall then have liability under this Agreement for costs for such remedial measures above $750,000; provided, however, Seller's liability under this Agreement for such remedial measures shall be considered to be an indemnification obligation that is counted toward and subject to the $12 million limitation on Seller's indemnification obligations as set forth in Article XI and Buyer's liability under this Agreement for such remedial measures shall be considered to be an indemnification obligation that is counted toward and subject to the $14 million limitation on Buyer's indemnification obligations as set forth in Article XI. Neither Seller's nor Buyer's liability under this Section _.__________ shall be subject to the $250,000 deductible otherwise applicable to indemnification obligations as set forth in Article XI.

(C) Notwithstanding anything to the contrary in Section _.__________, if the cumulative total costs to bring the Assets into substantial compliance with Applicable Environmental Laws, as estimated by the independent consultant retained by Buyer, exceeds $12 million, Buyer may elect to give Seller a notice of termination in accordance with Section 12.1 within twenty (20) business days after Buyer's receipt of the estimate by its independent consultant.

(D) Notwithstanding anything to the contrary contained in this Agreement, Buyer's obligation to pay or reimburse sums hereunder for remedial measures shall not be, and shall not be deemed or construed to be, an admission of liability by Buyer for any environmental noncompliance or violation.

(E) Buyer shall be timely advised about all remediation measures undertaken pursuant to this Section _.__.

(vi) Any election under Section _.________ shall be binding.

(vii) If Seller makes the election provided to it in Section _.___________, or Buyer makes the election provided to it in Section _.___________, Closing shall proceed as otherwise provided in this Agreement (with the Seller diligently pursuing and completing the remedial work either before or after the Closing Date). If Buyer fails timely to make the election provided to it in Section _.___________ after Seller's notice of termination, this Agreement shall be terminated immediately.

******

Representation and Warranties

******

Section _._. Environmental Audit Process. Buyer and Seller shall have completed in timely fashion the environmental investigation and/or audit process and made the elections available to them under Section _.______ hereof.

******

Covenants

******

Section __.__. Environmental Audit Process. Buyer and Seller shall have completed in timely fashion the environmental investigation and/or audit process and made the elections available to them under Section _.______ hereof.

******

Form 2.6
Phase I Audit as Condition Precedent to Closing Loan Transaction

ENVIRONMENTAL AUDIT. A Phase I environmental audit or assessment report of the Property, at Borrower's expense, performed to the Lender's satisfaction by an acceptable and qualified environmental consultant or independent engineer verifying that no unremediated environmental contamination, except as disclosed in said Environmental Audit on the Property has occurred or is imminent and containing adequately supported and documented conclusions (which conclusions must be satisfactory to the

Lender, in its sole discretion) which evaluate (i) whether any hazardous or toxic substances, hazardous wastes, pollutants, contaminants or any other environmental hazards are present in the soil, surface water or groundwater at or adjacent to the Property; (ii) whether operations at the Property are in compliance with all federal, state and local air quality and water quality regulations and other applicable environmental laws and (iii) whether there are any other potential or actual environmental concerns from current or prior ownership and uses of the Property. The contract between the Borrower and consultant for the conduct of an environmental audit or assessment shall contain a provision which expressly states that the Lender is an intended beneficiary of the contract and is entitled to rely on any report of findings or conclusions or the results of the environmental audit. In order to evaluate the environmental risks associated with this transaction, the Borrower will, upon request, give permission to the Lender and its agents and contractors to enter the Property for the purpose of conducting its own environmental assessment or hydrogeologic study of the Property including soil borings, installation of piezometers, the collection of soil and surface water samples, geophysical or geotechnical testing, soil vapor surveys and the installation and sampling of groundwater monitoring wells. All such assessments, studies, inspections and investigations deemed necessary by the Lender shall be conducted at Borrower's expense.

Form 2.7
Lender's Requirement for Periodic Environmental Audit

Environmental Audits. From time to time, as reasonably requested by Lender, at the sole expense of Borrower, Borrower shall provide Lender, or its designee, complete access to all of Borrower's facilities for the purpose of conducting an environmental audit of such facilities as Lender or its designees may deem necessary. Borrower agrees to cooperate with Lender with respect to any environmental audit conducted by Lender or its designee pursuant to this Section -.—.

Form 2.8
Seller's Acknowledgment of Purchaser's Right to Conduct Due Diligence

Purchaser's Due Diligence Investigation.

Seller acknowledges that Purchaser has conducted, and intends to conduct, at Purchaser's expense, investigations concerning Seller and its assets, liabilities, financial condition, business and affairs, the Transferred Assets and all other matters reasonably related to the transactions contemplated by this Agreement, including, without limitation, (i) one or more physical inspections of the Stores, ____________________ Equipment and Inventory, (ii) a review of all books, records, agreements, documents and other materials and information reasonably related to the transactions contemplated by this Agreement, including, without limitation, all Store Leases, the __________ Leases and Store Contracts (collectively, the "Due Diligence Materials"), (iii) an environmental audit of the Stores and the Store Real Property regarding the presence of unlawful hazardous materials under or within the Stores, the Store Real Property or the common areas of the Shopping Centers (to the extent permitted by the Store Leases), (iv) an environmental audit of the Stores and the Store Real Property regarding the presence of asbestos under or within the Stores, the Store Real Property or the common areas of the Shopping Centers (to the extent permitted by the Store Leases), and (v) a determination that Purchaser will not have any "successor liability" for any debts, obligations and liabilities of Seller (other than the Assumed Obligations) (collectively, the "Due Diligence Investigation").

Form 2.9
Lenders Right to Access and Right to Require Periodic Environmental Audits—Alternate 1

Access. Borrower shall provide access to Agent and any of its officers, employees and agents, or cause to be provided access to Agent and any of its officers, employees or agents, exercisable as frequently as Agent determines to be appropriate, upon reasonable advance notice (unless a Default shall have occurred and be continuing, in which event no notice shall be required and Agent shall have access at any and all times), during normal business hours (or at such other times as may reasonably be requested by Agent), to inspect the properties and facilities of Borrower and to inspect, audit and make extracts from all of Borrower's records, files and books of account. Borrower shall make available to Agent and its counsel, as quickly as practicable under the circumstances, originals or copies of all books, records, board minutes, contracts, insurance policies, environmental audits, business plans, files, financial statements (actual and pro forma), filings with Federal, state and local regulatory agencies, and other instruments and documents which Agent may request. Borrower shall deliver any document or instrument reasonably necessary for Agent, as it may from time to time request, to obtain records from any service bureau maintaining records for Borrower, and shall maintain duplicate records or supporting documentation on media, including computer tapes and discs owned by Borrower. Borrower shall instruct its banking and other financial institutions to make available to Agent such information and records as Agent may reasonably request.

Form 2.10
Lenders Right to Access and Right to Require Periodic Environmental Audits—Alternate 2 (with cost provisions)

Access. (a) Borrower shall provide full access during normal business hours, from time to time upon five (5) Business Days' prior notice, to Agent and any of its officers, employees and agents, up to four times a year (unless a Default or Event of Default shall have occurred and be continuing, in which event Agent and its officers, employees, designees, agents and representatives shall have access at any and all times, without limitation as to frequency per year, and without any advance notice), to the properties, facilities, books, records, suppliers, customers, advisors and employees (including officers) of Holdings, Borrower and its Subsidiaries, to the Collateral, to the accountants (including ___________) of Holdings, Borrower and the Subsidiaries thereof and to the work papers of such accountants, with Borrower agreeing to use reasonable best efforts to provide Agent access to the accountants and their work papers. Without limiting the generality of the foregoing, Borrower shall (i) permit Agent, and any of its officers, employees, agents and representatives, to inspect, audit and make extracts from all of Holdings', Borrower's and its Subsidiaries' records, files and books of account and (ii) permit Agent, and any of its officers, employees, agents and representatives, to inspect, review and evaluate the Collateral at Borrower's and its Subsidiaries' locations and at premises not owned by or leased to Borrower or any Subsidiary of Borrower. Borrower shall make available to Agent and its counsel, as quickly as is possible under the circumstances, originals or copies of all books, records, board minutes, contracts, insurance policies, environmental audits, business plans, files, financial statements (actual and pro forma), filings with federal, state and local regulatory agencies, and other instruments and documents which Agent may request. Borrower shall deliver any document or instrument necessary for Agent, as it may from time to time request, to obtain records from any service bureau or other Person which maintains records for Borrower, and shall maintain duplicate records or supporting documentation on media, including computer tapes and discs owned by Borrower. Borrower shall instruct its certified public accountants to make available to Agent such information and records as Agent may request. Notwithstanding the foregoing, Borrower shall not be required to provide access to any of the foregoing information which in the reasonable written opinion of counsel to Borrower, in form and substance reasonably satisfactory to Agent and its counsel, is subject to an attorney-client privilege and which cannot be disclosed (after giving effect to any possible disclosure method or protective agreement which may be requested by Agent or its counsel) without waiver of such privilege; provided, however, that in no event shall any information delivered or requested to be delivered pursuant to Schedule F or G be deemed to be protected by an attorney-client privilege.

(b) A fee of $500 per day per individual (plus all reasonable out-of-pocket costs and expenses) in connection with Agent's field examinations permitted under Section -.—(a) above and Section 4(c) of the Security Agreement shall be charged against the Revolving Credit Facility in connection with each field audit conducted after the Closing Date.

Form 2.11
Guidelines for Hazardous Materials Report—Scope of Work

The Real Estate Security will be surveyed by an environmental consultant approved by ________. The objective of the survey is to determine whether there is any environmental contamination present on the Real Estate Security, or whether such contamination is likely to occur in the future because of on-site or nearby activities or problems. The survey should indicate the extent of any problems and describe possible remedial measures. In addition, the consultant shall acknowledge in its report that The _______ Insurance Company of America will be relying on the survey in consideration of financing the Real Estate Security.

PHASE I - SITE INVESTIGATION

A. Historical Review

Identify any prior uses or activities at or near the site which might have created any environmental problems. Sources of information and supporting documentation may include, but are not necessarily limited to:

1. Area Plat Book
2. Title Search or assessor records
3. Topographic Maps
4. Aerial Photos
5. Construction Documents
6. Environmental Inquiries
    a. Federal, State and Local Environmental Authorities
        - record of environmental permits (owner and tenants)
        - record of environmental violations or other incidents
        - status of any past or present environmental issues or remedial plans
        - record of all of the above for all abutting properties
        - groundwater sampling data near the project site
        - location of sites undergoing or otherwise subject to remediation programs within one mile of the subject site. This shall include reference to the Federal CERCLIS and NPL lists, and appropriate state lists.
        - environmental regulations or requirements that may be relevant to this transaction (e.g., ISRA in New Jersey)
        - locations of nearby landfill or hazardous waste disposal sites - general groundwater flow direction and condition in the vicinity of the sites.

Identify any "regional" groundwater problems.

- possible environmental constraints on the use of the property including but not limited to:

  wetlands
  floodplain
  coastal zone
  sole source aquifer
  aquifer recharge area
  nearby environmentally sensitive sites

b. Local Fire Officials
   - record of underground tank installation, removals, ruptures, leaks or other incidents
   - record of fires involving toxic substances

c. Local Health Authorities
   - record of any health-related environmental issues pertaining to the property

d. OSHA
   - record of any worker or workplace-related environmental incidents

e. Local Water and Sewer Authorities
   - source of water for the Property
   - location of nearby water supply wells indicate whether the groundwater is used as a source of potable water
   - record of any drinking water contamination problems, either at the Property or the immediate vicinity
   - method of sewage disposal

B. Site Inspection

1. Interview Property owner, Property manager, maintenance personnel and tenants, and review available files for:
   a. any past or present environmental incidents, violation notices, or environmental litigation.
   b. any past or present use, storage, handling and disposal of any hazardous materials including petroleum, chemicals, pesticides and asbestos. Provide a copy of the "right to know" (hazard communication) and SARA Title III filings.
   c. record of required environmental permits by owner and/or tenants.
   d. waste generated and disposal methods used.

   e. waste water generated and disposal methods used. Specify the presence of sanitary sewer connections, septic systems, dry wells etc.
2. Visual Inspection for:
   a. storage or handling of hazardous materials in tanks, drums or containers.
   b. evidence of illegal dumping of hazardous materials, debris or construction materials.
   c. extensive use of fill from source of unknown origin.
   d. evidence of soil, surface water and/or groundwater contamination (*e.g.*, staining, distressed vegetation).
   e. potential for contaminated storm water drainage from adjacent properties.
   f. evidence of underground waste disposal (sumps, floor drains).
   g. any environmental concerns surfaced as the result of the Historical Review.
   h. any environmental concerns resulting from the handling of hazardous materials of nearby properties (*e.g.*, gas stations, manufacturing plants) - consider at least a 1/4 mile radius.
3. Underground and Above-Ground Storage Tanks and Other Equipment
   a. Provide an inventory of all storage tanks, including size, location, installation date, contents, tank material, etc.
   b. Provide a record of any tank leak tests and results.
   c. Verify all tank permits and registrations.
   d. Conduct physical inspection of above-ground tanks, process equipment and piping systems to the extent possible.
   e. Document the presence of any cathodic protection system, leak detection system or other means of equipment protection, and compare with regulatory requirements.
   f. Describe and evaluate any spill prevention or containment measures including SPCC plan if available.
   g. Determine all prior tank removals or repairs. Explain the reasons for the removal or repair, and whether or not any contaminated soil or groundwater was removed or remediated. The absence of post-excavation sampling may require a phase II investigation.
   h. Indicate any plans for future tank removals or repairs.
   i. Tank testing or subsurface investigation may be required at ________'s discretion, based on available data.

4. Polychlorinated Biphenyls (PCBs)
    a. Document the location and owner of all transformers or capacitors that contain PCBs.
    b. Verify that all PCB equipment (*i.e.*, contains any PCBs), including all transformers present on the property, is in compliance with all federal, state and local laws and regulations.
    c. Visually inspect PCB equipment and determine whether there are any leaks or other hazards.
    d. Determine whether all oil-filled transformers have been classified by PCB testing (mandatory for non-utility owned transformers) or other acceptable alternative.
    e. Verify that the PCB transformers located near (provide distance) commercial buildings have been properly registered with the building owner and are subject to an appropriate inspection program.

5. Radon

Conduct radon testing at all residential properties. Test other properties as appropriate, determined by the type of property, and the likelihood (regional experience) of the presence of radon.

6. Abutting and/or Nearby Upgradient Properties

To the extent feasible and possible, consider the potential that current or prior off-site activities may result in an environmental impact. This shall include:

a. abutting properties. The presence of a gasoline service station or facility handling a significant quantity of hazardous materials adjacent to the property shall normally constitute a basis for a phase II investigation.
b. nearby (at least within 1/4 mile) upgradient properties handling hazardous materials.
c. properties with known problems within one mile upgradient of the property.

7. Asbestos
    a. Determine whether the presence of asbestos is suspected or confirmed at the Property (evaluate any available asbestos survey).
    b. Examine available building records for an indication of the presence of asbestos.
    c. Determine whether there is an Operations and Maintenance plan (O&M) in effect.

d. If an asbestos bulk survey is required it shall be conducted consistent with the ________ Realty Group Asbestos Bulk Survey Scope of Work. A "phase I" or "limited" survey is not acceptable.

C. Report Requirements

1. The consultant shall acknowledge that ________ will be relying on this report in consideration of financing of the subject property. Alternatively, the survey report shall be addressed to ________, as well as the borrower.

2. Upon completion of Phase I, an evaluation of the environmental issues associated with the Property and a recommendation regarding the need for a subsurface investigation shall be made by the consultant. Conclusions and recommendations shall be summarized in an executive summary at the front of the report.

3. The consultant's final phase I report shall also include but not be limited to:

a. narrative description of the Property and the surrounding area
b. site plans and location maps indicating all referenced locations
c. site photographs
d. a discussion of all environmental issues identified
e. a description of all applicable state and local regulatory requirements
f. limitations of the report with explanations (*e.g.*, limited access)

4. Any recommendation for a subsurface investigation shall include, as a minimum:

a. reasons leading to the recommendation
b. recommended method(s) of subsurface exploration
c. number, depth, and location of borings and wells required
d. any remote sensing techniques (*i.e.*, soil gas survey, ground penetrating radar, etc.) to be utilized, and potential limitations of results so obtained
e. substances and parameters to be tested for by laboratory analysis and the rationale for this decision (*e.g.*, products handled on this or adjacent property)
f. projected time schedule for submission of the phase II results (*e.g.*, cost estimate for all work recommended)

PHASE II - SUBSURFACE INVESTIGATION

A. General Comments

1. The requirement for a subsurface investigation shall be established by ________ ___ staff.

2. The subsurface investigation will normally include at least three monitoring wells so that the direction of groundwater flow can be identified. If it is concluded that the direction of flow can be determined by other means (*e.g.*, existing data on a nearby site), or is not necessary for the analysis, then this requirement can be removed or reduced.

3. Analysis of groundwater samples shall be based on the Phase I results, but at least one sample shall normally be analyzed for priority pollutants with additional library search unless waived by ___ staff.

4. All analyses of metals will provide both total and dissolved results unless otherwise authorized by ___ staff.

5. The use of composite sampling must be specifically authorized by ___ staff.

6. The laboratory used for sample analysis shall be certified by any regulatory agency having jurisdiction, if applicable.

7. All residual materials from the subsurface investigation shall be disposed of by the consultant in an appropriate manner, consistent with all regulatory requirements.

B. The consultant's final Phase II report shall include:

1. Conclusions about environmental condition of the site.
2. Site plans and location maps indicating all referenced locations, including soil boring and monitoring wells
3. Site photographs if not provided in the Phase I report
4. Documentation and support data
5. Evolution of all data. This shall be related to state and local regulatory requirements and any actual, "suggested," or "guidance" action levels for reporting or cleaning up soil and groundwater contamination. Any state groundwater quality standards must be specifically referenced.
6. Groundwater level and direction of flow
7. Uncertainties relating to available data
8. Recommendations for additional site investigation with projected costs
9. Recommendations for remedial measures, if required
10. Estimated length of time and cost of remediation

Form 2.12
ASBESTOS BULK SURVEY

SURVEY OF WORK

The Real Estate Security will be surveyed by a qualified asbestos consultant approved by ________. The objective of the survey is to determine whether there are any asbestos-containing materials present on the Real Estate Security. The survey should indicate the extent of any such materials and describe possible remedial measures. In addition, the consultant shall acknowledge in its report that ________ will be relying on the survey in consideration of financing the Real Estate Security.

I. SAMPLING

A. The inspector conducting the sampling shall be identified, and the details of his training and experience shall be submitted to ________. The inspector must be accredited by all agencies asserting jurisdiction.

B. Define homogeneous areas for purposes of sampling. In defining a homogeneous area, uniformity in color and texture are key criteria. The consultant may consider functional space as criteria but this may not always be necessary or appropriate. Discretion shall be exercised when defining homogeneous areas, giving consideration to all factors which could reduce the number of samples required. This might include construction practices utilized, available building specifications and as-built drawings, and the type of materials under consideration.

C. Areas of specific items not accessible for bulk sampling because of limited access or requirement for destructive sampling shall be explained in the survey report.

D. Surfacing material (including sprayed on fireproofing and textured paint) shall be sampled in a statistically random manner, that is representative of each homogeneous area, as follows:

1. At least 3 bulk samples shall be collected from each homogeneous area that is 1,000 sq. ft. or less.
2. At least 5 bulk samples shall be collected from each homogeneous area that is greater than 1,000 sq. ft. but less than or equal to 5,000 sq. ft.

3. At least 7 bulk samples shall be collected from each homogeneous area that is greater than 5,000 sq. ft.

E. Thermal system insulation shall be sampled as follows:

1. Except as provided in paragraphs 2 through 4 below, collect, in a randomly distributed manner, at least three bulk samples from each homogeneous area of thermal system insulation.
2. Collect at least one bulk sample from each homogeneous area of patched thermal system insulation if the patched section is less than 6 linear or square feet (patched areas greater than 6 feet should be treated as separate homogeneous areas).
3. In a manner sufficient to determine whether the material is ACM or not ACM, collect bulk samples from each insulated mechanical system where cement or plaster is used on fittings such as tees, elbows or valves.
4. Bulk samples are not required to be collected from any homogeneous area where the accredited inspector has determined that the thermal system insulation is fiberglass, rubber or other material which does not contain asbestos.

F. Miscellaneous material suspected of containing asbestos shall be sampled as follows:

In a manner sufficient to determine whether material is ACM or not ACM, collect bulk samples from each homogeneous area of miscellaneous material suspected of containing asbestos. The sampling protocol used must be justified.

G. If it is anticipated that more than 250 samples will be required, the sampling plan must be reviewed with ________ .

H. If there is evidence of significant disturbance to suspected ACM, or if ACM is located in the air handling system, representative airborne asbestos concentrations may be required after consultation with the Property owner and ________ staff.

I. Unless specifically directed by ________, sampling of roofing material is not included in this scope of work.

II. ANALYSIS

A. If the project schedule allows, a portion of the samples can be retained for a second round of analysis to reduce costs. This should only be done if the material within a homogeneous area is expected to be ACM. A final

determination that any material is not ACM requires analysis of all samples.
B. Bulk samples shall be analyzed for asbestos using NVLAP certified laboratories. AIHA certification is recommended.
C. Bulk samples shall be analyzed by PLM (polarized light microscopy), using the "Interim Method for the Determination of Asbestos in Bulk Insulation Samples" found at Appendix A to Subpart F in 40 CFR Part 763. Samples shall not be composited for analysis.
D. Use TEM analysis for 10% (but at least one) of the samples from each homogeneous area of vinyl tile if otherwise indicated by PLM not to contain greater than 1% asbestos. The analysis of vinyl tile shall be done using the Chatfield method currently under consideration by ASTM. No other bulk sampling shall be conducted by TEM or SEM without the approval of the ____ staff of the ________.
E. Provide quantitative results of analyses for samples found to be ACM (*i.e.*, greater than one percent asbestos by weight). Results of analyses for other samples shall be reported as "less than one percent" unless the state or local regulatory requirements in the project area require otherwise.
F. The name and address of each laboratory performing an analysis, the date of analysis, and the name and signature of the person performing the analysis shall be submitted.

III. ASSESSMENT

Provide a written assessment of all ACM on the property indicating:
A. Location, description and results of all samples and the assumptions made regarding homogeneous areas.
B. Delineation of survey limitations, including inaccessible areas, and an evaluation of the significance of these uncertainties to the overall assessment.
C. Location, type and amount of the asbestos containing materials.
D. Condition of the materials, specifying:
   1. Friability
   2. Type of damage (*e.g.*, flaking, blistering, water damage or other signs of physical damage)
   3. Severity of damage (*e.g.*, major flaking, severely torn jackets, as opposed to occasional flaking, minor tears to jackets)
   4. Extent or spread of damage over large areas or large percentage of the homogeneous area

E. Whether the material is accessible for removal.

F. The potential that the material may be disturbed. If possible, this should reference a discussion with building management regarding maintenance practices and possible renovations.

G. Known or suspected causes of damage (*e.g.*, air erosion, vandalism, vibration, water).

H. Preventive measures which might eliminate the reasonable likelihood that undamaged ACM might become damaged.

I. Any immediate hazard presented by the ACM.

J. Any hazards that might result from the continued presence of any ACM in a building.

K. Results of air sampling if available.

L. The estimated costs for removal and an Operation and Maintenance (O&M) plan. For properties that have ACM and are proposed to be subject to a program of encapsulation, enclosure or other acceptable form of abatement, the cost of that program should also be provided. Cost estimates should be qualified with regard to the level of detail and margin of error.

M. Estimated costs should also be provided (separately) for restoration to pre-removal conditions. Cost estimates should be qualified with regard to the level of detail and margin of error.

N. Estimated time required for abatement.

O. State or local asbestos requirement affecting decisions regarding abatement.

IV. DEVIATIONS

All deviations from this scope of work must have prior approval from the ________ Realty Group Architects and Engineers.

Chapter 3

Environmental Compliance Audits

§ 3.1. Overview

Companies, public entities, and federal facilities have employed a variety of environmental auditing practices in various degrees for many years. Several hundred major firms in diverse industries now have environmental auditing programs, although they are often known by other names such as assessment, survey, surveillance, review, or appraisal.

The following documents typically define the environmental audit program of a public company:

- ✓ Annual report;
- ✓ Corporate organization chart;

✓ Written environmental policies;
✓ Audit program description and charter;
✓ Protocols and procedures;
✓ Employee training manuals;
✓ Audit reports and follow-up reports/action plans;
✓ Periodic status report;
✓ Corporate hazard ranking system and notification protocol.[1]

Environmental auditing[2] includes a variety of compliance assessment techniques that may go well beyond those legally required and are used to identify actual and potential environmental problems. EPA currently endorses the practice of environmental auditing and supports its accelerated use by regulated entities to help meet the goals of federal, state, and local environmental requirements because EPA believes that effective environmental auditing can lead to higher levels of overall compliance and reduced risk to human health and the environment. The existence of an auditing program does not, however, create any defense to, or otherwise limit, the responsibility of any regulated entity to comply with applicable regulatory requirements.[3] Audits can be designed to accomplish any or all of the following:

---

[1] *See* JACQUELINE SHIELD, AIR EMISSION, BASELINES & ENVIRONMENTAL AUDITING, at 48-9 (Van Nostrand Reinhold 1993).

[2] EPA defined environmental auditing in 1986 as "a systematic, documented, periodic and objective review by regulated entities of facility operations and practices related to meeting environmental requirements." 51 Fed. Reg. 25,004 (1986). "Regulated entities" include private firms and public agencies with facilities subject to environmental regulation. Public agencies can include federal, state, or local agencies as well as special-purpose organizations such as regional sewage commissions. *Id.*

[3] 51 Fed. Reg. 25,004 (1986). Environmental audits evaluate, and are not a substitute for, direct compliance activities such as obtaining permits, installing controls, monitoring compliance, reporting violations, and keeping records. Environmental auditing may verify but does not include activities required by law, regulation, or permit (*e.g.*, continuous emissions monitoring, composite correction plans at wastewater treatment plants). Audits do not in any way replace regulatory agency inspections. Environmental audits can, however, improve compliance by complementing conventional federal, state and local oversight. *Id.*

✓ verify compliance with environmental requirements;
✓ evaluate the effectiveness of environmental management systems already in place;
✓ assess risks from regulated and unregulated materials and practices;
✓ serve as a quality assurance check to help improve the effectiveness of basic environmental management by verifying that management practices are in place, functioning, and adequate.

Environmental auditing has developed not just to deter regulatory enforcement efforts but for sound business reasons. Auditing has become a means by which regulated entities manage pollution control affirmatively over time instead of reacting to crises. Auditing generally results in

✓ improving facility environmental performance,
✓ helping to communicate effective solutions to common environmental problems,
✓ focusing the facility manager's attention on current and upcoming regulatory requirements, and
✓ generating protocols and checklists that help facilities better manage themselves.

For these reasons, auditing often results in better-integrated management of overall health and environmental hazards, since auditors frequently identify environmental liabilities that go beyond regulatory compliance.

In order to conduct a compliance audit, any or all of the following should be reviewed:[4]

Waste management permits	Sanitary sewage discharge permits Stormwater discharge permits Steam and lake discharge permits Air emission permits
On-site environmental records review	Chemical inventory records Material safety data sheets Community right-to-know (SARA Title III) Transport manifests

[4] *See generally* KATHLEEN HESS, ENVIRONMENTAL SITE ASSESSMENT—PHASE I: A BASIC GUIDE, at 90-104 (Lewis 1993).

On-site records review—Other	Process flow charts Safety and health audits Purchasing records Reference manuals/documents
Off-site records	Commercial directories Newspapers Local library "vertical files" Industrial process references
Mining/natural resources records review	Mining Oil/gas wells Water wells

Although there are many different manners in which to conduct a compliance audit, there are certain discernible elements necessary for the environmental auditing system to be considered effective.[5] These include six critical elements:

✓ Explicit top management support;
✓ Adequate team staffing;
✓ State express objectives;
✓ Document periodic reviews;
✓ Prompt corrective action;
✓ Quality assurance to prevent repeat problems.

First, there must be explicit top management support for environmental auditing and commitment to follow-up on audit findings. This may be demonstrated by a written policy articulating upper management support for the auditing program and for compliance with all pertinent requirements. It also may be manifest by written corporate policies concerning environmental management and the importance of adhering to permit requirements as well as to federal, state, and local statutes and regulations.

Management support for the auditing program also should be demonstrated by an explicit written commitment to follow up on audit

[5] *See, e.g.*, WALTER WILLBORN, COMPENDIUM OF AUDIT STANDARDS (American Society for Quality Control, 1983); *Standards for the Professional Practice of Internal Auditing* (1981, The Institute of Internal Auditors, Inc.); 51 Fed. Reg. 25,004 (1986).

findings to correct identified problems and prevent their recurrence. This would necessarily include providing financial mechanisms to complete the required environmental work.

Second, the company must provide adequate team staffing and auditor training. Environmental auditors used by the company should possess or have ready access to the knowledge, skills, and disciplines necessary to accomplish audit objectives. Each individual auditor should comply with the company's professional standards of conduct. Auditors, whether full-time or part-time, should maintain their technical and analytical competence through continuing education and training, which will probably need to be offered at the employer's expense.

Third, the audit program must explicitly state objectives, scope, resources, and frequency. At a minimum, audit objectives should include assessing compliance with applicable environmental laws and evaluating the adequacy of internal compliance policies, procedures, and personnel training programs to ensure continued compliance. Audits should be based on a process that provides auditors all corporate policies, permits, and federal,[6] state, and local regulations pertinent to the facility and checklists or protocols addressing specific features that should be evaluated by auditors. Explicit written audit procedures generally should be used for planning audits, establishing audit scope, examining and evaluating audit findings, communicating audit results, and follow-up.

---

[6] Comprehensive assessment of compliance with federal environmental regulations requires an analysis of facility performance against numerous environmental statutes and implementing regulations, including

- ✓ Resource Conservation and Recovery Act
- ✓ Federal Water Pollution Control Act
- ✓ Clean Air Act
- ✓ Hazardous Materials Transportation Act
- ✓ Toxic Substances Control Act
- ✓ Comprehensive Environmental Response, Compensation and Liability Act
- ✓ Safe Drinking Water Act
- ✓ Federal Insecticide, Fungicide and Rodenticide Act
- ✓ Marine Protection, Research and Sanctuaries Act
- ✓ Uranium Mill Tailings Radiation Control Act

In addition, consideration must be given to state and local environmental laws. Many states have been delegated authority to administer federal programs. Many local governments' building, fire, safety, and health codes also have environmental requirements relevant to an audit evaluation.

Fourth, the audit process must collect, analyze, interpret, and document information sufficient to achieve audit objectives. Information should be collected before and during an on-site visit regarding environmental compliance, environmental management effectiveness, and other matters related to audit objectives and scope. This information should be sufficient,[7] reliable,[8] relevant,[9] and useful[10] to provide a sound basis for audit findings and recommendations.

The audit process should include a periodic review of the reliability and integrity of this information and the means used to identify, measure, classify, and report it. Audit procedures, including the testing and sampling techniques employed, should be selected in advance, to the extent practical, and expanded or altered if circumstances warrant. The process of collecting, analyzing, interpreting, and documenting information should provide reasonable assurance that audit objectivity is maintained and audit goals are met.

Fifth, the audit process must include specific procedures to promptly prepare candid, clear, and appropriate written reports on audit findings, corrective actions, and schedules for implementation. Procedures should be in place to ensure that such information is communicated to managers, including facility and corporate management, who can evaluate the information and ensure correction of identified problems. Procedures also should be in place for determining what internal findings are reportable to state or federal agencies.

Finally, the audit process must include quality assurance procedures to ensure the accuracy and thoroughness of environmental audits. This quality assurance may be accomplished through supervision, independent internal reviews, external reviews, or a combination of these approaches.

---

[7] EPA defined "*sufficient* information" as "factual, adequate and convincing so that a prudent, informed person would be likely to reach the same conclusions as the auditor." 51 Fed. Reg. 25,004 (1986).

[8] EPA defined "*reliable* information" as the "best attainable through use of appropriate audit techniques." 51 Fed. Reg. 25,004 (1986).

[9] EPA defined "*relevant* information" as supporting "audit findings and recommendations and is consistent with the objectives for the audit." 51 Fed. Reg. 25,004 (1986).

[10] EPA defined "*useful* information" as that which "helps the organization meet its goals." 51 Fed. Reg. 25,004 (1986).

An environmental audit could go well beyond the type of compliance assessment normally conducted during regulatory inspections, for example, by evaluating policies and practices, regardless of whether they are part of the environmental system or the operating and maintenance procedures. Specifically, audits should evaluate the extent to which systems or procedures:

1. **Develop organizational environmental policies** that
   a. implement regulatory requirements and
   b. provide management guidance for environmental hazards not specifically addressed in regulations;
2. **Train and motivate facility personnel** to work in an environmentally acceptable manner and to understand and comply with government regulations and the entity's environmental policy;
3. **Communicate relevant environmental developments** expeditiously to facility and other personnel;
4. **Communicate effectively with government** and the public regarding serious environmental incidents;
5. **Require that third parties** working for, with, or on behalf of the organization follow its environmental procedures;
6. **Make proficient personnel available** at all times to carry out environmental (especially emergency) procedures;
7. **Incorporate environmental protection into written operating procedures;**
8. **Apply best management practices** and operating procedures, including "good housekeeping" techniques;
9. **Institute preventive and corrective maintenance** systems to minimize actual and potential environmental harm;
10. **Utilize best available process and control technologies;**
11. **Use the most effective sampling** and monitoring techniques, test methods, recordkeeping systems, or reporting protocols (beyond minimum legal requirements);
12. **Evaluate causes** behind any serious environmental incidents and establish procedures to avoid recurrence;
13. **Exploit source reduction, recycle, and reuse** potential wherever practical; and

14. **Substitute materials or processes** to allow use of the least hazardous substances feasible.[11]

The precise actions necessary for an effective program to prevent and detect environmental problems or violations of law will depend upon a number of factors. Among the relevant factors are

1. the size of the organization;[12]
2. the likelihood that certain offenses may occur because of the nature of its business;[13] and
3. the prior history of the organization.[14]

An organization's failure to incorporate and follow applicable industry practice or the standards called for by an applicable governmental regulation weighs against a finding of an effective program to prevent and detect violations of law or environmental problems. EPA and the Department of Justice consider the "corporate attitude" of a company when making enforcement decisions. These include: voluntary compliance with pending environmental requirements significantly before such compliance is actually required; additional voluntary environmental cleanup; or pollution prevention or reduction measures performed, above and beyond those required by environmental statutes or regulations.

[11] 51 Fed. Reg. 25,004 (1986). *Accord* 56 Fed. Reg. 64,785 (1991). (The corporate attitude, policies, practices, and procedures of a listed facility's owner, operator, or supervisor will always be relevant when a facility that has been listed as the result of a criminal conviction requests removal from the List. How significant a factor the corporate attitude, policies, practices, and procedures will be depends upon the degree of intent involved in the violation at issue.)

[12] *Id.* The requisite degree of formality of a program to prevent and detect violations of law or environmental problems will vary with the size of the organization; the larger the organization, the more formal the program should typically be.

[13] If, because of the nature of an organization's business, there is a substantial risk that certain types of offenses or violations may occur, the organization must have taken steps to prevent and detect those types of offenses or violations. For example, if an organization handles toxic substances, it must have established standards and procedures designed to ensure that those substances are handled properly at all times.

[14] An organization's prior history may indicate types of offenses or violations that it should have taken actions to prevent. Recurrence of misconduct similar to that which an organization has previously committed casts doubt on whether it took all reasonable steps to prevent such misconduct.

§ 3.2 EPA Policy

§ 3.2.1 EPA's Historical Policy

In 1985, EPA decided to clarify its position regarding auditing in order to help encourage regulated entities to establish audit programs or upgrade systems already in place.[15] After notice and comment, EPA issued its final policy statement on audits, effective July 9, 1986.[16]

The 1986 policy statement specifically:

✓ Encouraged regulated entities to develop, implement, and upgrade an environmental auditing program;

✓ Discussed when the EPA may or may not request audit reports;

✓ Explained how EPA's inspection and enforcement activities may respond to regulated entities' efforts to ensure compliance through auditing;

✓ Endorsed environmental auditing at federal facilities;

✓ Encouraged state and local environmental auditing initiatives;[17] and

✓ Outlined elements of effective audit programs.

Certain very important points emerged from this early EPA enunciated policy.[18] First, although EPA sought to encourage regulated entities to modify environmental practices, EPA did not intend to mandate auditing (though in certain instances EPA would include provisions for environmental auditing as part of settlement agreements). According to EPA, auditing is—and has always been—a voluntary activity.

---

[15] 50 Fed. Reg. 46,504 (Nov. 8, 1985). EPA first published its auditing policy as interim guidance.

[16] 51 Fed. Reg. 25,004 (July 9, 1986).

[17] This includes encouraging states to adopt either its policy or similar and equally effective policies in order to advance the use of environmental auditing on a consistent, nationwide basis.

[18] EPA reiterated its historic policy on auditing again in 1991, when it articulated what it would consider an adequate change in "corporate attitude" for purposes of removing facilities from the EPA list of violating facilities following a criminal conviction. 56 Fed. Reg. 64,785 (Dec. 12, 1991). In that policy statement, EPA explained that the "hallmark of an effective program is that the organization exercises due diligence in seeking to prevent and detect environmental problems or violations or criminal conduct." *Id.*

Second, EPA said it would not routinely request environmental audit reports.[19] The agency recognized that to do so could inhibit the quality of audits conducted. EPA would request an audit report, or relevant portions thereof, only where the agency determined it needed to accomplish a statutory mission, or where the government considered it material to a criminal investigation.[20]

Third, EPA's 1986 auditing policy did not alter regulated entities' existing or future obligations to monitor, record, or report information required under environmental statutes, regulations, or permits, or to allow EPA access to that information. EPA retained its authority to request and receive any relevant information—including that contained in audit reports—under various environmental statutes *(e.g.,* Clean Water Act section 308, Clean Air Act sections 114 and 208) or in other administrative or judicial proceedings.[21]

Fourth, EPA continued to inspect self-audited facilities, although facilities with a good compliance history could be subject to fewer inspections. Recently, EPA expanded its position on voluntary auditing by creating what it called an "Environmental Leadership Program," which provides "special recognition to companies and facilities that reflect a

---

[19] An "environmental audit report" is a written report that candidly and thoroughly presents findings from a review conducted as part of an environmental audit of facility environmental performance and practices. An audit report is not a substitute for compliance monitoring reports or other reports or records that may be required by EPA or other regulatory agencies. 51 Fed. Reg. 25,004 (1986).

[20] Such requests "are limited, focused on particular information needs and are usually made only where the information needed cannot be obtained from monitoring, reporting or other data otherwise available to the Agency." 51 Fed. Reg. 25,004 (1986). Examples included situations where

1. audits are conducted under consent decrees or other settlement agreements;
2. a company has placed its management practices at issue by raising them as a defense; or
3. state of mind or intent are a relevant element of inquiry, such as during a criminal investigation.

[21] Audit findings are often required by law to be reported to government agencies. EPA encouraged regulated entities to notify appropriate state or federal officials of findings that suggest significant environmental or public health risks, even when not specifically required to do so. 51 Fed. Reg. 25,004 (1986).

CEO-level commitment to prevention-oriented environmental management that goes beyond mere compliance with regulations."[22]

Finally, in fashioning enforcement responses to violations, EPA policy since 1986 was to take into account, on a case-by-case basis, "the honest and genuine efforts of regulated entities to avoid and promptly correct violations and underlying environmental problems."[23] Evidence of such activity included auditing. When regulated entities took reasonable precautions to avoid noncompliance, expeditiously corrected underlying environmental problems discovered through audits or other means, and implemented measures to prevent their recurrence, EPA could exercise its prosecutorial discretion.[24] Leniency is especially appropriate under the 1986 policy when a regulated entity promptly reports violations or compliance data that otherwise were not required to be recorded or reported to EPA.

---

[22] 58 Fed. Reg. 4,802 (Jan. 15, 1993). For example, EPA is considering asking participating companies to publish a plan for pollution prevention that has been endorsed by the company's chief executive officer. Such a plan would describe how the company intends to reduce discharges to the environment by developing better recycling, treatment, and disposal programs. *Id.*

[23] *Id.*

[24] In 1990, EPA formally recognized that the condition leading to a conviction under section 309(c) of the CWA or section 113(c) of the CAA could include a convicted environmental violator's corporate attitude, policies, practices, and procedures regarding environmental compliance. In the Matter of Valmont Industries, Inc. (ML Docket No. 07-89-L068, Jan. 12, 1990) ("Valmont"). In *Valmont,* the decisions of both the Assistant Administrator for Enforcement (AA) and the EPA Case Examiner established the principle that the presence of a poor corporate attitude regarding compliance with environmental standards, thus creating a climate facilitating the likelihood of a violation, may be part of the condition giving rise to the condition that must be corrected prior to removal of the facility from the List. 40 C.F.R. § 15.20.

Valmont was convicted of crimes of falsification and deception. The AA determined that not only was Valmont required to correct the physical conditions that led to its conviction but that it also was required to demonstrate that it had implemented appropriate corporate policies, practices, and procedures designed to ensure that the mere appearance of compliance with environmental standards was not put above actual compliance with those standards. The Case Examiner later affirmed the use of the corporate attitude standard in determining whether the condition leading to listing has been corrected.

Following *Valmont,* EPA has applied the "corporate attitude test" in other cases where facilities have requested removal from the List, including cases involving knowing or negligent conduct, not those involving deliberate deception. *See* Colorado River Sewage System Joint Venture (ML Docket No. 09-89-L047, August 20, 1991); Zarcon Corp. (ML Docket No. 09-89-L058, Aug. 1, 1990); Sellen Construction Co. (ML Docket No. 10-89-L073, June 13, 1990).

For purposes of evaluating whether an organization has met its due diligence obligation, EPA defined due diligence as requiring, at a minimum, that the organization has taken at least the following types of steps to ensure compliance with environmental requirements.[25] First, the organization must have written policies defining the standards and procedures to be followed by its agents or employees, and these standards must have been consistently enforced through appropriate disciplinary mechanisms. Second, the organization must have specific high-level persons, not reporting to production managers, who have authority to ensure compliance with those standards and procedures. Third, the organization must have effectively communicated its standards and procedures to agents and employees.[26] Fourth, the organization must establish or have established an effective program for enforcing its standards.[27] Fifth, after an offense or a violation has been detected, the organization must have immediately taken appropriate steps to correct the condition and must also take all reasonable steps to prevent further similar offenses or violations. This includes notifying appropriate authorities of such offenses or violations, making any necessary modifications to the organization's program to prevent and detect environmental problems or violations of law, and disciplining individuals responsible for the offense or violation.[28]

§ 3.2.2 The 1995 Interim Audit Policy

On March 31, 1995, EPA announced as part of its interim guidance "incentives for those who take responsibility for voluntarily evaluating, disclosing and correcting violations."[29] EPA explained that "One of the

---

[25] 58 Fed. Reg. 4,802 (Jan. 15, 1993).

[26] *Id.* For example, by requiring participation in training programs and by the dissemination of publications.

[27] *Id.* For example, monitoring and auditing systems designed to prevent or detect noncompliance; and a well-publicized system, under which agents and employees are encouraged to report, without fear of retaliation, evidence of environmental problems or violations, or criminal conduct within the organization.

[28] *Id.* This may include conducting an independent environmental audit to ensure that there are no other environmental problems or violations at the facility.

[29] 60 Fed. Reg. 16,875 (April 3, 1995). *See generally* ENFORCEMENT *New Policy Creates Uncertainty, Obstacles To Audits, Industry Counsel Says,* BNA ENVIRONMENT DAILY

Environmental Protection Agency's most important responsibilities is obtaining compliance with federal laws that protect public health and safeguard the environment."[30] EPA acknowledged that the goal of obtaining optimal compliance "can be achieved only with the voluntary cooperation of thousands of businesses and other regulated entities subject to these requirements."[31] The policy is designed to "promote candid and thorough self-policing by providing greater certainty as to how EPA will exercise its criminal investigative discretion to encourage voluntary disclosure and prompt correction by regulated entities."[32]

The interim policy applies to violations under all of the federal environmental statutes that EPA administers and supersedes (unless otherwise noted) any conflicting or inconsistent provisions in the media-specific penalty or enforcement response policies and EPA's 1986 Environmental Auditing Policy Statement.[33]

EPA's interim policy was three pronged:

1. EPA "will completely eliminate gravity-based (or 'punitive')[34] penalties for companies or public agencies that voluntarily identify, disclose and correct violations according to the conditions outlined in this policy;"[35]
2. EPA will not recommend to the Department of Justice that criminal charges be brought against a company acting in good faith to identify, disclose, and correct violations, so long as no serious actual harm has occurred; and
3. EPA will not request voluntary environmental audits to trigger enforcement investigations.[36]

EPA caveated its policy by stating that "The incentives offered in this policy have been structured above all to protect human health and the

---

(Apr. 4, 1995); Enforcement *EPA Policy Offers No Audit Privilege; Lack Of Prosecution, Punitive Fines Possible, BNA* Environment Daily (Apr. 3, 1995).

[30] 60 Fed. Reg. 16,875 (April 3, 1995).

[31] *Id.*

[32] *Id.* at 16,879.

[33] *Id.* at 16,878.

[34] *Id.* at 16,875.

[35] *Id.* at 16,875.

[36] *Id.* at 16,875.

environment."[37] Thus, even where the conditions for mitigated enforcement are met, EPA specifically reserved the right to collect full civil penalties for criminal conduct, violations that present an imminent and substantial endangerment or result in serious actual harm, or repeat violations. This right to revoke its stated guidance is premised on two overriding concepts that have historically guided agency use of audits:

✓ Sources will not be allowed to gain an economic advantage over their competitors by delaying their investment in compliance.

✓ EPA will not hesitate to bring a criminal action against individuals responsible for criminal conduct.[38]

EPA set out seven conditions for reducing civil penalties and not making criminal referrals. These are as follows:

1. Voluntary self-policing[39]
2. Voluntary disclosure[40]
3. Prompt correction[41]
4. Remediation of imminent and substantial endangerment[42]
5. Remediation of harm and prevention of repeat violations[43]
6. No lack of appropriate preventive measures[44]
7. Cooperation[45]

---

[37] *Id.* at 16,875.

[38] *Id.*

[39] *Id.* at 16,877. The regulated entity discovers a violation through a voluntary environmental audit or voluntary self-evaluation appropriate to the size and nature of the regulated entity.

[40] 60 Fed. Reg. at 16,877. The regulated entity fully and voluntarily discloses the violation in writing to all appropriate federal, state, and local agencies as soon as it is discovered (including a reasonable time to determine that a violation exists), and prior to (1) the commencement of a federal, state, or local agency inspection, investigation or information request; (2) notice of a citizen suit; (3) legal complaint by a third party; or (4) the regulated entity's knowledge that the discovery of the violation by a regulatory agency or third party was imminent.

[41] *Id.* The regulated entity corrects the violation either within 60 days of discovering the violation or, if more time is needed, as expeditiously as practicable.

[42] *Id.* The regulated entity expeditiously remedies any condition that has created or may create an imminent and substantial endangerment to human health or the environment.

[43] *Id.* The regulated entity implements appropriate measures to remedy any environmental harm due to the violation and to prevent a recurrence of the violation.

[44] *Id.* The violation does not indicate that the regulated entity has failed to take appropriate steps to avoid repeat or recurring violations.

[45] The regulated entity cooperates as required by EPA and provides such information

EPA will retain its full discretion to recover any economic benefit gained as a result of noncompliance to preserve a "level playing field" in which violators do not gain a competitive advantage through noncompliance. EPA may, however, forgive the entire penalty for violations that meet the enumerated conditions but, in EPA's discretion, do not merit any penalty due to the insignificant amount of any economic benefit.[46]

Where appropriate, EPA may require that to satisfy any of these conditions, a regulated entity must enter into a written agreement, administrative consent order, or judicial consent decree, particularly where compliance or remedial measures are complex or a lengthy schedule for attaining and maintaining compliance or remediating harm is required.[47]

The "interim policy is intended to promote environmental compliance by providing greater certainty as to EPA's enforcement response to voluntary self-evaluations, and voluntary disclosure and prompt correction of violations."[48] The policy is further designed to provide guidance for states and local authorities in encouraging voluntary compliance initiatives among regulated entities.[49]

---

as is reasonably necessary and required by EPA to determine applicability of this policy. Cooperation may include providing all requested documents and access to employees and assistance in any further investigations into the violation. 60 Fed. Reg. at 16,877.

[46] EPA will not recommend to the Department of Justice that criminal charges be brought against a regulated entity where EPA determines that the seven enumerated conditions for reduction of civil penalties are met, and the violation does not demonstrate or involve (1) a prevalent corporate management philosophy or practice that concealed or condoned environmental violations; (2) high-level corporate officials' or managers' conscious involvement in or willful blindness to the violation; or (3) serious actual harm to human health or the environment.

EPA's policy does not apply to criminal acts of individual managers or employees. Where EPA determines that a criminal referral to the Department of Justice is unwarranted, EPA may nonetheless proceed with civil enforcement or other applicable enforcement response and penalty policies. *Id.* at 16,878.

[47] *Id.* at 16,877.

[48] *Id.*

[49] For purposes of the interim policy, the following definitions were developed:
"Environmental auditing" utilizes the definition given in EPA's 1986 policy on environmental auditing, *i.e.*, "a systematic, documented, periodic and objective review by regulated entities of facility operations and practices related to meeting environmental requirements."

Federal laws and regulations set minimum standards for protecting human health and achieving environmental protection goals such as clean air and clean water. EPA will continue to uphold these laws through vigorous enforcement actions that appropriately penalize violators. Penalties help ensure a level playing field by ensuring that violators do not obtain an unfair economic advantage over their competitors who made the necessary investment in compliance. Penalties also promote protection of the environment and public health by encouraging adoption of pollution prevention and recycling practices that limit exposure to liability for pollutant discharges and deter future violations by the violator and others.

At the same time, EPA recognizes that the agency cannot achieve maximum compliance without the cooperation of a regulated community willing to act responsibly by detecting, disclosing, and correcting violations. Already, regulated entities have many compelling incentives to implement environmental management/auditing systems. Recent surveys show that the vast majority of large companies engage in environmental auditing and/or have environmental management systems in place. EPA has concluded that the additional incentives will further promote the regulated community's commitment to adopting systems for maximizing compliance.[50] Seven principles guide EPA's policy for those engaged in voluntary compliance activities:

1. Self-policing by regulated entities can play a crucial role in finding, fixing, and preventing violations.
2. Violations discovered through self-policing should be disclosed and promptly corrected.

"Environmental audit report" means all documentation of information relating to an environmental audit, but not including the factual information underlying or testimonial evidence relating to such information.

"Regulated entity" means any entity, including a federal, state, and municipal facility, regulated under the federal environmental laws that EPA administers.

"Self-evaluation" means an assessment, not necessarily meeting all the criteria of a full environmental audit, by a regulated entity of its compliance with one or more environmental requirements.

"Voluntary" means not required by statute, regulation, permit, order, or agreement.

60 Fed. Reg. at 16,877.

[50] 60 Fed. Reg. 16,875–76.

3. Regulated entities that self-police and that voluntarily disclose and self-correct violations in accordance with this policy should be assessed penalties that are consistently and predictably lower than penalties for those who do not.
4. Regulated entities that self-police and voluntarily disclose and self-correct violations in accordance with this policy should also not be recommended for criminal prosecution.
5. Providing predictable incentives for voluntary disclosure and correction of violations identified through self-policing offers a positive alternative to across-the-board privileges and immunities that could be used to shield criminal misconduct, drive up litigation costs, and create an atmosphere of distrust among regulators, industry, and local communities.
6. EPA should not seek voluntary environmental audit information to trigger an investigation of a civil or criminal violation of environmental laws.
7. To preserve a level playing field, EPA should recover any economic benefit realized from violations of environmental law.

EPA also recognizes the development of and growing reliance on international voluntary environmental management standards in the U.S. and other countries.[51] "These standards, if properly crafted and implemented, can provide a powerful tool for organizations to improve their overall compliance with environmental requirements and move beyond compliance through innovative approaches to pollution prevention."[52]

EPA believes that its interim policy "offers a positive alternative to across-the-board privileges and immunities that could be used to shield criminal misconduct, drive up litigation costs and create an atmosphere of distrust between regulators, industry and local communities."[53]

In adopting the policy, EPA considered and expressly rejected the penalty immunity policies enacted or urged in many state legislatures. EPA explained that

---

[51] *Id.* at 16,877.
[52] *Id.* at 16,878.
[53] *Id.* at 16,876.

> Penalty immunity provisions for voluntary disclosures of violations can give lawbreakers an economic advantage over their law-abiding competitors. It makes sense to give substantial penalty reductions for those who come forward with their violations and promptly correct them, but to maintain a level playing field, the federal and state governments must be able to recoup the economic benefit of violations.

According to EPA, "A principal rationale for environmental audit privileges and penalty immunities for voluntary disclosures is to reduce the exposure of regulated entities that conduct self-evaluations and act on the findings by immediately correcting violations." EPA believes the interim policy adequately addresses those concerns without instituting a penalty immunity policy.

Even more significant, EPA determined that "Privilege runs counter to efforts to open up environmental decisionmaking and encourage public participation in matters that affect people's homes, workplaces and communities." The agency feared that an "environmental audit privilege could be misused to shield bad actors or to frustrate access to crucial factual information."[54] In addition, EPA asserted that "environmental audit privileges and penalty immunities could encourage increased litigation as opposing lawyers battle over what is privileged or immune from penalties and what is not."[55]

While recognizing "that states are important partners in federal enforcement, and that it is desirable to create a climate in which states can be innovative,"[56] EPA stated that it is required "to establish a certain

---

[54] *Id.* at 16,878.

[55] *Id.* ("Litigation over the scope of the privileges and immunities could burden our already taxed judicial system, drain government and private resources, and in some cases prevent quick action to address environmental emergencies.") The EPA further explained:

> The Supreme Court has noted, "privileges are not lightly created nor expansively construed for they are in derogation of the search for the truth." United States v. Nixon, 418 U.S. 683, 710 (1974). Moreover, the self-evaluation privilege has regularly and uniformly been rejected by the courts in cases where documents were sought by a governmental agency.

60 Fed. Reg. 16, 878.

[56] *Id.* at 16,878.

minimum consistency in federal enforcement, so that the sanctions a business faces for violating federal law do not depend on where the business is located."[57]

Accordingly, to maintain national consistency, EPA again decreed that "it will scrutinize enforcement more closely in states with audit privilege and/or penalty immunity laws." EPA said it may find it necessary to increase federal enforcement where environmental self-evaluation privileges or penalty immunities prevent a state from obtaining

1. information needed to establish criminal liability;
2. facts needed to establish the nature and extent of a violation;
3. appropriate penalties for imminent and substantial endangerment or serious harm to human health or the environment, or from recovering economic benefit;
4. appropriate sanctions or penalties for criminal conduct and repeat violations; or
5. prompt correction of violations, and expeditious remediation of those that involve imminent and substantial endangerment to human health or the environment.[58]

EPA will bring to the state's attention any provisions of state audit privilege and/or penalty immunity statutes that raise any of the concerns outlined above, and will work with the state to address those concerns and ensure that federal requirements are satisfied.[59]

In addition to the paper tiger threat to the states, a serious limitation is further placed on the potential usefulness of EPA's interim policy. The interim policy "is not final agency action, but is intended solely as guidance." Deviations from the guidelines, where merited, are authorized as long as the reasons for the deviations are documented.[60] As EPA expressly stated, the EPA interim auditing policy "is not intended, nor can it be relied upon, to create any rights enforceable by any party in litigation with the United States." Moreover, the policy may be revoked by EPA at any time without public notice or comment.[61]

---

[57] *Id.*

[58] *Id.*

[59] *Id.* at 16,879.

[60] *Id.*

[61] *Id.*

§ 3.2.3 EPA's Current Audit Policy

After holding numerous hearings addressing the question of whether or not EPA should alter its policy on environmental audits, on March 31, 1995, EPA announced its interim guidance "incentives for those who take responsibility for voluntarily evaluating, disclosing and correcting violations."[62] In that preliminary policy, EPA explained that "One of the Environmental Protection Agency's most important responsibilities is obtaining compliance with federal laws that protect public health and safeguard the environment."[63] EPA acknowledged that the goal of obtaining optimal compliance "can be achieved only with the voluntary cooperation of thousands of businesses and other regulated entities subject to these requirements."[64]

On December 22, 1995, EPA issued its final policy "to enhance protection of human health and the environment by encouraging regulated entities to voluntarily discover[65] and disclose and correct[66] violations of

---

[62] *Id.* at 16,875.

[63] *Id.*

[64] *Id.*

[65] The final policy generally applies to any violation that is voluntarily discovered, regardless of whether the violation is required to be reported. In order to be "voluntary," the violation must be identified and disclosed by the regulated entity prior to the commencement of a federal state or local agency inspection, investigation, or information request; notice of a citizen suit; legal complaint by a third party; the reporting of the violation to EPA by a "whistleblower" employee; and imminent discovery of the violation by a regulatory agency. Thus, the regulated entities must have taken the initiative to find violations and promptly report them, rather than reacting to knowledge of a pending enforcement action or third-party complaint.

The EPA policy gives three specific examples of discovery that would not be voluntary, and therefore would not be eligible for penalty mitigation:

> emissions violations detected through a required continuous emissions monitor, violations of NPDES discharge limits found through prescribed monitoring, and violations discovered through a compliance audit required to be performed by the terms of a consent order or settlement agreement.

[66] In order to receive the penalty mitigation benefits available, the regulated entity not only voluntarily discovers and promptly discloses a violation, but

1. expeditiously corrects it,
2. remedies any harm caused by that violation (including responding to any spill and carrying out any removal or remedial action required by law), and
3. expeditiously certifies in writing to appropriate state, local and EPA authorities that violations have been corrected.

environmental requirements."[67] The policy was developed in close consultation with thc U.S. Department of Justice, the states, public interest groups, and the regulated community. EPA has promised to apply the policy "uniformly by the Agency's enforcement programs."[68] Incentives in the new policy include

- ✓ eliminating or substantially reducing the gravity component of civil penalties and
- ✓ not recommending cases for criminal prosecution where specified conditions are met, for those who voluntarily self-disclose and promptly correct violations.[69]

Under the new EPA policy, "where violations are found through voluntary environmental audits or efforts that reflect a regulated entity's due diligence, and are promptly disclosed and expeditiously corrected, EPA will not seek gravity-based (*i.e.*, non-economic benefit) penalties[70] and will generally not recommend criminal prosecution against the regulated entity.[71] EPA will reduce gravity-based penalties by 75 percent for

---

The final policy requires that the violation be corrected within 60 days, or that the regulated entity provide written notice where violations may take longer to correct. EPA recognizes that some violations can and should be corrected immediately, while others (*e.g.*, where capital expenditures are involved), may take longer than 60 days to correct. In all cases, the regulated entity will be expected to do its utmost to achieve or return to compliance as expeditiously as possible.

Where correction of the violation depends upon issuance of a permit that has been applied for but not issued by federal or state authorities, the agency will, where appropriate, make reasonable efforts to secure timely review of the permit.

[67] 60 Fed. Reg. 66,706 (Dec. 22, 1995).

[68] *Id.*

[69] Criminal prosecution will almost always be recommended where the violation demonstrates or involves

(i) a prevalent management philosophy or practice that concealed or condoned environmental violations; or

(ii) high-level corporate officials' or managers' conscious involvement in, or willful blindness to, the violations.

[70] EPA defined "Gravity-based penalties" as "that portion of a penalty over and above the economic benefit., *i.e.*, the punitive portion of the penalty, rather than that portion representing a defendant's economic gain from non-compliance."

[71] Whether or not EPA refers the regulated entity for criminal prosecution, EPA expressly reserved the right to recommend prosecution for the criminal acts of individual managers or employees under other existing EPA and DOJ policies guiding the exercise of enforcement discretion.

violations that are voluntarily discovered, and are promptly disclosed and corrected, even if not found through a formal audit or due diligence.'"[72] EPA explained:

> The policy appropriately limits the complete waiver of gravity-based civil penalties to companies that meet the higher standard of environmental auditing or systematic compliance management. However, to provide additional encouragement for the kind of self-policing that benefits the public, gravity-based penalties will be reduced by 75% for a violation that is voluntarily discovered, promptly disclosed and expeditiously corrected, even if it was not found through an environmental audit and the company cannot document due diligence. EPA expects that this will encourage companies to come forward and work with the Agency to resolve environmental problems and begin to develop an effective compliance management program.[73]

EPA may require as a condition of penalty mitigation that a description of the regulated entity's due diligence efforts be made publicly available.

EPA's policy was designed to balance the concerns of business not to be penalized for their regulatory compliance efforts against the public's need for transparency and disclosure of events taking place in their community. Accordingly, the policy included important "safeguards to deter irresponsible behavior and protect the public and environment.'"[74] In addition to prompt disclosure and expeditious correction, "the policy requires companies to act to prevent recurrence of the violation and to remedy any environmental harm which may have occurred." Thus, repeated violations or those that result in actual harm or may present imminent and substantial endangerment are not eligible for relief or penalty mitigation.

Moreover, EPA has retained its policy of maintaining "an equal playing field." No company will be allowed to gain an economic advantage over competitors by delaying their investment in environmental compliance. Economic benefits may be waived, however, where the EPA determines that it is insignificant. EPA explained:

---

[72] 60 Fed. Reg. 66,706.

[73] *Id.*

[74] *Id.*

> After considering public comment, EPA has decided to retain the discretion to recover economic benefit for two reasons. First, it provides an incentive to comply on time. Taxpayers expect to pay interest or a penalty fee if their tax payments are late; the same principle should apply to corporations that have delayed their investment in compliance. Second, it is fair because it protects responsible companies from being undercut by their noncomplying competitors, thereby preserving a level playing field. The concept of recovering economic benefit was supported in public comments by many stakeholders, including industry representatives.[75]

Perhaps most significant, corporations remain criminally liable for violations that result from conscious disregard of their obligations under the law. Private individuals are similarly liable for criminal misconduct. EPA tells the public that the agency "has never recommended criminal prosecution of a regulated entity based on voluntary disclosure of violations discovered through audits and disclosed to the government before an investigation was already under way."[76] The new policy will not modify EPA's practice. Thus, EPA generally will not recommend criminal prosecution for a regulated entity that uncovers violations through environmental audits or due diligence, promptly discloses, and expeditiously corrects those violations.

To benefit from EPA's audit policy, the regulated entity must meet nine criteria:

1. Discovery of the violation through an environmental audit[77] or

---

[75] *Id.*

[76] *Id.*

[77] For purposes of the EPA policy, the term "Environmental Audit" has the definition given to it in EPA's 1986 audit policy on environmental auditing, *i.e.*, "a systematic, documented, periodic and objective review by regulated entities of facility operations and practices related to meeting environmental requirements."

Environmental audit report" means "the analysis, conclusions, and recommendations resulting from an environmental audit, but does not include data obtained in, or testimonial evidence concerning, the environmental audit."

due diligence,[78] or other systematic discovery[79]

2. Voluntary discovery[80]

---

[78] For purposes of the EPA policy, the term "due diligence" encompasses the regulated entity's systematic efforts, appropriate to the size and nature of its business, to prevent, detect, and correct violations through all of the following:

(a) Compliance policies, standards, and procedures that identify how employees and agents are to meet the requirements of laws, regulations, permits and other sources of authority for environmental requirements;

(b) Assignment of overall responsibility for overseeing compliance with policies, standards, and procedures, and assignment of specific responsibility for ensuring compliance at each facility or operation;

(c) Mechanisms for systematically assuring that compliance policies, standards and procedures are being carried out, including monitoring and auditing systems reasonably designed to detect and correct violations, periodic evaluation of the overall performance of the compliance management system, and a means for employees or agents to report violations of environmental requirements without fear of retaliation;

(d) Efforts to communicate effectively the regulated entity's standards and procedures to all employees and other agents;

(e) Appropriate incentives to managers and employees to perform in accordance with the compliance policies, standards, and procedures, including consistent enforcement through appropriate disciplinary mechanisms; and

(f) Procedures for the prompt and appropriate correction of any violations, and any necessary modifications to the regulated entity's program to prevent future violations.

60 Fed. Reg. at 66,711-12.

[79] The violation was discovered through:

(a) an environmental audit; or

(b) an objective, documented, systematic procedure or practice reflecting the regulated entity's due diligence in preventing, detecting, and correcting violations. The regulated entity must provide accurate and complete documentation to the Agency as to how it exercises due diligence to prevent, detect and correct violations according to the criteria for due diligence outlined in Section B. EPA may require as a condition of penalty mitigation that a description of the regulated entity's due diligence efforts be made publicly available.

[80] The violation was identified voluntarily, and not through a legally mandated monitoring or sampling requirement prescribed by statute, regulation, permit, judicial or administrative order, or consent agreement. For example, the policy does not apply to

(a) emissions violations detected through a continuous emissions monitor (or alternative monitor established in a permit) where any such monitoring is required;

(b) violations of National Pollutant Discharge Elimination System (NPDES) discharge limits detected through required sampling or monitoring;

(c) violations discovered through a compliance audit required to be performed by the terms of a consent order or settlement agreement.

3. Prompt disclosure[81]
4. Discovery and disclosure independent of government or third party plaintiff[82]
5. Correction and remediation[83]
6. Prevent recurrence[84]
7. No repeat violations[85]

---

[81] The regulated entity fully discloses a specific violation within 10 days (or such shorter period provided by law) after it has discovered that the violation has occurred, or may have occurred, in writing to EPA.

[82] The violation must also be identified and disclosed by the regulated entity prior to:
(a) the commencement of a federal, state, or local agency inspection or investigation, or the issuance by such agency of an information request to the regulated entity;
(b) notice of a citizen suit;
(c) the filing of a complaint by a third party;
(d) the reporting of the violation to EPA (or other government agency) by a "whistleblower" employee, rather than by one authorized to speak on behalf of the regulated entity; or
(e) imminent discovery of the violation by a regulatory agency.

[83] The regulated entity corrects the violation within 60 days, certifies in writing that violations have been corrected, and takes appropriate measures as determined by EPA to remedy any environmental or human harm due to the violation. If more than 60 days will be needed to correct the violation(s), the regulated entity must so notify EPA in writing before the 60-day period has passed. Where appropriate, EPA may require that to satisfy this condition, a regulated entity must enter into a publicly available written agreement, administrative consent order, or judicial consent decree, particularly where compliance or remedial measures are complex or a lengthy schedule for attaining and maintaining compliance or remediating harm is required.

[84] The regulated entity agrees in writing to take steps to prevent a recurrence of the violation, which may include improvements to its environmental auditing or due diligence efforts. 60 Fed. Reg. at 66,712.

[85] The specific violation (or closely related violation) has not occurred previously within the past three years at the same facility, or is not part of a pattern of federal, state, or local violations by the facility's parent organization (if any), which have occurred within the past five years. For the purposes of this section, a violation is
(a) any violation of federal, state, or local environmental law identified in a judicial or administrative order, consent agreement or order, complaint, or notice of violation, conviction, or plea agreement; or
(b) any act or omission for which the regulated entity has previously received penalty mitigation from EPA or a state or local agency.

8. Other violations excluded[86]
9. Cooperation[87]

EPA, thus, retains *inter alia* the proviso that the entity must "cooperate" with the agency before it will decide not to prosecute criminally.

EPA's policy, which became effective January 22, 1996, restates the agency's long-standing practice of not **routinely** requesting voluntary audit reports to trigger enforcement investigations. Like earlier policies, however, the most recent policy retains the agency's right to request an audit:

> If the Agency has independent evidence of a violation, it may seek information needed to establish the extent and nature of the problem and the degree of culpability. In general, however, an audit which results in prompt correction clearly will reduce liability, not expand it. Furthermore, a review of the criminal docket did not reveal a single criminal prosecution for violations discovered as a result of an audit self-disclosed to the government.[88]

EPA has promised to study the effects of the new audit policy over the next three years to determine the effectiveness of the new approach. EPA continues to oppose state environmental audit privilege legislation.[89]

---

[86] The violation is not one that

> (i) resulted in serious actual harm, or may have presented an imminent and substantial endangerment, to human health or the environment, or
>
> (ii) violates the specific terms of any judicial or administrative order, or consent agreement.

[87] The regulated entity cooperates as requested by EPA and provides such information as is necessary and requested by EPA to determine applicability of this policy. Cooperation includes, at a minimum, providing all requested documents and access to employees and assistance in investigating the violation, any noncompliance problems related to the disclosure, and any environmental consequences related to the violations.

[88] 60 Fed. Reg. 66,712.

[89] EPA said:

> As always, states are encouraged to experiment with different approaches that do not jeopardize the fundamental national interest in assuring that violations of federal law do not threaten the public health or the environment, or make it profitable not to comply. The Agency remains opposed to state legislation that does not include these basic protections, and reserves its right to bring independent action against regulated entities for violations of federal law that threaten human health or the environment, reflect criminal conduct or repeated

Specifically, EPA will examine

a) changes in compliance behavior within the regulated community, including improved compliance rates;
b) prompt disclosure and correction of violations, including timely and accurate compliance with reporting requirements;
c) corporate compliance programs that are successful in preventing violations, improving environmental performance, and promoting public disclosure;
d) consistency among state programs that provide incentives for voluntary compliance.

The EPA study will be made available to the public.

§ 3.2.4 Audit Provisions as Remedies in Enforcement Actions

EPA often proposes environmental auditing provisions in consent decrees and in other settlement negotiations because use of such audit programs is intended to reduce the likelihood of similar problems recurring in the future.[90] Environmental auditing provisions are most likely to be proposed in settlement negotiations where

- a pattern of violations can be attributed, at least in part, to the absence or poor functioning of an environmental management system; or
- the type or nature of violations indicates a likelihood that similar noncompliance problems may exist or occur elsewhere in the facility or at other facilities operated by the regulated entity.[91]

> noncompliance, or allow one company to make a substantial profit at the expense of its law-abiding competitors. Where a state has obtained appropriate sanctions needed to deter such misconduct, there is no need for EPA action. [*Id.*]

[90] EPA developed internal guidance for use by agency negotiators in structuring appropriate environmental audit provisions for consent decrees and other settlement negotiations. For an example of a consent decree based on this guidance, see United States v. Browning-Ferris Indus. Chem. Serv., Inc., 704 F. Supp. 1355 (M.D. La. 1988).

[91] 51 Fed. Reg. 25,004 (1986). *See also* 56 Fed. Reg. 64,785 (1991) (clarifying EPA's position concerning the role of corporate attitude, policies, practices, and procedures in determining whether, in mandatory contractor cases, the condition giving rise to a criminal conviction has been corrected).

EPA policy typically does not require automatic submission of environmental audits conducted under such settlement decrees to the agency. Rather, EPA will require only submission of notice that the audit was conducted. Moreover, in many cases EPA will expressly agree not to use the audit as evidence of any environmental violations, provided that the violations are corrected within a prearranged timetable (usually 60 to 90 days).[92]

§ 3.3 Justice Department Policy

On July 1, 1991, the Justice Department issued a statement indicating its "Factors in Decisions on Criminal Prosecutions for Environmental Violations in the Context of Significant Voluntary Compliance or Disclosure Efforts by the Violator."[93]

In that statement Justice stated that, "it is the policy of the Department of Justice to encourage self-auditing, self-policing and voluntary disclosure of environmental violations by the regulated community by indicating that these activities are viewed as mitigating factors in the Department's exercise of criminal environmental enforcement discretion." The document was designed to give the regulated community a sense of how the federal government exercises its criminal prosecutorial discretion with respect to such factors as the defendant's voluntary disclosure of violations, cooperation with the government in investigating the violations, use of environmental audits and other procedures to ensure compliance with all applicable environmental laws and regulations, and use of measures to remedy expeditiously and completely any violations and the harm caused thereby.[94]

---

[92] *E.g.*, United States v. Browning-Ferris Indus. Chem. Serv., Inc., 704 F. Supp. 1355 (M.D. La. 1988).

[93] *See* "FACTORS IN DECISIONS ON CRIMINAL PROSECUTIONS FOR ENVIRONMENTAL VIOLATIONS IN THE CONTEXT OF SIGNIFICANT VOLUNTARY COMPLIANCE OR DISCLOSURE EFFORTS BY THE VIOLATOR." U.S. Department of Justice, July 1, 1991.

[94] *See generally* Robert Darnell, *Environmental Criminal Enforcement and Corporate Environmental Auditing: Time for a Compromise?*, 31 AM. CRIM. L. REV. 123-143 (1993).

§ 3.3.2 Department of Justice (DOJ) Factors to Be Considered

The Justice Department stated that it will particularly consider the following factors in determining whether and how to prosecute: degree of cooperation, whether the disclosure was voluntary, and the presence or absence of preventative measures and compliance programs. Any one factor will not be dispositive in any given case. All relevant factors are considered on a case-by-case basis.[95]

Voluntary Disclosure. First, the Justice Department will consider whether the person[96] made a voluntary, timely, and complete disclosure of the matter under investigation. Consideration is given to whether the person came forward promptly after discovering the noncompliance and to the quantity and quality of information provided. Particular consideration is also given to whether the disclosure substantially aided the government's investigation process and whether it occurred before a law enforcement or regulatory authority (federal, state, or local authority) had already obtained knowledge regarding noncompliance. A disclosure is not considered to be "voluntary" if that disclosure is already specifically required by law, regulation, or permit.[97]

Cooperation. Second, the Justice Department will consider the degree and timeliness of cooperation by the person. The Justice Department states that "full and prompt cooperation is essential, whether in the context of a voluntary disclosure or after the government has independently learned of violation." Consideration is also given to the Justice Department's analysis of the violator's assistance to the government's investigation.

---

[95] *See* U.S. Dept. of Justice, FEDERAL PRINCIPLES OF PROSECUTION, Comment to Part A.2; Part B.3 (1980).

[96] As used in this document, the terms "person" and "violator" are intended to refer to business and nonprofit entities as well as individuals.

[97] For example, any person in charge of a vessel or of any onshore facility or an offshore facility is required to notify the appropriate agency of the United States Government of any discharge of oil or hazardous substance into or upon *inter alia* the navigable waters of the United States. § 311(b)(5) of the Clean Water Act, 33 U.S.C. § 1321(b)(5), as amended by the Oil Pollution Act of 1990, Pub. L. 101-380, § 4301(a), 104 Stat. 485 533 (1990).

Preventive Measures and Compliance Programs. Third, the Justice Department considers the existence and scope of any regularized, intensive, and comprehensive environmental compliance program, including an environmental compliance or management audit. Particular consideration is given to whether the compliance or audit program includes sufficient measures to identify and prevent future noncompliance and whether the program was adopted in good faith in a timely manner.

The Justice Department has no specific criteria covering what constitutes a "good" policy, but evaluates any program by considering the following questions:

1. Was there a strong institutional policy to comply with all environmental requirements?
2. Had safeguards beyond those required by existing law been developed and implemented to prevent noncompliance from occurring?
3. Were there regular procedures, including internal or external compliance and management audits, to evaluate, detect, prevent, and remedy circumstances like those that led to the noncompliance?
4. Were there procedures and safeguards to ensure the integrity of any audit conducted?
5. Did the audit evaluate all sources of pollution (*i.e.*, all media), including the possibility of cross-media transfers of pollutants?
6. Were the auditor's recommendations implemented in a timely fashion?
7. Were adequate resources committed to the auditing program and to implementing its recommendations?
8. Was environmental compliance a standard by which employee and corporate departmental performance were judged?

Pervasiveness of Noncompliance. Fourth, the Justice Department considers the pervasiveness of noncompliance. As the Department explained:

> Pervasive noncompliance may indicate systemic or repeated participation in or condonation of criminal behavior. It may also indicate the lack of a meaningful compliance program.

In evaluating this factor, the Department will consider, among other things, the number and level of employees participating in the unlawful activities and the obviousness, seriousness, duration, history, and frequency of noncompliance.

Internal Disciplinary Action. Fifth, the Justice Department will consider whether there was an effective system of discipline for employees who violated company environmental compliance policies. The Justice Department considers effective internal disciplinary action crucial to any compliance program. In evaluating this criteria, the Department queries "whether the disciplinary system was designed to or did establish an awareness in other employees that unlawful conduct would not be condoned by the company?"

Subsequent Compliance Efforts. Sixth, the Department will consider the extent of any efforts to remedy any ongoing noncompliance. The promptness and completeness of any action taken to remove the source of the noncompliance and to lessen the environmental harm resulting from the noncompliance is also considered. Considerable weight is given to prompt, good-faith efforts to reach environmental compliance agreements with federal or state authorities, or both, and full compliance with such agreement is a significant factor in any decision whether to prosecute.

§ 3.3.3 Application of These Factors to Hypothetical Examples[98]

The Justice Department has articulated certain "best case" and "worst case" examples intended to assist federal prosecutors in evaluating environmental cases.

§ 3.3.4 DOJ's Best Case

The Justice Department stated that this was the "ideal case in terms of criteria satisfaction and consequent prosecution leniency":

1. Company A regularly conducts a comprehensive audit of its compliance with environmental requirements.
2. The audit uncovers information about employees' disposing of hazardous wastes by dumping them in an unpermitted location.

[98] While this policy applies to both individuals and organizational violators, these examples focus particularly upon situations involving organizations.

3. An internal company investigation confirms the audit information. (Depending upon the nature of the audit, this follow-up investigation may be unnecessary.)
4. Prior to the violations the company had a sound compliance program, which included clear policies, employee training, and a hotline for suspected violations.
5. As soon as the company confirms the violations, it discloses all pertinent information to the appropriate government agency, it undertakes compliance planning with that agency, and it carries out satisfactory remediation measures.
6. The company also undertakes to correct any false information previously submitted to the government in relation to the violations.
7. Internally the company disciplines the employees actually involved in the violations, including any supervisor who was lax in preventing or detecting the activity. Also, the company reviews its compliance program to determine how the violations slipped by, and corrects the weaknesses found by that review.
8. The company discloses to the government the names of the employees actually responsible for the violations, and it cooperates with the government by providing documentation necessary to the investigation of those persons.

The Justice Department said that, "Under these circumstances Company A would stand a good chance of being favorably considered for prosecutorial leniency, to the extent of not being criminally prosecuted at all." However, Justice warned that the degree of any leniency may turn upon other relevant factors not specifically dealt with in these guidelines.[99]

§ 3.3.5 DOJ's Worst Case

At the opposite end of the scale, Justice described Company Z, which meets few of the criteria. In this worst case, the likelihood of prosecutorial leniency was considered remote. According to Justice, Company Z's circumstances may include any of the following:

---

[99] For example, if the company had a long history of noncompliance, the compliance audit was done only under pressure from regulators, and a timely audit would have ended the violations much sooner, those circumstances would be considered.

1. Because an employee has threatened to report a violation to federal authorities, the company is afraid that investigators may begin looking at it. An audit is undertaken, but it focuses only upon the particular violation, ignoring the possibility that the violation may be indicative of widespread activities in the organization.
2. After completing the audit, Company Z reports the violations discovered to the government.
3. The company has a compliance program, but it is effectively no more than a collection of paper. No effort is made to disseminate its content, impress upon employees its significance, train employees in its application, or oversee its implementation.
4. Even after "discovery" of the violation the company makes no effort to strengthen its compliance procedures.
5. The company makes no effort to come to terms with regulators regarding its violations. It resists any remedial work and refuses to pay any monetary sanctions.
6. Because of the noncompliance, information submitted to regulators over the years has been materially inaccurate, painting a substantially false picture of the company's true compliance situation. The company fails to take any steps to correct that inaccuracy.
7. The company does not cooperate with prosecutors in identifying those employees (including managers) who actually were involved in the violation, and it resists disclosure of any documents relating either to the violations or to the responsible employees.

Justice explained that leniency is unlikely, because the only positive action taken by Company Z was the audit, which was "so narrowly focused as to be of questionable value, and it was undertaken only to head off a possible criminal investigation." Company Z otherwise demonstrated no good faith either in terms of compliance efforts or in assisting the government in obtaining a full understanding of the violation and discovering its sources.

Justice stated that these factors do not ensure a criminal prosecution of Company Z. As with Company A, other circumstances may be present that affect the balance struck by prosecutors. For example, the effect of the violation (because of substance, duration, or amount) may be such that prosecutors would not consider it to be an appropriate criminal case. Administrative or civil proceedings may be considered a more appropriate response.

§ 3.3.6 Other DOJ examples

Between these extremes there are a range of possibilities. The presence, absence, or degree of any criterion may affect the prosecution's exercise of discretion. For example, the Justice guidance said that if a company performs an audit that is very limited in scope and reflects no more than an effort to avoid prosecution, but is otherwise cooperative in terms of both bringing itself into compliance and providing information regarding the crime and its perpetrators, "the result could be any of a number of outcomes, including prosecution of a lesser charge or a decision to prosecute the individuals rather than the company." If, on the other hand, the company refuses to reveal any information regarding the individual violators, the Justice guidance indicates that the likelihood of the government's prosecuting the company is substantially increased.

If the company chooses to "sit on" the audit and take corrective action without telling the government and the government learns of the situation months or years after the fact, consideration would be given by prosecutors for remedial efforts; hence, prosecution of fewer or lesser charges might result. However, if the company's silence adversely affected others who are entitled to fair regulatory treatment or if the company deprived those legally responsible for evaluating cleanup needs of the ability to carry out their functions, the likelihood of its totally escaping criminal prosecution would be significantly reduced.[100] Justice explained that

---

[100] Justice gave the following example. Company D failed to report discharges of X contaminant into a municipal sewer system, discharges that were terminated as a result of an audit. The sewer authority, though, knowing only that there have been excessive loadings of X, but not knowing that Company D was a source, tightens limitations upon all known sources of X. Thus, all of those sources incur additional treatment expenses, but Company D is unaffected. Had Company D revealed its audit results, the other companies would not have suffered unnecessary expenses.

In some situations, moreover, failure to report is a crime. *See, e.g.*, 33 U.S.C. § 1321(b)(5) and 42 U.S.C. § 9603(b). To illustrate the effect of this factor, consider Company

> A complicating fact here is that environmental regulatory programs are self policing: they include a substantial number of reporting requirements. If reports which in fact presented false information are allowed to stand uncorrected, the reliability of this system is undermined. They also may lead to adverse and unfair impacts upon other members of the regulated community.
>
> If the company concentrates upon correcting only the easier, less expensive, less significant among the violations discovered in the audit, then its lackadaisical approach to correction will not make it a strong candidate for leniency.

If the company performs an audit and finds violations involving failure to comply with reporting requirements, and the company undertakes a program of gradually correcting its violations, but does not bring them to the government's attention before the government learns of the situation, the company "could receive some consideration for its efforts, but its failure to disclose and the slowness of its remedial work probably mean that it cannot expect a substantial degree of leniency."

Although Justice considered comprehensive audits to be positive efforts toward good faith compliance, such audits are not indispensable to enforcement leniency. If a company does not have a formal audit program but, as a part of its efforts to ensure compliance, does realize that it is committing an environmental violation, and takes steps to come into compliance and cooperate with the government, then it may be "a likely candidate for leniency, including possibly no criminal prosecution."

In sum, the Justice Department clearly stated that mitigating efforts made by the regulated community will be recognized and evaluated. The greater the showing of good faith, the more likely it will be met with leniency. Conversely, the less good faith shown, the less likely that prosecutorial discretion will tend toward leniency.

E, which conducts a thorough audit and finds that hazardous wastes have been disposed of by dumping them on the ground. The company cleans up the area and tightens up its compliance program, but does not reveal the situation to regulators. Assuming that a reportable quantity of a hazardous substance was released, the company was under a legal obligation under 42 U.S.C. § 9603(b) to report that release as soon as it had knowledge of it, thereby allowing regulators the opportunity to ensure proper clean up. Company E's knowing failure to report the release upon learning of it is itself a felony.

§ 3.3.7 Limitation of the DOJ Guidance

The decision to prosecute "generally rests entirely in [the prosecutor's] discretion."[101] This discretion is especially firmly held by the criminal prosecutor.[102] The Justice Department's policy, although made available to the public, was intended only as internal guidance to Department of Justice attorneys. The guidance clearly states that it is not intended to and "may not be relied upon to create a right or benefit, substantive or procedural, enforceable at law by a party to litigation with the United States, nor do they in any way limit the lawful litigation prerogatives, including civil enforcement actions, of the Department of Justice or the Environmental Protection Agency." The guidance is provided only to guide the effective use of limited enforcement resources, and does not derive from, find its basis in, or constitute any legal requirement, whether constitutional, statutory, or otherwise, to forgo or modify any enforcement action or the use of any evidentiary material.[103] Thus, although such guidance aids the private attorney in counseling the client, it is not binding, and as such, cannot be relied upon in fashioning corporate posture.

---

[101] Bordenkircher v. Hayes, 434 U.S. 357, 364 (1978), *reh'g denied,* 435 U.S. 918 (1978). Although some statutes have occasionally been held to require civil enforcement actions, *see, e.g.*, Dunlop v. Bachowski, 421 U.S. 560 (1975), those are unusual cases, and the general rule is that both civil and criminal enforcement is at the enforcement agency's discretion where not prescribed by law. Heckler v. Chaney, 470 U.S. 821, 830-35 (1985); Cutler v. Hayes, 818 F.2d 879, 893 (D.C. Cir. 1987) (decisions not to enforce are not reviewable unless the statute provides an "inflexible mandate").

[102] *See* Newman v. United States, 382 F.2d 479, 480 (D.C. Cir. 1967).

[103] *See* PRINCIPLES OF FEDERAL PROSECUTION (U.S. Dept. of Justice, 1980), p. 4; UNITED STATES ATTORNEY'S MANUAL (U.S. Dept. of Justice, 1986) 1-1.000.

§ 3.4 International Standards Organization (ISO)

§ 3.4.1 The ISO 9000 Quality Standards

Private international environmental management systems (EMSs) began in earnest with the development and adoption of the International Standards Organization (ISO) 9000 Quality Standards,[104] which grew out of the European Union's (EU) efforts to form a single internal market. The EU developed a conformity assessment program in its efforts to create a single internal market. To do so, physical, technical, and fiscal barriers to trade were removed. The EU assessment program covered all product processes: testing and certification, quality system registration, and standards and laboratory practices.

The ISO 9000 Standards is a set of generic standards that provide quality management guidance and identify generic quality system elements necessary to achieve quality assurance. The standards are also independent of any specific industry. Their fundamental objective is to make quality assurance an integral part of the way companies do business. The series contains five standards: ISO 9000-1, ISO 9001, ISO 9002, ISO 9003, and ISO 9004-1.

ISO 9000-1 and 9004-1 provide guidance for internal quality management purposes. Specifically, ISO 9000-1 lays out those elements that will enable an organization to meet stakeholder expectations, in general, and those of customers, in particular.[105] This standard states, among other things, that an organization should achieve, maintain, and continually improve the quality of its products, and improve the quality of its own

---

[104] ISO refers to the International Organization for Standardization, founded in 1946 to develop a common set of manufacturing, trade, and communication standards. The Geneva, Switzerland, based organization is composed of 92 member countries. The American National Standards Institute (ANSI) is the U.S. member body to ISO. ISO is structured into 180 technical committees that draft standards. Member nations form technical advisory groups (TAGs) that provide input into the standards development process. All drafts are subject to comment by all interested parties prior to promulgation. All standards are voluntary, unless nations and industries attach legal obligations to them, thereby making them mandatory.

[105] Stakeholders under ISO 9000 are customers, employees, owners, suppliers, and society.

operations to meet the needs of stakeholders. Each increment in the organization's work should be viewed as a process step—a transformation that adds value. Each process step is then identified, organized, and managed into an integrated quality system. Then it is recommended all processes be documented to allow the organization to evaluate systems, achieve improvement, and maintain those improvements.

ISO 9004-1 provides guidance for internal quality management without regard to external contractual requirements (*i.e.,* customer expectations). The objective of this standard is to ensure the technical, administrative, and human factors affecting the quality of its products. Thus a company should develop a system that is suitable for its needs since the aim of the standard is to reduce, eliminate, and prevent quality imperfections.[106]

ISO 9001, ISO 9002, and ISO 9003 provide guidance for external quality assurance. ISO 9001 is the most comprehensive and is used primarily when a supplier must ensure product conformance to specified needs throughout the entire product cycle. It most commonly applies to the manufacturing and processing industries and the construction, architecture, and engineering trades.[107] ISO 9002 applies to industries whose work is

---

[106] ISO 9004-1 specifically addresses the following: management responsibility, quality system elements, financial considerations, quality in marketing, specification and design, purchasing and processes, control of processes, product verification, control of inspection, control of nonconforming product, corrective action, personnel, and product safety.

[107] ISO 9001 includes all the quality assurance requirements listed in ISO 9002 and ISO 9003. Those requirements are

- ✓ management responsibility;
- ✓ quality system;
- ✓ contract review;
- ✓ design control;
- ✓ document and data control;
- ✓ purchasing;
- ✓ control of customer-supplied product;
- ✓ product identification and traceability;
- ✓ process control;
- ✓ inspection and testing;
- ✓ control of inspection;
- ✓ inspection status;
- ✓ control of nonconforming product;
- ✓ corrective and preventive action;

based on technical designs and specifications provided by their customers. It is relevant for products that do not involve a design aspect. ISO 9003 applies to organizations whose products or services can be sufficiently assessed by testing and inspection. Thus, the standard addresses the requirements for detection and control of problems only during final inspection and testing.

Related, complementary standards to the ISO 9000 Standards are the ISO 10011-1, ISO 10011-2, and ISO 10011-3 auditing standards. ISO 10011-1 establishes guidelines for implementing an audit system.[108] ISO 10011-2 establishes the minimum criteria required to qualify auditors.[109] ISO 10011-3 establishes guidelines for the management of ongoing audit programs.[110] Once companies have implemented the ISO 9000 Standards, the next step is to register the company as ISO 9000 certified. Once a registrar is selected and the registration process complete,[111] a company is either approved, conditionally approved, or disapproved.

To date, there are 7,600 ISO 9000 registrations in the U.S., and of this number 80% are individual companies. Reasons to attain ISO 9000 registration are driven primarily by market demands and gaining a competitive advantage.[112] In addition the greatest benefits of registration

---

- ✓ handling, storage, packaging, preservation, and delivery;
- ✓ control of quality records;
- ✓ internal quality audits;
- ✓ training; and
- ✓ servicing.

[108] The part provides guidelines for actually conducting an audit of an organization's quality system. It provides guidelines for establishing, carrying out, and documenting audit systems.

[109] The criteria upon which an auditor is judged include education and training, experience, personal attributes, management capabilities, and competency.

[110] Guidelines include the applicable standards, staff qualifications, the monitoring and maintenance of auditor performance, operational factors, and a code of ethics.

[111] The registration process has six basic steps. They are application, document review, preassessment, assessment, registration, and surveillance.

[112] A joint *Quality Systems Update*/Deloitte & Touche survey in 1993 revealed that the top five reasons for gaining ISO 9000 registration are customer demand at 27.4 percent; quality benefits at 21.8 percent; market advantage at 15.6 percent; 9.0 percent requirement of EC regulations; and 8.9 percent corporate mandate/part of larger strategy. The following industries had the highest response rate: chemicals and allied products at 17.7 percent; fabricated metal products at 14 percent; and industrial and commercial equipment and computer equipment at 17.4 percent.

were greater quality awareness, better documentation, and positive cultural change.[113] Finally, the annual average savings for ISO 9000 companies was $179,000.

The success of ISO 9000 is clear. Customers demand consistently made quality products, and in turn, manufacturers need to produce consistently made quality products. In addition, government regulators need to see consistently made safe products. ISO 9000 registration has produced what all stakeholders want—consistency. This consistency has been achieved by making quality assurance a routine part of the business ethic. The success of the ISO 9000 series led to the development of the ISO 14000 EMS standards.

§ 3.4.2 The ISO 14000 Environmental Management Standards

The driving force behind the EMS movement is a mix that includes the desire to come into compliance with environmental regulations because of civil and criminal sanctions, public disclosure requirements, reduction in liability, pollution prevention and waste reduction, the "green" products market, improved internal management, shareholder pressure, and community goodwill. Additionally, in the EU, the formation of a single market made the harmonization of standards an urgent task. Thus the International Organization of Standards under Technical Committee (TC) 207 set out to establish generic environmental standards for any type or size business relating to organizational evaluation and product evaluation.[114]

---

[113] The top three answers for external benefits of registration are higher perceived quality at 33.5 percent; improved customer satisfaction at 26.6 percent; and competitive edge at 21.5 percent. The top three answers for internal benefits of registration are better documentation at 32.4 percent; greater quality awareness at 25.6 percent; positive cultural change at 15.0 percent.

[114] The ISO Technical Committee 207 (TC 207) has been working to develop international standards for EMSs, which draw their members from 57 countries. The American National Standards Institute (ANSI) is the U.S. representative to the TC 207. It organized The U.S. Technical Advisory Group (U.S. TAG), which is responsible for U.S. input on these emerging standards to reflect U.S. interests. There are 310 members in the U.S. TAG, which includes government agencies like U.S. EPA, corporations like AT&T, and public interest groups like National Wildlife Federation, to name a few.

The organization evaluation standards include

ISO 14004: EMS—General Guidelines;

ISO 14001: EMS—Specifications;

ISO 14010: Guidelines for Environmental Auditing—General Principles;

ISO 14011/1: Guidelines for Environmental Auditing—Audit Procedures—Part 1: Auditing of EMS;

ISO 14012: Guidelines for Environmental Auditing—Qualification Criteria for Environmental Auditors;

ISO 14031: Evaluation of the Environmental Performance of the Management System and Its Relationship to the Environment;

ISO 1401x: Environmental Site Assessment; and

ISO 1403x: Evaluation of the Environmental Performance of the Operational System and Its Relationship to the Environment.

Of these eight standards only ISO 14004, ISO 14001, ISO 14011/1, and ISO 14012 are draft international standards (DIS), which means companies can begin the implementation and audit process with little concern that substantial changes will be made to them.

The product evaluation standards include

ISO 14020: Goal and Principles of All Environmental Labeling;

ISO 14021: Self Declaration Environmental Claims;

ISO 14024: Environmental Labeling—Guiding Principles, Practices, and Criteria for Multiple Criteria-Based Practitioner Programmes (Type I)—Guide for Certification Procedures;

ISO 14040: Environmental Management—Life Cycle Assessment—Principles and Practice;

ISO 14041: Environmental Management—Life Cycle Assessment—Life Cycle Inventory Analysis;

ISO 14042: Environmental Management—Life Cycle Assessment—Life Cycle Impact Assessment;

ISO 14043: Environmental Management—Life Cycle Assessment—Life Cycle Improvement Assessment;

ISO 14060: Guide for the Inclusion of Environmental Aspects in Product Standards; and

ISO 1405x: Terms and Definitions—Guide on the Principles for ISO/TC 207/SC6 Terminology Work.

None of the product evaluation standards is in DIS form. To date it is uncertain when any of these standards will "go DIS" or finally be approved.

§ 3.4.3 The ISO 14000 Environmental Auditing Standards

There is a three-part series of standards in the field of environmental auditing. They are:

ISO 14010: Guidelines for Environmental Auditing—General Principles of Environmental Auditing;

ISO 14011: Guidelines for Environmental Auditing—Audit Procedures—Auditing of EMSs; and

ISO 14012: Guidelines for Environmental Auditing—Qualification Criteria for Environmental Auditors.

ISO 14010: General Audit Principles. This standard states that an environmental audit should focus on clearly defined and documented subject matter. There should also be sufficient and appropriate information about the subject matter of the audit, adequate resources to support the audit process, and adequate cooperation from the auditee. In addition, ISO standard 14010 provides seven general principles of environmental auditing that are applicable to all types of environmental audits.

1. *Objectives and scope.* The audit should be based on objectives defined by the client. The scope is determined by the lead auditor in consultation with the client in order to meet these objectives.
2. *Objectivity, independence, and competence.* To ensure objectivity of the audit process, the audit team should be independent of the activities it audits. Furthermore, the audit team should possess the appropriate combination of knowledge, skills, and experience to carry out audit responsibilities.
3. *Due professional care.* Auditors should use care, diligence, skill, and judgment in carrying out the audit. Also, the audit team/client relationship should be one of confidentiality and discretion.
4. *Systematic procedures.* The environmental audit should be conducted in accordance with these general principles and any guidelines developed for the appropriate type of environmental audit (*i.e.*, ISO 14011).
5. *Audit criteria, evidence, and findings.* Criteria should be agreed to at the outset of the audit process. Appropriate information should be collected, analyzed, interpreted, and documented to be used as audit evidence. The evidence should be of such quality and quantity that independent auditors would reach similar findings from the same evidence against the same criteria.

6. *Reliability of audit findings and conclusions.* The environmental auditing process should be designed to provide the client and auditor with the desired level of confidence in the reliability of the audit findings and any audit conclusions.
7. *Reporting.* The audit findings or a summary should be communicated to the client in a written report.

These audit principles can be used under a regulatory reform program as the type of principles that must be demonstrated when auditing an EMS for certification and admittance into a regulatory reform program. The principles of defining the scope of the audit, objectivity in conducting the audit, due professional care, systematic audit procedures, criteria, reliability, and reporting are essentially universally accepted principles that make them uncontroversial. Furthermore, these broad principles give an audit team ample discretion in drawing up their own guidance.

ISO 14011: Audit Procedures for EMSs. ISO 14011 provides generic procedures for the conduct of EMS audits. There are three broad elements under this standard:

1. *EMS audit objectives, roles, and responsibilities.* The first element, EMS audit objectives, roles, and responsibilities, has two steps: audit objectives; and roles, responsibilities, and activities. Under the first step, audit objectives the standard states that EMS audits should have defined objectives. A typical objective would be to determine whether the auditee's EMS has been properly implemented and maintained. Under the second step, roles, responsibilities, and activities, the standard sets forth detailed responsibilities and activities for the lead auditor, auditor, audit team, client, and auditee. For example, the lead auditor is responsible for completion of the audit within the agreed-upon audit scope and plan. The auditor supports the lead auditor. In addition, the audit team should possess the overall experience and expertise needed to carry out the audit. The client determines the need for the audit. And the auditee works with the audit team to ensure that the audit objectives are achieved.
2. *Auditing* has four steps:
   - ✓ initiating the audit;
   - ✓ preparing the audit;
   - ✓ executing the audit; and
   - ✓ audit reports and records.

Under the first step, initiating the audit, the audit scope should be defined. The scope describes the extent and boundaries of the audit in terms of factors such as physical location and organizational activities. In addition, a preliminary document review should be done on the organization's environmental policy statements, programs, records, or manuals for meeting EMS requirements.

Under the second step, preparing the audit, an audit plan should be defined. The audit plan should be flexible to permit scope changes based on audit information and to permit effective use of resources. Audit team members should be assigned specific activities to audit. And working documents (*i.e.*, checklists) should be maintained to facilitate the auditor's investigation.

Under the third step, executing the audit, an opening meeting should take place to review the scope, objectives, and audit plan. Sufficient audit evidence should be collected to determine whether the auditee's EMS conforms to the EMS audit criteria. Audit findings should be reviewed with the responsible auditee manager with a view to obtaining acknowledgment of the factual basis of all findings of nonconformity. A closing meeting should be held to present audit findings to the auditee to obtain its understanding and acknowledgment of the factual basis of the findings.

Under the fourth step, audit reports and records, the audit report is prepared under the direction of the lead auditor, who is responsible for its accuracy and completeness. The audit report should contain the audit findings or a summary thereof with reference to supporting evidence. In addition, the audit report should be sent to the client, whose distribution of the report should be determined by the client in accordance with the audit plan. Finally, all documents pertaining to the audit should be retained in accordance with any applicable agreements or requirements.

3. *Audit completion.* The audit is completed once all activities defined in the audit plan have been concluded. As with the audit principles, the audit standards could be adopted under a regulatory reform program. First, these standards were drawn up by industry and government agencies in a cooperative spirit.[115]

---

[115] *See US Gets ISO 14001 Language Changes; Initial Review Sought,* 2 INTERNATIONAL ENVIRONMENTAL SYSTEMS UPDATE 1 (March 1995), where EPA sought and got pollution prevention and compliance auditing language included in the standards, or else refused to support them.

And second, adopting these principles saves a government agency the time and effort of drawing up its own duplicative standards. Additionally, the standards are broad enough so that they provide an audit team with the discretion it needs to conduct its audits at different facilities.

ISO 14012: Qualification Criteria for Auditors. This standard sets guidance on qualification criteria for environmental system auditors to support the application of EMSs and environmental auditing. This standard is applicable to both internal and external auditors. There are eight elements to this standard. They are

1. education and work experience;
2. auditor training;
3. objective evidence of education, experience, and training;
4. personal attributes and skills;
5. lead auditor;
6. maintenance of competence;
7. due professional care; and
8. language.

The education and work experience element indicates that auditors should have completed at least secondary education, or its equivalent. Auditors should also have work experience, which contributes to the development of skills and understanding in areas like environmental science, environmental law, or audit procedures, to name a few.

The auditor training element states that in addition to the criteria set forth in the first element, auditors should have completed both formal training and on-the-job training to develop competence in carrying out environmental audits. Formal training should address environmental science, environmental law, and audit procedures, to name a few. However, this training can be waived if competency can be demonstrated through accredited exams or relevant professional qualifications. In addition, an auditor should have completed a period of on-the-job training for a total of twenty equivalent workdays of environmental auditing, and for a minimum of four environmental audits.

The objective evidence of education, experience, and training element states that individuals should keep objective evidence of their education, experience, and training. This presumably means documents like training certificates and diplomas, etc.

The personal attributes and skills element states that auditors should possess certain personal attributes and skills. Some of these include interpersonal skills conducive to performance of the audit, such as diplomacy, tact, and the ability to listen and reach sound judgment based on objective evidence. Others include competence in clearly expressing concepts and ideas orally and in writing.

The lead auditor element states that the lead auditor should have the necessary leadership and management skills to ensure an effective and efficient audit process. The standard also sets forth additional criteria that a lead auditor should meet in addition to those skills already highlighted above.

The maintenance of competence element states that auditors should maintain their competence by ensuring the currency of their knowledge of environmental science, environmental law, and audit processes, to name a few. In addition, auditors should ensure that their experience in the execution of audits is current, and also participate in refresher training as necessary.

The due professional care element states that auditors should exercise due professional care, as addressed in ISO 14010, and adhere to an appropriate code of ethics.

The language element states that no auditor should participate unsupported in an audit when unable to communicate effectively in the language necessary for performing his or her responsibilities. When necessary, support should be obtained from a person with the necessary language skills.

As with the other two standards, the auditor standard can be used under a regulatory reform program to provide support and credibility to audit results. The standard covers broad qualification aspects like education and/or training, management and interpersonal skills, and ethics. Auditors need to demonstrate these specific skills and abilities objectively prior to their own certification. Moreover, the private sector is now providing certification for auditors that the government can use and take advantage of when initiating a regulatory reform program.[116]

---

[116] The American National Standards Institute (ANSI) announced in New York on October 31, 1995, that it will establish a national program to accredit ISO 14000 EMS registrars, auditor certifiers, and training providers. This private sector development means

The ISO 14000 standards cover all aspects of the development and implementation of an EMS. The standards also covers those aspects relating to auditing an EMS and auditors qualifications. However, these standards cannot provide enhanced environmental protection unless they are implemented industrywide.

§ 3.4.4 Industry Incentives for Implementing the ISO 14000 Standards

The type of economic activity a business engages in will, in part, determine the usefulness of establishing an EMS, in general, and an ISO 14000 certified EMS, in particular.[117] Chemical and textile manufacturers, along with mining and forestry operations, are just a few of the businesses that manage numerous environmental aspects of their economic activities. In addition, they are also heavily regulated and face tremendous compliance pressures. An ISO 14000 certified EMS can help manage these aspects more effectively, and relieve transactional and enforcement pressures.[118] Fewer environmental impacts mean fewer liabilities and fewer disclosure obligations.[119] For example, an EMS designed to reduce the generation of hazardous wastes can reduce off-site shipments and/or on-site treatment. This can lead to reduced environmental impacts, a more cost-effective process, fewer regulatory requirements, and lower insurance costs.[120]

---

that regulatory agencies can have a high degree of confidence in any third party certification of an ISO 14000 EMS. This confidence can be derived from the fact that an independent entity can objectively train those third parties that facilities will seek as auditors of their EMSs.

[117] The Standard Industrial Classification (SIC) Codes are a practical way to view industrial activity labels.

[118] EPA estimates that the average cost of cleaning up a Superfund site exceeds $30 million. *See* National Priorities List for Uncontrolled Hazardous Waste Sites, 60 Fed. Reg. 20,330 (1995).

[119] The SEC scrutinizes whether companies subject to SEC reporting requirements have disclosed their environmental problems adequately. *See* Lisa Sotto, *"Companies That Fail to Make Adequate Disclosures of Potential Environmental Liabilities Under Superfund Have Become the Objectives of Increased SEC Scrutiny,"* The National Law Journal, December 4, 1995, at B5.

[120] Fluke Corporation, a vertically integrated electronics manufacturer, incorporated a pollution prevention element in its ISO 14000 EMS to reduce off-site hazardous waste

Conversely, wholesale trade, retail trade, finance, and real estate have fewer environmental aspects associated with their economic activities, and therefore face fewer regulatory pressures. These businesses may, however, face consumer pressures to offer environmentally sound products and services. An ISO 14000 certified EMS is one way of demonstrating environmentally sound products and services. For example, service industries can contract with their suppliers to deliver products from ISO 14000 certified manufacturers.[121] Real estate developers could pledge to use environmentally sound land use planning techniques, and reduce impacts to agricultural lands or historic sites. Additionally, the European Union is considering using the ISO 14000 series as part of its environmental, eco-management auditing regulations. Companies that do international business and have overseas sites will need to implement an ISO 14000 certified EMS as part of the cost of doing business in many overseas markets.

ISO 9000 certified companies are well placed to develop and implement ISO 14000 certified EMSs.[122] Many ISO 9000 certified companies also manage numerous environmental aspects, and consequentially have environmental impacts and regulatory pressures.[123]

---

shipments from 1,000 tons per year to 12 tons per year. *See* International Standards Initiative (ISI) Update, October 10, 1995 (Issaquah, WA).

[121] L. L. Bean and Levi Strauss purchase denim from Cone Mills of Greensboro, NC, which is the world's largest producer of denim. Both purchasers conduct environmental assessments to see how "green" their bluejeans supplier is. Cone Mills is on the E3 "honor roll." All three companies try to anticipate regulatory changes rather than react to them by taking environmental responsibility into account when doing business. *See "The Environment's New Clothes"*, THE CHRISTIAN SCIENCE MONITOR, August 19, 1993, at 10-11.

[122] *See also* ManiMoli Dinesh, *Experience With BS 7750 Foreshadows ISO 14001*, 2 INTERNATIONAL ENVIRONMENTAL SYSTEMS UPDATE 12 (March 1995); David Jones, *Responsible Care Seen as Road to ISO 14001*, 2 INTERNATIONAL ENVIRONMENTAL SYSTEMS UPDATE 14 (April 1995).

[123] A recent article entitled *Quality Systems/Deloitte & Touche Survey of Companies by Industry*, reprinted in THE ISO 9000 HANDBOOK (1995), states that the chemicals, computers, and metals industries have the highest ISO 9000 registrations. These industry sectors also have high environmental aspects, and conversely high environmental impacts. Since both ISO 9000 and ISO 14000 have many of the same elements, arguably organizations with an ISO 9000 system can easily integrate ISO 14000 into their current operations.

These companies sought ISO 9000 registration primarily due to customer demands/expectations and quality benefits.[124] These are some of the same pressures companies face in seeking ISO 14000 certification. They also understand the barriers to successful registration, the associated costs, and savings derived from the ISO 9000 process. In addition, companies that are ISO 9000 certified believe ISO 14000 certification is the next logical step in industrial management.[125] Whether a business is ISO 9000 certified or not, the ISO 14000 Standards may become important to the future business success of companies. An assessment of North American businesses recently surveyed by Arthur D. Little[126] found that businesses believe the ISO 14000 Standards are important for competitive advantage in the marketplace, as well as important to the development of a strategic proactive regulatory plan.[127]

---

[124] *Id.* Customer demands/expectation was the primary reason at 25 percent, and quality benefits was the secondary reason at 20 percent.

[125] Deb Reese, R/QA Engineer and ISO 9000 Director for Matsushita Semiconductor Corporation of America has stated that ISO 14000 is a natural expansion of the ISO 9000 quality series. ISO 14000 expands the scope of the quality systems in manufacturing. *See* David A. Jones, *Responsible Care Seen as Road to ISO 14001,* 2 INTERNATIONAL ENVIRONMENTAL SYSTEMS UPDATE 14 (April 1995). Those companies that adhere to industry codes like CMA's Responsible Care or API's STEPs are well placed to meet ISO 14001 requirements.

[126] 18 Int'l Env't Rep. (BNA) 869 (November 15, 1995) (the Arthur D. Little survey is not available to the public).

[127] *See First Impressions, ROI Key To Top Commitment,* 2 INTERNATIONAL ENVIRONMENTAL SYSTEMS UPDATE 11 (March 1995), where for ISO 14000 to work it must be related to cost-savings in the areas of, but not limited to, waste management and energy reduction as well as provide for future savings possibilities in lower insurance rates and better access to capital. In short, return on investment (ROI) is the bottom line when it comes to convincing management that a certified EMS is worth pursuing. *See also* an unpublished informal survey conducted by the U.S. TAG from July through November 1995, where 24 percent of respondents stated competitive advantage as the best reason for certifying under ISO 14000, and where 26 percent stated customer demand as the best reason for certifying under ISO 14000.

To purchase copies of the ISO 14000 draft standards or to find out more information about their development, contact:

Name/Address	**Phone/Fax**
American National Standards Institute (ANSI) US Delegate to TC 207 11 West 42nd Street New York, New York 10036	212-642-4900 212-642-4969 (fax)
American Society for Testing and Materials (ASTM) Administer of US TAG 1916 Race Street Philadelphia, PA 19103	215-299-5400 215-299-2630 (fax)
NSF International (NSFI) P.O. Box 130140 Ann Arbor, MI 48113-0410	313-769-8010 313-769-0109(fax)

§ 3.5 Provisional American Society for Testing and Materials (ASTM) Standards for Organization's Environmental Management Systems

The American Society for Testing and Materials (ASTM) recently published two new provisional standards concerning environmental compliance audits:

- ✓ Provisional Standard 11-95 deals with environmental compliance audits.
- ✓ Provisional Standard 12-95 deals with evaluation of a company's environmental management.

These provisional standards are currently under review.[128]

[128] The standard on evaluating an organization's environmental management system is not directly related to either a compliance audit or a transactional (Phase I) audit. An environmental management systems audit determines whether a company has a sufficient environmental system in place to comply with various environmental laws. The actual compliance audit will check if there is compliance. The environmental management system audit is mainly concerned with a facility's management processes.

Under the proposed ASTM guidelines, the relevant concerns would include

✓ the facility or company's maintenance of a written environmental policy;

✓ the existence of an environmental manager, the qualifications and independence of the environmental manager;

✓ managerial support for the environmental manager in his/her duties throughout the facility;

✓ training programs;

✓ written policies;

✓ record-retention policies on all applicable media regulations, internal observation and enforcement of environmental laws; and

✓ a quality-assurance mechanism to ensure the quality and independence of the company's environmental management system.

§ 3.6 Practical Advice for Conducting a Compliance Audit

First, when preparing to conduct an environmental compliance audit, identify both the facilities to be audited and the auditors. If several facilities are to be audited, the order and priority among them should be determined. Decide whether the audits will be conducted by in-house auditors, outside auditors, or a combination. Despite whether the audit is to be conducted by in-house or by outside consultants, the auditors should consist of a small team that has technical expertise, some knowledge of the facilities to be audited, and knowledge about environmental laws and regulations. The team should make special efforts to conduct the audit informally, creating a good working relationship with plant personnel.

Decisions on how to conduct the audit need to be made before the audit process has begun. For example, the environmental audit can consist of one or more of the following:

✓ review of the company's environmental files, such as permits, reports to environmental agencies, Material Safety Data Sheets (MSDS) records, waste manifests, etc.;

✓ tours of the company's facilities;

✓ interviews with company managers and other personnel;

✓ selective sampling and analysis of wastewaters, emissions, wastes, or soil;

✓ review of federal and state environmental agency files on the facility; and

✓ discussions with federal and state environmental agency personnel on company's compliance with environmental requirements.

At the outset, the company will need to decide which of these actions will be included in its environmental compliance survey.

Developing and screening a list of all chemicals used, stored, and disposed of in the plant is a practical place to begin an audit. This list can be important for conducting the environmental audit so that the auditors can be knowledgeable about the chemicals in each department of the plant. This step may also include collection of material safety data sheets and determinations of whether the chemicals are on various lists of hazardous substances.

Conducting an environmental regulation review is also an important and useful step for senior management and department heads and assistants. These briefing sessions can help "set the atmosphere" for the cooperation and full disclosure needed for the audit's success. Therefore, they should include

1. A summary of the reasons and planning for the audit.
2. A survey of key environmental laws and regulations that apply to the plant.
3. A discussion of the potential liability of the company (and its responsible officers) for violations of environmental laws or for pollution.
4. Outlines of information on
   - wastewater discharges;
   - air emissions;
   - solid and hazardous wastes;
   - Superfund liability and hazardous substance release reporting;
   - other laws and rules applicable to the plant.

Conducting a full environmental audit is, of course, the most comprehensive course of conduct. When scheduling an audit, it is best to give advance notice to all those involved and set a realistic schedule. Advance notice also provides each department with an opportunity to clean up its area, check its files, and generally "prepare for inspection."

For each audit session, the auditors should prepare by briefly discussing any areas of operation or environmental matters on which they should focus their attention and any special environmental concerns with that department or section. The auditors would then meet with the department or section manager or assistant manager. As an example, each session could consist of four parts:

Sessions	**Description**
1. Introduction	Varies with circumstances
2. Sit-down discussion concerning:	(a) The process, equipment, and areas for which that department or section was responsible, and (b) The "environmental" aspects of the department or section, including (i) wastewater operation and flow patterns; (ii) air pollution emissions; (iii) solid and hazardous waste generation and handling; (iv) chemicals used; and (v) anything else, such as spill prevention or response.
3. Tour of the department's or section's facilities	Site specific.
4. A limited review of files, logs, or other records concerning "environmental" compliance	As specified. Not necessary for all facilities.

At the beginning of each session, the auditors should explain that the purposes of the audit are to determine whether the company is complying with environmental requirements and to identify and present potential problems and possible solutions to those problems. It should also be clear that the report on the audit would address these purposes and not "point fingers at anyone"; that the attorney is participating in the audit because of his "legal expertise" concerning environmental laws and regulations and because his presence could help preserve the confidentiality of information learned and generated during the audit; and that the auditors will conduct a postaudit review with the manager of each department or section to discuss the draft report on the audit before making the report final.

The discussion may generally follow the outline set forth above, but the auditors should listen very carefully and be especially alert to ask questions about any matter raised in the discussion.

During tours of the area and walks around the facility, the auditors can observe any sources of wastewater, emissions, or fugitive dust; any drums, tanks, or other containers for storage or use of chemicals; and the direction that any runoff or spill would flow. The auditors should be particularly interested in containment areas around storage tanks, leaks or potential leaks, and similar "common sense" concerns.

To complete the audit, the auditors should review the files concerning environmental matters, with particular focus on any violations or malfunctions during the last two years; in-house summaries of environmental matters; and compliance with environmental requirements, permits, and permit applications. The auditors should also review the organization and effectiveness of the filing system and the mechanisms set up to provide timely reminders of deadlines for reports and other requirements. The auditors should then examine the environmental statutes, regulations, and other materials available in the company library and environmental coordinator's office and the methods for updating those materials.

Finally, the auditors should question the status of the company's environmental compliance and present or future problems based on existing conditions; on any possible modifications to or expansions of the plant; and on future, more stringent regulatory requirements.

When designing a pollution prevention audit, begin by identifying high priority waste streams through a review of documents, discussions with plant personnel, and a tour of the facility. Then, analyze each waste stream for potential reduction strategies, such as

✓ material conservation;

✓ material substitution;

✓ improved inventory management, including spill and leak prevention;

✓ recycling;

✓ process modifications.

Next, the audit should assess the costs and benefits of each potential reduction strategy, including assessment of the company's long-term options for waste disposal.

Conducting an environmental audit is, of course, useless unless there is adequate follow-up. For that reason, following the audit sessions the auditors (preferably the attorney, to maintain confidentiality) should prepare a report. As an example, the report can be written with the following format:

✓ an introduction summarizing how the audit was conducted;

✓ an executive summary of the overall findings and recommendations;

✓ a section on each of the major production departments and the other sections that were audited; and

✓ detailed findings and recommendations.

The section on each department can generally describe the environmental aspects of that department's operations; identify favorable "environmental" performance or procedures of the department; list present or potential problems; and propose solutions to those problems. In most cases, the auditors should send the draft report to the company managers who were audited and request corrections, changes, and suggestions. The final report may include recommendations for capital improvements that might enhance environmental performance; operational changes that would improve environmental performance; reaffirmation of existing operational procedures; reexamination, in some cases, by engineering or legal personnel of applicable environmental laws or regulations; and other approaches.

After submission of the report, senior management will generally review and discuss the report with the auditors. Following management review of the report, there will probably be follow-up actions to put some or all of the recommendations into effect.

Forms to Be Used in Compliance Audits

Form 3.1
Environmental Audit Requirement as Condition of Consent Decree

ENVIRONMENTAL AUDIT

The Defendants shall initiate an Environmental Audit as set forth in Attachment __, which is incorporated by reference into this Consent Decree.

Form 3.2
Model Attachment to Consent Decree

Attachment ___

ENVIRONMENTAL AUDIT

I. SCOPE OF WORK

In order to ensure full compliance with applicable environmental laws, regulations, and permits by the Defendants' Facility in an efficient and coordinated manner, the Defendants shall undertake the following auditing tasks, under which: (a) An independent auditor retained by and paid for by the Defendants and approved by EPA and the State Agency, shall, subject to EPA and the State Agency oversight, audit the Facility and prepare a report on their assessment of the Defendants' compliance with RCRA, CWA, and CAA laws, regulations, and permits (and equivalent state laws, regulations, and permits); (b) The independent auditor shall perform an analysis of the Defendants' environmental management systems, practices and policies, at the Facility and in the Defendants' corporate offices having responsibility for supervision of environmental compliance activities at the Facility.

The performance standard for each audit is to complete a detailed and professional investigation, as set forth in the Audit Work Plan (described below), of recordkeeping and operating practices and environmental management operations at Defendants' Facility and corporate offices having responsibility for the supervision of environmental compliance activities at the Facility.

II. SELECTION OF AUDIT FIRM

Within forty-five (45) days after entry of this Consent Decree, the Defendants shall propose an Audit Firm to EPA and the State Agency for approval. The Audit Firm shall be an expert in environmental auditing, environmental management systems, and RCRA, CWA, and CAA compliance. The Audit Firm must be capable of exercising the same independent judgment that a Certified Public Accounting firm would be expected to exercise in auditing a publicly held corporation. In addition, the Defendants shall require the proposed Audit Firm to describe: (a) any stock ownership in the Company, any parent, subsidiary, or affiliated corporation; (b) all contractual agreements the Audit Firm has or has had with the Defendants in regard to the development of recordkeeping or operational procedures, the designing of equipment or structures, or the construction of equipment or structures at any of the Defendants' facilities. Such descriptions shall be included in the proposal submitted to EPA and the State Agency within forty-five (45) days after entry of this Consent Decree.

If EPA and the State Agency reject the proposed audit firm, then the Defendants shall propose an alternate audit firm for EPA and the State Agency approval.

III. AUDIT WORK PLAN

Within forty-five (45) days after approval of the Audit Firm, the Defendants shall submit a proposed Audit Work Plan to EPA and the State Agency for approval. The proposed Audit Work Plan shall include auditing protocols to be used by the Audit Firm, a schedule for conducting the facility audit and for the completion of all tasks set forth in the Scope of Work, and the names and resumes of those Audit Firm employees who will be primarily responsible for performance of the tasks set forth in the Scope of Work.

The portion of the Audit Work Plan dealing with the Facility Compliance Audit shall include, but not be limited to: (1) waste pre-acceptance and acceptance records and procedures currently utilized by the facility to ensure compliance with the federal and state land disposal restrictions; (2) waste pre-acceptance and acceptance records and procedures utilized by the facility to ensure compliance with federal and state restrictions regarding disposal of liquid wastes in landfills, including test procedures implemented by the facility to confirm the absence of free liquids; (3) waste pre-acceptance and acceptance records and procedures to ensure ignitable and reactive wastes are treated, stored, or disposed of at the facility in accordance with the applicable regulations, including test procedures implemented by the facility to ensure that after treatment, the resulting waste, mixture or dissolution or material no longer meets the definition of ignitable or reactive waste; and (4) waste stream and waste load pre-acceptance procedures in order to ensure that detailed chemical and physical analyses of all wastes proposed for acceptance at the facility are reviewed and approved by the facility technical manager prior to treatment, storage, or disposal.

The Audit Work Plan, or the Audit Work Plan as modified by EPA and the State Agency, shall be initiated within thirty (30) days of notification by EPA and the State Agency of their approval, in accordance with the schedule contained therein.

IV. AUDIT PROCEDURES

The Audit Firm shall have access to and may review any records (except attorney-client communications) which bear on determining compliance by the facility with applicable regulatory requirements. However, the Audit Firm shall focus on determining compliance with the applicable regulations during the previous 12-month period. Records to be reviewed during the audit may include, but not be limited to: (1) All records required by State, Federal, or local laws to be maintained by the Defendants; (2) Facility operating records, such as waste profile sheets containing waste pre-acceptance verification data, waste tracking data for intrafacility movement of received wastes or wastes generated on-site, waste storage data, waste treatment data, and data reflecting the disposition of received wastes, including landfill cell grid sheets; (3) Applications, licenses, permits and approvals, unit operation plans, operation and maintenance plans, or other regulatory documents pertaining to on-site activities at the facility; (4) Environmental monitoring plans, with specific attention to groundwater monitoring; (5) Waste treatability procedures, methods, and study results; (6) Manifests for wastes entering or leaving the facility; (7) Records of use, maintenance, and decommissioning of vehicles used on-site for transportation of hazardous wastes; (8) Vehicle washing records; (9) Facility inspection records and maintenance records for equipment and structures, including the groundwater monitoring wells; (10) Employee training records; (11) Contingency Plan and emergency procedures; (12) Corporate and facility guidelines, policies, and internal operating rules pertaining to facility operations, inspections, personnel training, recordkeeping practices, emergency response and communications, closure and post-closure activities (but not including internal corporate compliance audit materials); (13) Any effluent data, including effluent and intermediate discharges to surface water and discharge locations specified by federal or state permits; and (14) All other records bearing on RCRA, CWA, and CAA compliance kept at the Facility or Defendants' corporate headquarters.

The Audit Firm shall have access to all hazardous waste management units (HWMUs), equipment, and structures used at the facility to manage, treat, store, or dispose of wastes. The Audit Firm shall observe and evaluate actual operation and maintenance procedures for all HWMUs and other principal equipment and structures at the facility, including the observation and evaluation of sample collection at the groundwater monitoring wells.

The Audit Firm shall provide a minimum of three blind standards (*i.e.*, pre-analyzed samples containing known constituent concentrations) to each laboratory that routinely conducts preacceptance testing, surface water and groundwater analysis for the Defendants, and also to the Facility on-site laboratory. The results of the analyses of these samples will be used to evaluate the ability of each laboratory to quantify representative hazardous constituents in various media.

V. FACILITY AUDIT REPORT

The Facility Audit Report shall describe in detail: (1) The procedures followed during the audit, the facility itself, the regulatory history and current compliance status of the facility; (2) Each potential violation detected during the audit; (3) Any other information, which in the judgment of the Audit Firm, merits review by EPA or the State Agency. For each potential violation reported, the Facility Audit Report shall also indicate the relevant statutory or regulatory section; the particular area of the facility where the violation occurred (if appropriate); the dates during which the violation occurred or existed (if it can reasonably be determined); and any other relevant or appropriate information.

In addition, the Facility Audit Report shall include, as a minimum, a discussion of: (1) All hazardous wastes whose recorded location differs from their observed physical location, or whose physical location cannot be corroborated by facility records. (2) Whether the facility waste analysis plan (WAP) is adequate to sufficiently manage the hazardous waste generated, treated, stored, disposed of, or otherwise handled at the facility in accordance with the applicable federal or state laws and regulations, and whether the Defendants are in compliance with the procedures in the WAP for preacceptance and acceptance of hazardous wastes. (3) Any observed deviations from Defendants' written operating procedures, including documentation on any untimely response to repair of deteriorating or malfunctioning waste containers, structures (including groundwater monitoring wells), or equipment. (4) Recommendations as to potential significant improvements and/or modifications that should be made to Defendants' operating procedures to achieve compliance with all applicable regulatory requirements. (5) The results of the laboratory evaluations conducted pursuant to Section IV. (6) Adequacy of the facility's waste tracking system, including the computerized waste tracking system, when implemented.

VI. COMPLIANCE REPORT AND PLAN

The Defendants shall implement as soon as practicable procedures to correct any violation identified in the Facility Audit Report. No later than sixty (60) days after the Defendants receive the Facility Audit Report, the Defendants shall prepare a Compliance Report and Plan indicating how the facility has corrected or is correcting each deficiency. All compliance actions taken by the Defendants shall be undertaken in accordance with applicable Federal, State, and local laws.

The Compliance Report and Plan shall describe in detail every compliance action taken in response to the Facility Audit Report. In the case of violations which are not corrected within sixty (60) days of receipt of the Facility Audit Report, the Compliance Report and Plan shall describe actions to be taken in response to any violations or findings in the Facility Audit Report. The Defendants shall certify that the information presented in the Compliance Report and Plan is accurate.

Failure to commence or complete any specific work activity in response to the Facility Audit Report shall not be construed as a failure to meet a deadline within the meaning of Section X(A)(2) of the Consent Decree.

VII. CORPORATE MANAGEMENT SYSTEMS REPORT

No later than ninety (90) days after the Facility Audit Report is submitted to the Defendants, the Audit Firm shall submit to the Defendants a Corporate Management Systems Report which shall: (1) Identify and describe the existing facility environmental operations and the corporate offices responsible for overall company-wide environmental compliance and management systems, policies, and prevailing practices as they affect all applicable statutory and regulatory compliance; (2) Evaluate such operations and systems, practices, and policies and identify and describe fully the perceived weaknesses in such operations and systems, practices, and policies by comparing them, to the extent practicable, to: (a) their ability to promote compliance with all applicable statutory and regulatory requirements; (b) the existing practices, programs, and policies of other hazardous waste disposal corporations and other industrial corporations with similar operations operating within the United States, based upon available literature, public information, and Audit Firm's non-confidential experience pertinent to regulatory compliance programs, practices, and policies currently operative in the disposal industry in the United States; (c) the history of operations in terms of the facility's compliance programs, compliance record, and environmental management practices over the previous five years.

The auditor shall evaluate the foregoing information, using such factors as the auditor believes to be relevant and appropriate, which factors shall be stated in the report. Based on the evaluation required in paragraph 2 above, the auditor shall identify and describe fully with supporting rationales, the perceived areas, if any, where Defendants' environmental management systems, practices, and policies may be improved as they affect the facility's compliance obligations, listing specific options for any improvements at the facility in the following areas: (a) environmental compliance program management operation, staffing, education, and experience requirements; (b) lines of authority to Defendants' corporate offices responsible for overall company-wide environmental compliance and management systems, policies, and practices, and relationship to the operating facility manager; (c) personnel training for individual employee compliance obligations; (d) Operations and Maintenance (O & M) procedures for all Solid Waste Management Units' (SWMUs) waste leachate collection and sampling equipment; (e) evaluation of operations and pollution control equipment in terms of adequacy of design and compatibility with wastes being passed through such equipment; (f) quality and thoroughness of implementation of all waste and wastewater analysis plans for both incoming and outgoing waste streams directly discharged, emitted, released to the ambient environment, or conveyed offsite in bulk shipments; (g) preparation of Quality Assurance and Quality Control programs for sampling and analysis and for environmental testing procedures, including corporate and facility laboratories and contract laboratories for the facility; (h) preparation of records needed to provide the facility management with an adequate data base to accurately determine compliance with all applicable statutory and regulatory requirements, with particular attention to waste generation (including quantity and chemical composition), movements, treatment, and ultimate disposition by location of waste source, handling points, and final disposition; (i) preparation of self-monitoring reports required to be filed with the EPA and the State Agency; (j) preparation and review of Incident Reports evaluating causes of equipment malfunctions, improper waste handling, or breakdowns, with specific recommendations for corrective steps and preventive O & M, along with procedures for reporting these recommendations to corporate headquarters.

VIII. CORPORATE MANAGEMENT REPORT AND PLAN

No later than ninety (90) days after it has received the Corporate Management Systems Report, Defendants shall prepare a Corporate Management Systems Report and Plan describing in full what actions it has taken or will take to implement the recommendations of the Corporate Management Systems Report, or responding to the findings and recommendations and explaining why implementation of the recommendations is not necessary.

IX. SUBMISSION OF AUDIT REPORTS

All Audit Reports required to be prepared under this Consent Decree shall not be submitted to EPA or the State Agency unless they are specifically requested in writing from the Defendants or Audit Firm by the Director, Hazardous Waste Management Division, EPA—Region __, or the Secretary of the State Agency. If the reports are requested, the Defendants or Audit Firm shall provide the reports within ten (10) days of receipt of the request.

The Audit Firm shall notify EPA and the State Agency of the availability of the Facility Audit Report and the Corporate Management Systems Report simultaneously upon submission of the respective reports to the Defendants. The Defendants shall also notify EPA and the State Agency upon completion of the Compliance Report and Plan and the Corporate Management Report and Plan.

X. ADDITIONAL AUDIT REQUIREMENT

The Defendants, through the use of a different independent audit firm, shall conduct a second environmental audit of the Facility according to the procedures set forth above. The second environmental audit shall begin one year after submission of the Corporate Management Report and Plan by the Defendants, commencing with Section II—Selection of Audit Firm.

XI. USE OF AUDIT REPORTS BY EPA AND THE STATE AGENCY

The United States and the State of ___________ agree that the Facility Audit Report and the Compliance Report and Plan may not be used as direct evidence in a proceeding to assess a civil penalty under RCRA, CWA, or

CAA if the violation is corrected within the time allowed below after delivery of the Facility Audit Report to the facility. The Facility Audit Report and the Compliance Report and Plan may be used as direct evidence in a proceeding to assess a civil penalty under RCRA, CWA, or CAA if the violations remain uncorrected after the period specified below and may be used in any circumstances as direct evidence of the duration or degree of Defendants' knowledge of a RCRA, CWA, or CAA violation. Defendants do not waive any legal rights they may have to object to the admissibility of such evidence. Documents prepared by or for the Defendants pursuant to these audit provisions are not required by statute or regulation and shall remain confidential to the maximum extent permitted by law.

The restriction on use of the Audit Reports as direct evidence in proceedings to assess civil penalties under RCRA, CWA, or CAA shall not apply to any violations identified through reporting requirements imposed on the Defendants by statute or regulation. With regard to any RCRA, CWA, or CAA violation (or violation under equivalent state statutes or regulations) which occurred during the time period beginning with commencement of the audit period and ending with the time allowed for corrective action identified below, and which is identified specifically or by specific type in the Facility Audit Report or the Compliance Report and Plan, but not those specific types of violations which arise from a knowing or willful disregard of applicable legal standards, EPA and the State Agency will not seek civil penalties for such violations: (1) if corrected as soon as practicable, but not to exceed sixty (60) days after delivery of the Facility Audit Report; or (2) for any violation which cannot be corrected within sixty (60) days, if the Defendants commence correction as soon as practicable, but not more than sixty (60) days after receipt of the Facility Audit Report and progress expeditiously to completion of the correction.

EPA and the State Agency may seek civil penalties under RCRA, CWA, and CAA for violations identified by EPA or the State Agency through inspections, information requests, or any other means independent of the audit. The burden of proof that the evidence of the violation was not independently obtained is on the Defendants. EPA and the State Agency may seek civil penalties for violations identified through self-reporting requirements, including but not limited to information obtained from Discharge Monitoring Reports (DMRs) and Non-Compliance Reports (NCRs) or any such reports filed pursuant to RCRA, CWA, or CAA requirements.

Form 3.3

Facility Site Inspection Checklist for *RCRA Small Quantity Generators* (SQG)

Section A—GENERAL SITE INFORMATION

1. Are any solid waste facilities located in the 100-year floodplain?
   NO ___ YES ___
2. Describe land use within one mile ______________
3. Are there any closed or abandoned solid waste facilities?
   NO ___ YES ___

NOTE: Attach PLANT MAP or SKETCH showing site orientation, waste management facilities, and major topographic features.

Section B—UNAUTHORIZED DISCHARGES

1. Is there any evidence of fires and explosions or leaks and discharges to the environment from solid waste facilities or any other type of facility?
   NO ___ YES ___
2. If yes, have they been reported and remedied?
   N/A ___ YES ___ NO ___

Section C—ACCUMULATION TIME EXEMPTIONS

N/A ___

NOTE: Hazardous wastes may be accumulated in Containers or Tanks without a permit for up to 180 days for SQGs, or 270 days for SQGs that must transport waste >200 miles.

* 1. Is the beginning date of accumulation clearly indicated on each container?
N/A ___ YES ___ NO ___

* 2. Is each container or tank clearly labeled or marked "Hazardous Waste"?
N/A ___ YES ___ NO ___

**3. Did generator exceed the Accumulation Time limitation?
N/A ___ NO ___ YES ___

4. Did SQG or CE-SQG exceed Accumulation Quantity limitation?
N/A ___ NO ___ YES ___

NOTE: SQG: Total quantity of HW must never exceed 6000 kg.
CE-SQG: Total quantity of HW must never exceed 1000 kg.

* **Not applicable to Municipal and Industrial CE-SQGs.**

**** Prohibition found in 268.50(a)(1) applies also, *i.e.,* must accumulate solely for the purpose of necessity to facilitate proper recovery, treatment, or disposal and complies with the accumulation time requirement. Otherwise permit required violation applies.**

Section D—PREPAREDNESS and PREVENTION

1. Is the facility equipped with:

   a. Internal communication or alarm system within easy access?
   N/A ___ YES ___ NO ___

   b. Communication system to call off-site emergency assistance?
   N/A ___ YES ___ NO ___

   c. Fire, spill control, and decontamination equipment?
   N/A ___ YES ___ NO ___

   d. Adequate fire-water supply (volume and pressure)?
   YES ___ NO ___

2. Is the above-noted emergency equipment regularly tested?
   YES ___ NO ___

3. Is aisle space sufficient to allow unobstructed movement of personnel and equipment?
   N/A ___ YES ___ NO ___

NOTE: Measure or estimate aisle space: ________________

Section E—EMERGENCY PROCEDURES FOR SQGs

N/A ___

1. Is the following information posted by the telephone?

   a. Name and telephone no. of the emergency coordinator
   N/A ___ YES ___ NO ___

   b. Location of emergency equipment
   N/A ___ YES ___ NO ___

   c. Telephone no. of fire department
   N/A ___ YES ___ NO ___
   (Unless facility has direct connection)

Section F—CONTAINER STORAGE AREA REQUIREMENTS

N/A ___

Container Requirements

1. Are containers in good condition?
   N/A ___ YES ___ NO ___

2. Are the containers compatible with the wastes being stored?
   N/A ___ YES ___ NO ___

3. Are containers kept closed and stored in a safe manner?
   N/A ___ YES ___ NO ___
4. Are containers inspected weekly for leakage and deterioration?
   N/A ___ YES ___ NO ___
5. Are containers holding incompatible wastes separated by a physical barrier or sufficient distance ?
   N/A ___ YES ___ NO ___
6. Describe CSA(s).

LDR Container Storage Requirements

7. Have restricted wastes exceeding treatment standards been stored in the CSA for more than one year?
   N/A ___ YES ___ NO ___

If yes, continue:

Can the generator demonstrate storage was for the sole purpose of accumulating sufficient quantities to facilitate proper recovery, treatment, or disposal?

N/A ___ YES ___ NO ___

NOTE: Ques. no. 7 does not apply to Conditionally Exempt Small Quantity Generators.

Section G—SATELLITE ACCUMULATION AREAS

N/A ___

NOTE: Generators may accumulate HW in containers at or near the point-of-generation without a permit if they meet the following conditions.

1. Are containers in good condition?
   YES ___ NO ___
2. Is the waste compatible with the containers?
   YES ___ NO ___
3. Are containers kept closed (except when adding or removing waste)?
   YES ___ NO ___
4. Are containers marked "hazardous waste" or labeled to identify the contents?
   YES ___ NO ___
5. If waste accumulation has exceeded 55 gallons (or 1 qt. of acutely HW):
   a. Has container holding excess amount been marked with beginning date of excess accumulation?
   N/A ___ YES ___ NO ___

b. Have excess amounts remained in satellite area over 3 days?
N/A ___ YES ___ NO ___

6. Describe satellite accumulation area(s).

Section H—SQG Tank, Inspections

N/A ___

1. Where present, does the owner/operator inspect the following at least each operating day?
   a. Discharge control equipment
   N/A ___ YES ___ NO ___
   b. Data gathered from monitoring equipment to ensure operation within design
   N/A ___ YES ___ NO ___
   c. For uncovered tanks, freeboard level
   N/A ___ YES ___ NO ___
2. Are the following inspected at least weekly?
   a. Construction materials of the tank
   N/A ___ YES ___ NO ___
   b. Area immediately around the tank for signs of leakage or erosion
   N/A ___ YES ___ NO ___
3. For tanks which are continuously fed hazardous waste, is there a means to stop inflow (*i.e.*, waste feed cutoff or bypass to a stand-by tank)?
   N/A ___ YES ___ NO ___

*** NOTE—If SQG has Surface Impoundments, Landfills, or Groundwater Monitoring in progress, attach required checklists and check status of facility.**

A facility cannot be an SQG if it has active land disposal of hazardous waste.

Form 3.4
Model Corresponding Personnel and Responsibility Chart

ACCOUNTABILITY CHART

ENVIRONMENTAL RESPONSIBILITIES	KEY PERSONNEL RESPONSIBLE[1]
1. Establish environmental goals and values.	1. President, CEO, Board of Directors
2. Develop environmental policy.	2. President, CEO, Chief Environmental Officer, General Counsel
3. Monitor EMS performance.	3. Chief Environmental Officer
4. Ensure external regulatory compliance.	4. Operations Manager, General Counsel
5. Ensure internal EMS compliance.	5. All managers
6. Ensure continual improvement.	6. All managers
7. Identify customer expectations.	7. Sales and marketing staff
8. Identify suppliers expectations.	8. Procurement staff
9. Develop and maintain accounting procedures.	9. Accounting staff
10. Develop and maintain process and design procedures.	10. Process/ design engineering staff
11. Comply with all defined procedures.	11. All staff

[1] While personnel at Company have clearly delineated roles, each employee is empowered to ensure that the organization adheres to its environmental policy. To this end Company protects the anonymity and job security of all employees who bring operational discrepancies to the chief environmental manager. In addition any employee who contributes to the continual improvement of the EMS will be recognized as appropriate through the organization's reward and recognition system.

Chapter 4

Common Law Privilege of Environmental Audits

§ 4.1 Overview

A great deal of media as well as regulatory debate has centered on the degree to which environmental audits can be privileged from discovery in both civil as well as criminal enforcement and third-party litigation. Compliance oriented environmental audits have often been structured by attorneys to take advantage of the attorney-client privilege and the work product doctrine. Audits structured as such are typically labeled "privileged."

Recent legislative attempts to codify an environmental audit privilege extend a more recent self-critical evaluation privilege that has been adopted in a number of contexts—including environmental auditing—by a limited number of state and federal courts.

Below are further discussions of each of these common law doctrines.

§ 4.2 Attorney-Client Privilege

The attorney-client privilege ensures the confidentiality of communications between an attorney and client.[1] The privilege has been defined as follows:

---

[1] Other theories designed to protect confidentiality include the work product doctrine and the self-evaluation theory. Under the work product doctrine, any "work product" of an attorney (or, arguably, the client or consultants to the attorney) in preparation for threatened or pending litigation is protected from discovery and disclosure to persons outside the corporation.

The self-evaluation theory has been recognized, however, by only a few courts. Many commentators argue that environmental audits should be protected from discovery since disclosure of internal investigations conducted to discover and remedy problems does not further public interest of encouraging critical self-examination.

> (1) Where legal advice of any kind is sought, (2) from a professional legal advisor in his capacity as such, (3) the communications relating to that purpose, (4) made in confidence, (5) by the client, (6) are at [this] instance permanently protected, (7) from disclosure by himself or by the legal adviser, (8) except the protection may be waived.[2]

The privilege is usually, but not necessarily, destroyed by the presence of a third party.[3]

The leading case on the effect of third-party involvement on the privilege is *United States v. Kovel.*[4] That case held that the privilege was not automatically destroyed by communication of the client's confidences to a third-party accountant. The court compared the presence of an accountant to the presence of a translator: "Accounting concepts are a foreign language to some lawyers in almost all cases, and to almost all lawyers in some cases. . . . [T]he presence of the accountant is necessary, or at least useful, for the effective consultation between the client and the lawyer which the privilege is designed to permit."[5] Subsequent cases have followed *Kovel* and held that the privilege is not always destroyed by the presence of a third party.[6]

The key test for application of the privilege is "[t]hat the communication be made *in confidence* for the purpose of obtaining *legal* advice *from the lawyer.*"[7] However, the cases state that this element can be satisfied where the client seeks advice from the attorney's agent, as well, if it is for the purpose of helping the attorney render legal advice to the client.[8] The privilege may apply if the consultant can be classified as the attorney's agent or assistant.

---

[2] 8 J. Wigmore, Evidence, § 2292 (McNaughton Rev. 1961). *See also* United States v. United Shoe Machinery Corp., 89 F. Supp. 357, 358-59 (D. Mass. 1950).

[3] *See, e.g.*, In re Walsh, 623 F.2d 489, 495 (7th Cir. 1980), *cert. denied,* 449 U.S. 994 (1980).

[4] 296 F.2d 918 (2d Cir. 1961).

[5] *Id.*

[6] *See, e.g.*, Federal Trade Comm'n v. TRW, Inc., 628 F.2d 207 (D.C. Cir. 1980); United States v. Cote, 456 F.2d 142 (8th Cir. 1972); United States v. Judson, 322 F.2d 460 (9th Cir. 1963).

[7] Kovel, 296 F.2d at 922.

[8] *Id.* at 921, citing 8 Wigmore, Evidence § 2301.

Most of the cases that address the issue of protection from discovery of communications made to third parties involve communications made to an accountant regarding tax matters. An analogy can be drawn for purposes of application of the privilege between those communications and communications made to a technical consultant regarding environmental matters. First, the federal law of privilege controls in both cases. Tax matters are investigated under the Internal Revenue Code, and environmental matters are investigated under various federal environmental statutes. Both contexts involve a federal question; therefore, the privilege issue will be governed by federal common law under Rule 501 of the Federal Rules of Evidence. Second, both financial data and environmental data are complex. Interpretation of the data requires specialized knowledge possessed either by accountants or technical consultants, as the case may be. The presence of a consultant arguably is as necessary to an environmental review as the presence of an accountant is to a financial review. Therefore, the cases should apply to communications made to a technical consultant in the course of an environmental review.

Under this line of cases, in order to protect confidential information divulged to the consultant in the course of a review, all three elements of the test for applicability of the attorney-client privilege must be satisfied. The client must take precautions to keep the information confidential, must be seeking legal advice, and must be seeking advice from a lawyer acting as such. The technical consultant must act as an agent of the lawyer, not as an independent consultant, for the privilege to apply. It must be clear that the consultant is translating data **for the attorney** so that **the attorney** may better advise the client on legal matters.

It is unclear, under this progeny, whether the attorney has to formally hire the consultant in order for it to be clear that the consultant is aiding him in rendering legal advice. It should be sufficient for the attorney to advise the client to retain a consultant for the express purpose of aiding the attorney in the interpretation of data. In fact, the early cases seem to say that the formal relationship between the attorney, the client, and the consultant is irrelevant.[9] Subsequent cases have put more emphasis on who hires whom. The issue of who must formally employ the third party has not been

[9] *See* Kovel, 296 F.2d at 921.

directly addressed, but it has been a relevant fact in many of the decisions. Together, the cases suggest that the third party should actually be employed by the attorney.[10]

In *United States v. Chevron,*[11] Chevron claimed that its Corporate Environment Compliance Review Status Reports were protected by the attorney-client privilege. After noting that a party asserting the privilege has the burden of proving its existence, the court found that the privilege did not apply because the court found that "it is not enough to assert the privilege merely because an attorney was present in his or her capacity as a business advisor."[12] Rather, the court explained that "the communication's primary purpose must be to gain or provide legal assistance."[13] Because Chevron did not prove that the audit was conducted in order for the attorney to advise the client, the court held that the privilege did not apply.

Federal Trade Comm'n v. TRW, Inc.,[14] clouds the issue further. The case states that reports produced by outside consultants for the purpose of determining potential liabilities and solutions are "[o]n the frontier of the law, still not well defined, separating those independent outside services performed for an attorney that are discoverable, from those that are not."[15] The court did not reach a conclusion as to whether the reports were, in fact, discoverable. By throwing the reports into limbo, the opinion appears to increase the risk that a court will find that the information disclosed to the consultant is discoverable.

In summary, the law is not settled concerning attorney-client privilege, especially in light of *Federal Trade Comm'n v. TRW, Inc.*[16] In order to protect the information disclosed by the client to the technical consultant to the greatest extent possible, the consultant should be hired by the attorney—acting as an attorney and not as a general business advisor—

---

[10] *See, e.g.,* United States v. El Paso Co., 682 F.2d 530, 541 (5th Cir. 1982), *cert denied,* 466 U.S. 944 (1984); United States v. Pipkins, 528 F.2d 559 (5th Cir. 1976), *cert denied,* 426 U.S. 952 (1976); United States v. Brown, 478 F.2d 1038, 1040 (7th Cir. 1973); Bauer v. Orser, 258 F. Supp. 338, 342 (D.C.N.D. 1966).

[11] 1989 WL 121616 (E.D. Pa. Oct. 16, 1989).

[12] *Id.*

[13] *Id.*

[14] 628 F.2d 207, 212 (D.C. Cir. 1980).

[15] *Id.*

[16] *Id.*

preferably with a statement in the hiring letter stating that the audit will be conducted in order to gather factual data upon which the attorney will base his or her legal conclusions.[17] It must be clear that the consultant was retained for the purpose of interpreting data for the attorney so that the attorney can better advise the client. If the consultant is hired by the attorney and reports to the attorney, the consultant's role as the attorney's agent is more evident. It may be possible to establish a formal employment relationship between the attorney and the consultant and then have the consultant bill the client directly. However, the consultant should take direction from and report directly to the attorney. All possible precautions should be taken to convey the image of the consultant as the attorney's agent.

In order for the attorney-client privilege to extend to communications made to a technical consultant in the course of an environmental review, it must be clear that the consultant is aiding the attorney in advising the client. Even so, application of the privilege to communications made to outside consultants is uncertain; protection of information disclosed during an environmental review is not guaranteed. In order to further the perception of the consultant as the attorney's agent, thereby increasing the likelihood that the privilege will apply, the attorney should hire and supervise the consultant directly.[18]

Another dilemma presented in the context of evaluating the efficacy of the attorney-client privilege is the ethical dilemma that an environmental attorney may face under certain circumstances. Almost all the state and federal environmental statutes include criminal liability provisions. If an attorney acquires privileged information that a client is engaging in activity constituting an environmental crime, there may be a duty to disclose that information to authorities. Under the Model Rules of Professional Conduct and under many state codes of professional conduct, an attorney *may* reveal confidential information to the extent he or she believes is reasonably

[17] *See* Chevron, 1989 WL 121616 (E.D. Pa. Oct. 16, 1989).

[18] *Id.* 1989 WL 121616 (E.D. Pa. 1989) (Corporate Environmental Compliance Review Status Reports not protected by attorney-client privilege unless requested by an attorney in his or her capacity as an attorney, rather than as a business advisor).

necessary to prevent the client from committing a criminal act that the attorney believes is likely to result in imminent death or substantial bodily harm.[19] Likewise, under the Model Code of Professional Responsibility, an attorney may reveal the intention of his client to commit a crime and information necessary to prevent the crime.[20] Disclosure is not mandatory under the model code. Other jurisdictions, however, require an attorney to disclose the otherwise privileged information in such a circumstance. In these other jurisdictions, disclosure is not discretionary, but required.

In either case, the environmental attorney's judgment is likely to be taxed by questions resembling those that are fundamental to environmental law in general. The attorney will be faced with enormous moral, scientific, and factual questions. Determining when a risk resulting from an environmental crime is likely to result in imminent death or substantial bodily harm is more comfortably within the province of the scientist and not the environmental attorney.

§ 4.3 Work Product Doctrine

Another potential means of protecting an environmental audit from disclosure is through the work product doctrine first enunciated by *Hickman v. Taylor.*[21] In that case, the Supreme Court "recognized a qualified immunity from discovery for the 'work product of the lawyer'; such material could only be discovered upon a substantial showing of 'necessity or justification.'"[22] The holding of *Hickman* has largely been embodied in

---

[19] Model Rules of Professional Conduct Rule 1.6(b)(1) (1991).

[20] Model Code of Professional Responsibility DR 4-101(C)(3). ABA Opinion 314 (1965) indicates that an attorney must disclose the confidences of his or her client if facts in the attorney's possession indicate beyond reasonable doubt that a crime will be committed. *See also* Washington, D.C., Rules of Professional Conduct Rule 1.6 (c)(1); New York Code of Professional Responsibility DR 4-101(C)(3) (1990).

[21] 329 U.S. 495, 510, 91 L. Ed. 451, 67 S. Ct. 385 (1947).

[22] F.T.C. v. Grolier Inc., 462 U.S. 19, 24, 76 L. Ed. 2d 387, 103 S. Ct. 2209 (1983).

Fed. R. Civ. P. 26(b)(3),[23] which outlines "the extent to which trial preparation materials are discoverable in federal court."[24]

The work product doctrine protects only those documents that are prepared for use in anticipated or ongoing litigation.[25] To be covered by the work product doctrine, documents must have been produced "pursuant to a clear decision, based on substantial bona fide reasons, to reject the claim and to litigate."[26] "Three conditions must be met to earn work product protection. The material must (1) be a document or tangible thing, (2) that was prepared in anticipation of litigation, and (3) was prepared by or for a party, or by or for his representative."[27] "Once these conditions are satisfied, the party seeking discovery must establish a substantial need for the material and a practical inability to secure the substantial equivalent thereof by alternate means."[28] The courts have taken a flexible approach in determining whether the work product doctrine is applicable, asking not whether litigation was a certainty, but whether the document was created "with an eye toward litigation."[29]

---

[23] FED. R. CIV. P. 26(b)(3) provides in pertinent part:

> a party may obtain discovery of documents and tangible things . . . prepared in anticipation of litigation or for trial by or for another party or by or for that other party's representative (including the other party's attorney, consultant, surety, indemnitor, insurer, or agent) only upon a showing that the party seeking discovery has substantial need of the materials in the preparation of his case and that the party is unable without undue hardship to obtain the substantial equivalent of the materials by other means.

The phrase "in anticipation of litigation" has not required work product to be case specific. *See* Fine v. Facet Aerospace Products Co., 133 F.R.D. 439, 445 (D.C. N.Y. 1990); *In Re* Grand Jury Proceedings, 73 F.R.D. 647 (M.D. Fla. 1977).

[24] *Id.* at 25. See Niagara Mohawk Power v. Stone & Webster, 125 F.R.D. 578, 585 (N.D.N.Y. 1989).

[25] FED. R. CIV. P. 26(b)(3). *See also* 68 F.R.D. at 580 ("material compiled by counsel in preparation for this lawsuit would be protected from discovery by the 'work product' doctrine, which exists independently of the attorney-client privilege"). *See generally* United States v. Nobles, 422 U.S. 225, 238 n. 11, 45 L. Ed. 2d 141, 95 S. Ct. 2160 (1975). Work product immunity extends to documents and tangible things that would reveal an attorney's mental processes in preparation for litigation. *See* Fed. R. Civ. P. 26(b)(3); Hickman v. Taylor, 329 U.S. 495, 508, 511, 91 L. Ed. 451, 67 S. Ct. 385 (1947).

[26] Fine v. Bellefonte Underwriters Ins. Co., 91 F.R.D. 420, 422 (S.D.N.Y. 1981).

[27] *In re* Grand Jury Subpoenas Dated Dec. 18, 1981, etc., 561 F. Supp. 1247, 1257 (E.D.N.Y. 1982).

[28] *Id.*

[29] Hickman, 347 U.S. at 511. The work product privilege draws no distinction between materials that are factual in nature and those that are deliberative. See Martin v. Office

By its terms, Rule 26(b)(3) makes clear that the work product doctrine protects work done in anticipation of litigation or trial by an attorney, as well as a party or other agents of the party or its representative.[30] The definition of the term "representative" in Fed. R. Civ. P. 26(b)(3) goes beyond attorney work product and includes the work product prepared on an attorney's behalf.[31]

The protection of the work product doctrine attaches to documents and other items produced by a party or its representative in anticipation of litigation.[32] The test to determine whether a document constitutes work

---

of Special Counsel, MSPB, 260 U.S. App. D.C. 382, 819 F.2d 1181, 1187 (D.C. Cir. 1987). Where documents withheld are privileged under the work product doctrine, it is irrelevant that they do not also fall within the scope of the deliberative process privilege.

[30] *See, e.g.*, United States v. Nobles, 422 U.S. at 238-39 (attorneys often must rely on investigators and work product doctrine applies to materials prepared by such agents); Sterling Drug Inc. v. Harris, 488 F. Supp. 1019 (S.D.N.Y.1980) (privilege extends to documents prepared by agent of attorney responsible for coordinating and gathering evidence in anticipation of hearing). *See also In re* Joint Eastern & Southern Dist. Asbestos Litigation, 119 F.R.D. 4, 6 (E.D. & S.D.N.Y. 1988), providing that:

> [A] court should weigh (1) any attempt at improper use of the work product doctrine to exceed the limits of preparation on the one hand and concealment on the other," (2) the degree to which the documents in question are composed of factual material rather than an attorney's legal analysis, and (3) whether the disclosure demand "constitutes a fishing expedition."

Memorandum & Order at 7, quoting *In re* Joint Eastern & Southern Dist. Asbestos Litigation, 119 F.R.D. at 6.

[31] *See* United States v. Nobles, 422 U.S. 225 (1976); Sprague v. Director of Office of Workers' Compensation, 688 F.2d 862, 869-70 (1st Cir. 1982);

[32] Federal Rule of Civil Procedure 26(b)(3) provides in pertinent part:

Trial Preparation: Materials. Subject to the provisions of subdivision (b)(4) of this rule, a party may obtain discovery of documents and tangible things otherwise discoverable under subdivision (b)(1) of this rule and prepared in anticipation of litigation or for trial by or for another party or by or for that other party's representative (including his attorney, consultant, surety, indemnitor, insurer, or agent) only upon a showing that the party seeking discovery has substantial need of the materials in the preparation of his case and that he is unable without undue hardship to obtain the substantial equivalent of the materials by other means. In ordering discovery of such materials when the required showing has been made, the court shall protect against disclosure of the mental impressions, conclusions, opinions, or legal theories of an attorney or other representative of a party concerning the litigation. . . . FED. R. CIV. P. 26(b)(3).

product is whether it was prepared by the party or its representative because of the prospect of litigation.[33] A party's representative includes its attorney, insurer, employee, or other agent, so long as they were working on behalf of the party and preparing the document with the prospect of litigation in mind.[34] Documents that were produced in anticipation of litigating one case remain protected in a separate case, at least where the cases are closely related.[35]

Commentators have referred to the work product doctrine as more of a qualified immunity than a privilege.[36] The protection is similar to qualified immunity because the party seeking discovery can obtain discovery of certain work product upon a showing of substantial need for the materials and necessity.[37] Plaintiff must show that he cannot obtain the substantial equivalent of the work product through other efforts, such as conducting his own investigations, and that he has a substantial need for the materials in prosecuting his case.[38] The rule, like its common law background,[39] meticulously protects against revealing an attorney's mental impressions, conclusions, opinions, and legal theories.[40] It is questionable whether any showing justifies disclosure of an attorney's mental impressions.[41] In

---

[33] *See* 8 WRIGHT & MILLER, FEDERAL PRACTICE AND PROCEDURE § 2024 at 198.

[34] Lott v. Seaboard S.R., Inc., 109 F.R.D. 554, 557 (S.D. Ga 1985).

[35] Fine v. Facet Aerospace Products Co., 133 F.R.D. 439, 445 (D.C. N.Y. 1990); *In Re* Grand Jury Proceedings, 73 F.R.D. 647 (M.D. Fla. 1977).

[36] 8 WRIGHT & MILLER, FEDERAL PRACTICE AND PROCEDURE § 2025 at 212. *See also Lott,* 109 F.R.D. at 557. Because the work product doctrine is not considered a substantive privilege, Fed. R. Evid. 501 does not require that state law be applied. Fine, 133 F.R.D. at 444.

[37] *See* FED. R. CIV. P. 26(b)(3).

[38] *Id.*

[39] The seminal case on the work product doctrine is Hickman v. Taylor, 329 U.S. 495, 67 S. Ct. 385, 91 L. Ed. 451 (1947).

[40] *Id.*

[41] *See* Upjohn Co. v. United States, 449 U.S. 383, 401, 101 S. Ct. 677, 66 L. Ed. 2d 584, 599 (1981); Duplan Corp. v. Moulinage et Retorderie de Chavanoz, 509 F.2d 730, 737 (4th Cir. 1974), *cert. denied,* 420 U.S. 997, 43 L. Ed. 2d 680, 95 S. Ct. 1438 (1975); FDIC v. Cherry, Bekaert & Holland, 131 F.R.D. 596 (M.D. Fla. 1990), *citing* Bd. of Trustees of Leland Stanford Jr. Univ. v. Coulter Corp., 118 F.R.D. 532, 534 (S.D. Fla. 1987).

addition, "the work product doctrine does not apply to the situation in which a client seeks access to documents or other tangible things created or amassed by its attorney during the course of the representation."[42]

There are numerous cases that deal with the application of the work product doctrine to environmental audits.[43] The first step in determining whether the work product doctrine will apply to environmental audit documents is determining whether the documents were prepared in anticipation of litigation. In *Bituminous Casualty,*[44] Tonka asserted that the documents in question were alternatively protected by the work product doctrine. Tonka argued that the documents were prepared in anticipation of litigation with the Minnesota Pollution Control Agency and that the insurers had not shown a substantial need. Tonka also argued that some of the documents were completely immune from discovery to the extent that they contained opinions and mental impressions of Tonka's counsel.[45] The insurers argued that the documents were not protected by work product since they were not prepared in anticipation of the instant litigation.[46]

---

[42] *See, e.g.*, Maxwell v. Florida, 479 U.S. 972, 976 n.2, 107 S.Ct. 474, 477 n.2, 93 L.Ed.2d 418 (1986) (Marshall, J., dissenting from denial of writ of certiorari); Gottlieb v. Wiles, 143 F.R.D. 241, 247 (D.Colo. 1992) ("An attorney may not withhold work product from his own client.").

[43] *See, e.g.*, Continental Casualty Co. v. United States Fire Ins. Co., 142 F.R.D. 471 (D. Colo. 1992); Metro Wastewater Reclamation Dist. v. Continental Casualty Co., 142 F.R.D. 471 (D. Colo. 1992); Chemcentral/Grand Rapids Corp. v. United States Environmental Protection Agency, 1992 WL 281322 (N.D. Ill. Oct 6, 1992), 23 Envtl. L. Rep. 20797; Arizona v. Ybarra, 777 P.2d 686 (Ariz. 1989); Bituminous Casualty Corp. v. Tonka Corp., 140 F.R.D. 381 (D. Minn. 1992); Kansas-Nebraska Natural Gas Co., Inc. v. Marathon Oil Co., 109 F.R.D. 12 (D. Neb. 1985); Olen Properties Corp. v. Sheldahl, Inc., 1994 WL 212135 (C.D. Cal. April 12, 1994); Continental Illinois Nat'l Bank and Trust Co. of Chicago v. Indem. Ins. Co. of North America, 1989 WL 135203 (N.D. Ill. 1989); Martin v. Bally's Park Place Hotel & Casino, 983 F.2d 1252 (3d Cir. 1993).

[44] 140 F.R.D. at 387.

[45] *Id.*

[46] *Id.* at 388.

The Court agreed that none of the documents that Tonka sought to protect had been prepared in anticipation of the instant litigation, but determined that, in the 8th Circuit, what must be determined is whether the documents were prepared in anticipation of some litigation.[47] The Court determined that Barr Engineering's preliminary draft documents were the fruits of counsel's investigative endeavors. The key issue was whether counsel's investigation and Tonka's subsequent remedial measures were conducted in the course of Tonka's ordinary business or in anticipation of litigation.[48] The Court wrote that "[a]ny action that a company takes in response to the MPCA's requests before and after designation of responsible party status are therefore made under the strong prospect of future litigation over issues such as the propriety of the designation of responsible party status, the scope of the clean up and the adequacy of the clean up."[49] The fact that Tonka was able to avoid actual litigation with the MPCA was of little consequence since it did not mean that the efforts undertaken by Tonka's counsel and its consultants were not done in anticipation of litigation. Thus, documents generated after the MPCA issued its Request for an Initial Investigation on February 5, 1985, were generated in anticipation of litigation and were therefore protected from discovery.[50]

The preparation in anticipation of litigation element was not met in *Metro Wastewater Reclamation District v. Continental Casualty Company*.[51] In that case, Metro Wastewater was notified by EPA that it might be a potentially responsible party (PRP) for the clean up of contamination at the Lowry Landfill Superfund site, since Metro Wastewater had deposited hazardous substances from processed sewage sludge near the landfill. After an EPA letter directed Metro Wastewater to participate in a Remedial Investigation and Feasibility Study, Metro Wastewater negotiated an administrative Consent Order with EPA along with other PRPs. Metro Wastewater's insurer rejected its demand for defense and indemnification, and Metro Wastewater sued. During this

[47] *Id.*

[48] Bituminous Casualty, 140 F.R.D. at 389.

[49] *Id.* at 390.

[50] *Id.*

[51] 142 F.R.D. 471 (D. Colo. 1992).

litigation, Metro Wastewater refused to produce thousands of documents based upon various privileges and immunities.[52] In particular, Metro Wastewater raised the work product doctrine as a bar to the production of in-house memoranda and correspondence or reports from its environmental consultants that were generated after the EPA proceedings were commenced and the Consent Order was agreed upon.[53]

The Court rejected Metro Wastewater's work product doctrine because the documents relating to the EPA Consent Order cannot be deemed to have been prepared in connection with litigation or in anticipation thereof. Once a settlement with EPA was agreed upon, activities undertaken to comply with the settlement terms do not relate to existing or anticipated litigation. Since the documents appeared to relate to the implementation of the Consent Order, they could not be protected by the work product doctrine.[54]

The initial focus of the inquiry into whether the report was prepared "in anticipation of litigation," is a person's unilateral belief that litigation is impending. That person's belief, however, must be objectively reasonable.[55] In *Martin v. Bally's Park Place Hotel & Casino,*[56] the Third Circuit was called to inquire into to what extent the OSHA records access rule[57] incorporated the work product doctrine, and whether the doctrine was properly invoked in refusing OSHA's request for the production of a consultant's report.[58] In that case, a bartender at a Bally's hotel telephoned an OSHA regional office complaining about skin, eye, and throat irritation. The source of this irritation was determined to have been caused by a chemical in a nearby dishwasher. After two letters from OSHA, one dated March 10 and another dated April 3, the Bally's Director of Labor Relations reported the matter to Bally's General Counsel. In an internal memorandum dated April 17, the General Counsel ordered the Director of Labor Relations to hire an expert to prepare a report for the General Counsel's

---

52 *Id.* at 475.
53 *Id.* at 477.
54 *Id.* at 477–78.
55 Martin v. Bally's Park Place Hotel & Casino, 983 F.2d 1252, 1260 (3d Cir. 1993).
56 *Id.*
57 29 C.F.R. § 1910.20 (1991).
58 Bally's Park Place Hotel & Casino, 983 F.2d at 1253.

exclusive use.[59] On July 27, OSHA served Bally's with a subpoena duces tecum demanding the consultant's report, which Bally objected to, based upon the work product doctrine. In response to Bally's refusal to produce the consultant's report, the Secretary of Labor (hereinafter Secretary) refuted the contention that the report was prepared in anticipation of litigation, and argued that Bally's General Counsel could not, under the circumstances, have anticipated litigation when he commissioned the report.[60]

The Court disagreed with the Secretary's contention. In none of OSHA's correspondence did it unequivocally commit to end its inquiry. Furthermore, even if OSHA had made such a commitment, it could not have guaranteed Bally's that the union or one or more of its employees would not initiate litigation. Especially probative in this regard was the fact that the bartender, whose complaint to OSHA initiated OSHA's response, had brought an action against Bally's.[61] Under these circumstances, Bally's General Counsel's belief in the likelihood of impending litigation was a reasonable one.

Even if the work product doctrine applies, only legal theories, opinions, and conclusions receive absolute protection. In *Arizona v. Ybarra*,[62] the Court inquired into what, in the environmental context, can be deemed a legal theory, opinion, or conclusion, and the effect of the availability of information to an opposing party. *Ybarra* involved Excel Industries, which, in preparation for impending criminal litigation, directed the engineering firm Western Technologies to conduct a soil test on property located behind Excel's Phoenix facility. Western Technologies completed the soil test and prepared a written report. This report described the chemical contaminants in the soil and their concentrations. It also noted which chemicals were considered to be hazardous wastes. Subsequently, Western Technologies was served with a grand jury subpoena duces tecum requiring it to provide certain information, which included the soil test and report. Western Technologies submitted the report and analysis to a special agent of the attorney general, who testified extensively about the analysis

---

[59] *Id.* at 1255.
[60] *Id.* at 1261.
[61] *Id.*
[62] 777 P.2d 686 (Ariz. 1989).

and report to the grand jury. In a motion for a new determination of probable cause, Excel Industries argued that some of the evidence presented to the grand jury was obtained in violation of the attorney-client privilege.

The Court was first called upon to determine whether Western Technologies was part of the defense counsel's "investigative staff."[63] The Court noted that experts retained to investigate and produce reports on technical aspects of specific litigation are part of counsel's investigative staff. With regard to Western Technologies, the work product doctrine could apply since Western Technologies prepared the report on the instruction of counsel to defend against specific criminal charges.[64] The Court noted that

> . . .this case is a perfect example of a situation in which a lawyer, trained and experienced in criminal law, would need the assistance of experts to prepare an effective legal defense. Gas chromatography by Dubowski's technique to analyze soil samples for chemicals that may fall within the statutory hazardous waste definition is a skill not usually learned in law school. Few lawyers can ever learn, much less litigate, such matters without expert advice. Assuming the adversary process is the best method of attaining the truth, we must encourage lawyers to make the necessary investigations without fear that their very diligence may eventually provide ammunition for the opponent.[65]

Next, the Court inquired into whether the contents of the report were protected by the work product doctrine.[66] Noting that such reports would be protected only to the extent that they contain the theories or conclusions of defense counsel or his or her legal or investigative staff, the Court wrote:

> . . .the determination that certain compounds or chemicals exist in the soil samples is a conclusion. Additionally, the determination that certain material in the soil samples is a hazardous waste can only be made by applying the Code of

---

63 *Id.* at 689.
64 *Id.* at 690.
65 *Id.*
66 *Id.* at 691.

> Federal Regulations. Thus, determining whether the substances in the soil samples were hazardous wastes requires the application of law to fact, and is, in the truest sense, a conclusion.[67]

Lastly, the Court considered the availability of the disputed documents in discovery.[68] The Court concluded that the soil samples were physical facts that both parties had equal access to. "In short, the state had equal access to the soil, could have taken its own samples, and could have hired its own expert."[69]

An environmental audit may be shielded under the work product doctrine if it is prepared in anticipation of litigation under the direction of an attorney. Audits so prepared should ideally be labeled "Attorney-Client Privilege; Work Product Doctrine." In addition, it is advisable that the correspondence hiring the consultant conducting the audit specifically state that the attorney is requiring the audit in order to render legal advice.

§ 4.4 Critical Self-Evaluation Privilege

The critical self-evaluation privilege is another means of potentially protecting an environmental audit from disclosure. The self-evaluation privilege was first recognized in *Bredice v. Doctors Hospital, Inc.*[70]

A basic issue is whether self-evaluation[71] of corporate actions should be encouraged and protected from discovery in litigated matters, as a matter of public policy.[72] The doctrine of "privileges" is dynamic in nature, not

---

[67] *Id.* at 691.

[68] *Id.* at 692.

[69] *Id. See also* Joseph M. Udall, Note, *Environmental Engineer's Report is Protected as Work Product—State v. Ybarra, 161 Ariz. 188, 777 P.2d 686 (1989),* 23 ARIZ. ST. L. J. 699 (1991).

[70] 50 F.R.D. 249 (D.D. C.), *aff'd,* 479 F.2d 920 (D.C. Cir. 1973) (holding that a medical malpractice plaintiff could not discover hospital peer review committee minutes and reports).

[71] The self-evaluation privilege, adopted in several federal jurisdictions, prevents the disclosure of confidential, critical, evaluative, and deliberative material whenever the public interest in confidentiality outweighs an individual's need for full discovery.

[72] *See generally* J. Murphy and R. Oyer, *The Self-Evaluation Privilege and Beyond,* INSIGHT, March 1993; *Self-Critical Analysis—An Emerging Privilege,* N.Y.L.J., Dec. 17, 1992; *Attorney Client Privilege: Killing Limited Waiver,* N.Y.L.J., Jan. 16, 1992, p.5;

static. A privilege develops when the "public need for disclosure" is outweighed by the "public need for confidentiality" of information.[73] Even though privileges ". . . are not lightly created nor expansively construed . . .,"[74] there are instances when protection of the "public need for confidentiality" must be acknowledged and developed.

Several federal courts have acknowledged the existence of a qualified privilege of self-examination or self-critical analysis.[75] The privilege prevents disclosure of confidential critical, evaluative, and/or deliberative material whenever the public interest in confidentiality outweighs an individual's need for full discovery.[76] Factual information, however, should not be protected by the qualified privilege.[77]

This privilege has been considered most significantly in conjunction with a line of Securities and Exchange Commission (SEC) investigations. In the early 1970s, the SEC implemented a voluntary disclosure program in order to facilitate investigations. In order to encourage compliance with the voluntary disclosure program, the court created a doctrine of limited waiver whereby the release of privileged documents to investigative governmental

---

Privileges in Enforcement Proceedings—The Doctrine of Selected Waiver: Self-Policing or Self-Destruction, INSIGHT, Aug. 1992, p.11; Note, *Privilege of Self-Critical Analysis,* 96 HARV. L. REV. 1083 (1983).

[73] *See, e.g.,* Hickman v. Taylor, 329 U.S. 495, 509-12, 67 S.Ct. 385, 392-394, 91 L.Ed. 451 (1947) (work-product doctrine); Upjohn v. United States, 449 U.S. 383, 389, 101 S.Ct. 677, 682, 66 L.Ed.2d 584 (1981) (attorney-client communications); Hague v. Williams, 37 N.J. 328, 336 (1962) (physician-patient communications).

[74] United States v. Nixon, 418 U.S. 683, 710, 94 S.Ct. 3090, 3108, 41 L.Ed.2d 1039 (1974).

[75] *See, e.g.,* Bredice v. Doctors Hospital, Inc., 50 F.R.D. 249 (D.D.C.), *aff'd on rehearing,* 51 F.R.D. 187 (D.D.C.1970), *aff'd* 479 F.2d 920 (D.C.Cir.1973); Banks v. Lockhead-Georgia Co., 53 F.R.D. 283 (N.D.Ga.1971); Rosario v. New York Times Co., 84 F.R.D. 626 (S.D.N.Y.1979).

[76] *See, e.g.,* Roberts v. National Detroit Corp., 87 F.R.D. 30 (E.D.Mich.1980); Gillman v. United States, 53 F.R.D. 316 (S.D.N.Y.1971).

[77] *See generally* Murphy, *The Self-Evaluation Privilege,* 7 J. CORP. L. 489 (1982); Crisman and Matthews, *Limited Waiver of Attorney-Client Privilege and Work-Product Doctrine In Internal Corporate Investigations: An Emerging Corporate 'Self-Evaluative' Privilege,* 21 AM. CRIM. L. REV. 123 (1983). See, however, Note, *The Privilege of Self-Critical Analysis,* 96 HARV. L. REV. 1083, 1093-96 (1983), which argues in favor of a qualified privilege for factual information contained in such a document.

agencies did not constitute a waiver of the privilege in subsequent proceedings or litigation.[78] The limited waiver was intended to encourage cooperation with the SEC's voluntary program without fearing a complete waiver of the attorney-client privileged materials.

State courts have recognized that privileges may be created in the absence of statutory authority when demanded by public policy.[79] Thus, when confronted with a request to create a qualified privilege for self-evaluation, state courts have weighed the public interest in confidentiality against a litigant's interest in disclosure.[80]

The information that is the subject of a production request must be the criticisms or evaluations or the product of an evaluation or critique conducted by the party opposing the production request.[81] Additionally, the "public need for confidentiality" of such analysis must be such that the unfettered internal availability of such information should be encouraged as

---

[78] *See* Diversified Industries v. Meredith, 572 F.2d 596 (8th Cir. 1977).

[79] *See, e.g.,* Hague v. Williams, 37 N.J. 328, 336 (1962) (in absence of statutory authority, the New Jersey Supreme Court acknowledged a qualified physician-patient privilege); Berst v. Chipman, 232 Kan. 180, 653 P.2d 107 (1982) (qualified privilege may exist independently of a specific statutory privilege).

[80] Garrow v. Elizabeth General Hospital and Dispensary, 79 N.J. 549 (1979) (good-faith statements concerning the qualifications of a candidate for staff appointment held qualifiedly privileged); Dade County Medical Ass'n v. Hlis, 372 So.2d 117 (Fla.App.1979) (interest in confidentiality of ethic committee records outweighs asserted grounds for discovery); Wesley Medical Center v. Clark, 234 Kan. 13, 669 P.2d 209 (1983) (interest in discovery of peer review committee records outweighs need for confidential communications); Oviatt v. Archbishop Bergan Mercy Hospital, 191 Neb. 224, 214 N.W.2d 490 (1974) (records of medical staff committee concerning defendant physician's suspension held privileged); Davison v. St. Paul Fire and Marine Insurance Co. 75 Wis.2d 190, 248 N.W.2d 433 (1977) (public interest in confidentiality of documents relating to the care of plaintiff by hospital held insufficient to prevent discovery).

[81] Rule 26(b)(1) of the Federal Rules of Civil Procedure provides that "[p]arties may obtain discovery regarding any matter, not privileged, which is relevant to the subject matter involved in the pending action."

Rule 501 of the Federal Rules of Evidence states that in cases in which federal law controls, privileges "shall be governed by the principles of the common law as they may be interpreted by the courts of the United States in the light of reason and experience." Rule 501 allows court "flexibility to develop rules of privilege on a cases by case basis." University of Pennsylvania v. EEOC, 493 U.S. 182, 189 (1990). Open discovery is, however, favored and privileges must be strictly construed. *See Id.*

a matter of public policy.[82] As a corollary to this second point, the analysis or evaluation or opinion must be of the character that would result in the termination of such self-evaluative inquiries or critical input in future situations if this information is subject to disclosure. In short, confidentiality and the "public need for confidentiality" are the sine qua non of effective internal self-critical analysis since such confidentiality will encourage open and frank criticism. Such criticism is essential in recognizing the cause of past problems and the elimination of future problems. If the public policy of safety improvements in practice and procedure is to be encouraged, the "public need for disclosure" for the individual must give way. This is especially so if such improvements are to continue and develop.

Valuable criticism can neither be sought nor obtained, nor generated in the shadow of potential or even possible public disclosure. It is not realistic to expect candid expressions of opinion or suggestions as to future policy or procedures in an air of apprehension that such statements may well be used against one's colleague or employer in a subsequent litigated matter. The purpose of an investigation intended to seek criticism, opinion or suggestion and form the basis of criticism of then existing policy or procedure is self-improvement. The value of the investigation is questionable if the input is not reliable. It is clear that the reliability of the input in this situation varies inversely with the risk of disclosure of the input or resulting criticisms.

Unless confidential safety investigations are encouraged, safety improvements in the workplace and in general will be stifled. Moreover, "[A]n uncertain privilege, or one which purports to be certain but results in widely varying applications by the courts, is little better than no privilege at all.'"[83] Accordingly, public policy demands that the "evaluative" portions of certain audits be protected by the privilege of self-critical analysis.[84]

---

[82] *See* Emerson Electric Co. v. Schlesinger, 609 F.2d 898 (8th Cir. 1979) (courts will look to whether the information resulted from an internal review involving confidential self-analysis and whether disclosure of the internal review will serve or harm the public interest).

[83] Upjohn v. United States, 449 U.S. at 393, 101 S.Ct. at 684.

[84] Wylie v. Mills, 195 N.J. Super. 332; 478 A.2d 1273 (1984), wherein the court applied the privilege to an in-house corporate report on an accident involving an employee. In arriving at this result, the court stressed that self-critical analysis was important to the recognition of past problems and the elimination of future ones. Without such protection, the

Nevertheless, factual information contained within the report may be required to be disclosed.[85]

Under Fed. R. Evid. 501, state law on privilege governs in a case such as this, brought under the court's jurisdiction in diversity. The "critical self-analysis" privilege has been recognized to protect documents that reflect an internal self-analysis of self-criticism. The purpose of this privilege is to encourage self-evaluation and the benefits that may flow therefrom, and to avoid the chilling effect upon such self-analysis that would result from complete disclosure. Only evaluative opinions, and not the facts disclosed in the course of self-evaluation, may be protected. This privilege applies to the opinions and reports of peer review committees.[86]

The critical self-evaluation privilege is not absolute.[87] Where the need for this information is substantial and disclosure would have little effect on self-analysis, discovery has been compelled. The privilege may be punctured by a showing of particularized need that outweighs the public interest in confidentiality. In determining whether a party has made the required showing under the privilege, courts balance the following three factors:

1. the extent to which the information may be available from other sources;
2. the degree of harm that the litigant will suffer from its unavailability; and

court indicated its concern that candid opinions and suggestions as to future policy would not be volunteered for fear that the statements might be used against an employer in subsequent litigation.

[85] *Id. See also* Todd v. South Jersey Hospital System et al., Civ. No. 91-4757 (D.N.J. Jan. 21, 1994) (The statutory utilization review privilege is absolute as to documents created by the utilization review committee; such documents are therefore not discoverable. However, plaintiffs' need for the information sought is compelling enough to overcome the common-law privilege of self-critical analysis.)

[86] The utilization review privilege is a separate statutory privilege created by N.J.S.A. 2A:84A-22.8. This matter provides, with few exceptions, that information and data obtained by utilization review committees cannot be disclosed. The privilege is extremely limited and applies only to utilization review committees. The privilege is intended to encourage participation on the utilization review committees and to aid in the implementation of Medicare.

[87] *See In re* Crazy Eddie Securities Litigation, 792 F.Supp. at 205; Lasky v. American Broadcasting Cos., 5 Fed. R. Serv. 3d 1366, 1369 (S.D.N.Y. 1986).

3. the possible prejudice to the agency's investigation.[88]

Courts have not favored the application of the self-evaluation privilege in the environmental context.[89]

[88] In determining whether to extend the self-evaluation privilege, the courts have given serious consideration to the public policy implications of full disclosure. The courts have employed a balancing test to determine if the information sought is so significant as to defeat the strong public policy encouraging candid internal investigations. Therefore, when the public policy in favor of self-evaluation outweighs the benefits possibly conferred by full disclosure, the courts will often provide immunity. See Marc K. Shaye, *Keeping Environmental Audit Results Confidential,* MICHIGAN LAWYERS WEEKLY 24A (Aug. 30, 1993).

[89] For instance, in State v. Cecos International, Inc., 66 Ohio App.3d 262, 583 N.E.2d 1118, (Oh. Ct. App. 1990), the court ruled that the self-critical analysis privilege did not apply to environmental reports generated by the operators of a hazardous waste treatment facility. The court determined that because CECOS is engaged in a potentially dangerous operation, and because the legislature mandated heavy regulation of the hazardous waste industry requiring disclosure of company records to state regulatory officials, the self-critical analysis privilege did not apply.

Similarly, in Artesian Water Co. v. New Castle County, 1981 WL 15606 (Del. Ch. 1981), the Delaware Chancery Court determined that the self-critical analysis privilege did not apply in the context of a claim related to groundwater contamination. That case considered whether the results of a round-table discussion among experts funded in part by USEPA could be subject to the self-critical analysis privilege. The court first questioned whether the privilege was recognized under Delaware law; it then concluded that even if the privilege was recognized in Delaware, the report would not be subject to it. The court reasoned that because the nature of the relief sought plaintiff—an injunction seeking to prevent future groundwater pollution of its water supply wells—the need for the data contained in the report outweighed any claim for privilege.

The court in United States v. Dexter Corp., 132 F.R.D. 8 (D. Conn. 1990) reached a like result. That case was an action to enforce the federal Clean Water Act in which the government sought to compel production of environmental audit reports. The court ruled that, because the official stated policy of the United States was that "there should be no discharge of oil or hazardous substances" into the water, application of a privilege of self-critical analysis would be against public policy and would hamper the ability of USEPA to administer the act.

The most recent case to consider the issue is Arkwright Mutual Insurance Co. v. National Union Fire Insurance Co., 1993 WL 14448 (S.D.N.Y. 1993). In that case, General Electric was issued a subpoena to produce records concerning the presence of contamination at a facility that was the subject of an action between a primary insurer of GE's and its reinsurer. The subpoena sought among other documents an environmental audit report prepared by GE concerning the site. GE sought to prevent production of the report based upon both the attorney-client privilege and the self-critical analysis privilege.

Recently, numerous states have either enacted or proposed legislation that would create an environmental self-evaluation privilege. This privilege would, in theory, create a presumption against the imposition of any administrative, civil, or criminal penalties for voluntary disclosures arising out of any environmental self-evaluation.[90] Whether a state statute could be used to effectively shield enforcement of a federal environmental statute is, however, an open question.

In response to these claims of privilege, the court directed GE to produce both redacted and unredacted copies of the report for in camera review. With one exception, the court determined that the redacted version of the report was all that was required to be produced. It is not clear from the decision whether the basis for the redaction was that the self-critical portions of the report were excised. To date, however, the Arkwright decision appears to be the only reported case where some portion of an environmental audit report was withheld from disclosure.

[90] *See, e.g.*, 1994 CO S.B. 139 (March 18, 1994) (introduced Jan. 28, 1994).

Chapter 5

State Environmental Audit Privilege Legislation

§ 5.1 Overview

State legislation encouraging environmental audits has been enacted using a variety of different legal mechanisms. State initiatives to encourage environmental audits have continued to catch the attention of the American public. Currently, over 30 states have enacted or are considering some form of environmental audit privilege legislation or other form of legislation designed to encourage audits. Below is a chart indicating the states that are considering or have enacted environmental audit privilege legislation:

State	**Audit Privilege Statutes**
ALABAMA	Legislation introduced February 1996.
ALASKA	Legislation introduced January 1996.
ARIZONA	
ARKANSAS	ARK. CODE ANN. §8-1-301 *et seq.*
CALIFORNIA	Legislation introduced in January and February 1996.
COLORADO	COLO. REV. STAT. §§ 13-25-126.5(3), 13-90-107(j), 25-1-114.5
CONNECTICUT	
DELAWARE	Legislation introduced June 1996.
FLORIDA	Legislation died in committees.
GEORGIA	
HAWAII	Legislation introduced January 1995.
IDAHO	IDAHO CODE § 9-801 *et seq.*
ILLINOIS	205 ILCS 5/52.2
INDIANA	IND. CODE 13-10-3 *et seq.*
IOWA	Introduced March 1995.
KANSAS	KAN. STAT. ANN. §60-3332 *et seq.*

State	**Audit Privilege Statutes**
KENTUCKY	KY. REV. STAT. ANN. § 224.01-040
LOUISIANA	
MAINE	
MARYLAND	Legislation introduced February 1996.
MASSACHUSETTS	Legislation introduced February 1995.
MICHIGAN	1996 Mich. Pub. Acts 132
MINNESOTA	1995 Minn. Laws 168, to be codified at MINN. STAT. ANN. § 115.074
MISSISSIPPI	MISS. COD. ANN. § 49-2-51
MISSOURI	Legislation introduced January 1996.
MONTANA	
NEBRASKA	
NEVADA	
NEW HAMPSHIRE	1996 N.H. Laws 4
NEW JERSEY	Legislation introduced January 1996.
NEW MEXICO	
NEW YORK	Legislation introduced January 1996 and March 1996.
NORTH CAROLINA	Legislation introduced May 1995.
NORTH DAKOTA	
OHIO	Legislation introduced April 1995.
OKLAHOMA	Legislation introduced Jamuary 1996.
OREGON	ORE. REV. STAT. §468.963
PENNSYLVANIA	Legislation introduced October 1995.
RHODE ISLAND	Legislation introduced February 1995.

State	Audit Privilege Statutes
SOUTH CAROLINA	1996 S.C. Acts 384
SOUTH DAKOTA	
TENNESSEE	Legislation introduced February 1995.
TEXAS	71 TEX. REV. CIV. STAT. ANN. ART. 4447cc-1 *et seq.*
UTAH	UTAH CODE ANN. § 19-7-101 *et seq.*
VERMONT	Introduced January 1996.
VIRGINIA	VA. CODE ANN. § 10.1-1193 *et seq.*
WASHINGTON	
WEST VIRGINIA	Legislation introduced in January 1996 and February 1996.
WISCONSIN	Legislation introduced March 1996 and May 1995.
WYOMING	WYO. STAT. § 144.765

In addition to environmental audit privilege legislation, many states are also considering legislation that encourages audits in other manners. For example, New York continues to encourage the state to conduct environmental audits in a number of different contexts.

§ 5.2 Issues Raised Concerning Audit Privilege Legislation

As states consider environmental audit privilege legislation certain themes continue to emerge and develop. These include:

1. How should the state define the terms "environmental audit" and "environmental report"?
2. Must the document be labeled "environmental audit" or "privileged" in order for the company to avail itself of a privilege?
3. To what extent should there be a testimonial privilege protecting the data contained in the audit report?
4. What elements and what conditions must exist in order to overcome the privilege?
5. What is the burden of proof? Should it be a preponderance of the evidence standard? What is necessary to establish a prima facie case?

6. To what extent will an immunity be granted for voluntary disclosure? And more significantly, what disclosure is considered voluntary?
7. Should "bad actors" be excluded from the privilege?

The following table describes how the certain states enacting environmental audit privilege legislation have addressed each of these issues:

Legislative Issues	Ark.	Colo.	Idaho	Ill.	Ind.	Ky.	Ore.	Utah	Va.	Wyo.
Effective date of legislation	2/95	6/94	7/95	1/95	7/94	7/94	1991	3/95	7/95	7/95
The environmental audit report must be labeled	Yes	No	Yes	Yes	Yes	Yes	Yes	No	No	Yes
Penalty reduction or civil immunity for voluntary disclosure	No	Yes	Yes	No	No	No	No	No	Yes	Yes
Criminal immunity for voluntary disclosure	No	Yes	Yes	No	No	No	No	No	No	No
Express waiver of privilege	Yes	Yes	Yes	Yes	Yes	Yes	Yes	Yes	Yes	Yes
Implied waiver	Yes	NC[1]	NC	NC	NC	Yes	Yes	NC	NC	Yes
Waiver by failing to file a petition for in camera review within specified time	Yes[2]	NC	NC	Yes[3]	Yes[4]	Yes[5]	Yes[6]	NC	NC	Yes[7]
Waiver by introduction of all or part of audit by party asserting privilege	No	NC	No	NC	NC	Yes	NC	No	NC	No
No privilege if asserted for fraudulent purpose	Yes	Yes	Yes	Yes	Yes	Yes	Yes	NC	Yes	Yes
No privilege if material is not subject to privilege	Yes	NC	Yes	Yes	Yes	Yes	Yes	Yes	Yes	Yes

---

[1] NC indicates that it is not yet clear, because the legislature did not specifically address the question.

[2] The time required is 30 days.

[3] The time required is 30 days.

[4] The time required is 30 days.

[5] The time required is 20 days.

[6] The time required is 30 days.

[7] The time required is 20 days.

Legislative Issues	Ark.	Colo.	Idaho	Ill.	Ind.	Ky.	Ore.	Utah	Va.	Wyo.
No privilege if material shows evidence of non-compliance efforts and compliance efforts were not prompt and pursued with reasonable diligence	Yes	Yes	NC	Yes	Yes	Yes	Yes	NC	NC	Yes
No privilege in a criminal proceeding if the prosecutor has a need requiring the information[8]	NC	Yes	NC	NC	Yes	Yes	Yes	NC	NC	Yes
Party asserting the privilege has the burden of proof for both privilege and demonstration of efforts toward compliance	Yes	Yes	Yes	No	Yes	Yes	Yes	Yes	Yes	Yes
Party seeking disclosure has burden of proving fraudulent purpose	Yes	Yes	Yes	Yes	Yes	Yes	Yes	NC	Yes	Yes
Legal official or party seeking disclosure has burden of proving other conditions allowing disclosure	Yes	Yes	NC	Yes	Yes	Yes	Yes	Yes	Yes	Yes
Partial disclosure of only issues in dispute allowed	Yes	NC	NC	Yes	Yes	Yes	Yes	Yes	Yes	Yes

§ 5.3 State-by-State Review

State legislation encouraging environmental audits has been enacted using a variety of different legal mechanisms. Brief descriptions of some of these statutes follow.

---

[8] The degree of need varies tremendously from statute to statute among the following:

- ✓ substantial need;
- ✓ compelling need; or
- ✓ compelling circumstance.

§ 5.3.1 Arkansas[9]

On February 17, 1995, the governor of Arkansas signed certain environmental audit privilege legislation into law.[10] The purpose of the law is to protect the environment. The legislature found that environmental protection "is enhanced by the public's voluntary compliance with environmental laws and that the public will benefit from incentives to identify and remedy environmental compliance issues."[11] Limited expansion of the protection against disclosure will encourage such voluntary compliance and improve environmental quality. The voluntary provisions of the act were designed so that they would "not inhibit the exercise of the regulatory authority by those entrusted with protecting our environment."[12]

The environmental audit privilege was created to protect the confidentiality of communications relating to voluntary internal environmental audits.[13] This privilege would encourage owners and operators of facilities and persons conducting other activities regulated under the environmental laws of the state of Arkansas, or its federal counterparts or extensions, both to conduct voluntary internal environmental audits of their compliance programs and management systems and to assess and improve compliance with statutory and regulatory requirements.[14]

---

[9] 1995 Ark. Acts 350.

[10] *Id.*

[11] ARK. CODE ANN. § 8-1-301 (Purpose).

[12] *Id.*

[13] *Id.* § 8-1-303(a).

[14] *Id.* § 8-1-303(a). For the environmental audit privilege, the following definitions were adopted, "unless the context otherwise requires":

1. "Commission" means the Arkansas Pollution Control and Ecology Commission.
2. "Director" means the Director of the Arkansas Department of Pollution Control and Ecology.
3. "Environmental audit" means a voluntary, internal and comprehensive evaluation of one (1) or more facilities or an activity at one (1) or more facilities regulated under this chapter, or federal regional or local counterparts or extensions thereof, or of management systems related to that facility or activity, that is designed to identify and prevent noncompliance and to improve compliance with statutory or regulatory requirements. An environmental audit may be conducted by the owner or operator, by the owner's or operator's employees, or by independent contractors . . .

Id. § 8-1-302 (Definitions).

The environmental audit report will be privileged and will not be admissible as evidence in any legal action in any civil, criminal, or administrative legal action including enforcement actions.[15] For purposes of the privilege, "Environmental audit report" means a set of documents prepared as a result of an environmental audit and labeled "ENVIRONMENTAL AUDIT REPORT; PRIVILEGED DOCUMENT" that may include:

(A) Field notes, records of observations, findings, opinions, suggestions, conclusions, drafts, memoranda, drawings, photographs, computer generated or electronically recorded information, maps, charts, graphs, and surveys collected or developed for the primary purpose of preparing an environmental audit; and

(B) An audit report prepared by the auditor that includes:
   1. The scope of the audit;
   2. The information gained in the audit;
   3. Conclusions and recommendations; and
   4. Exhibits and appendices.

(C) Memoranda and documents analyzing a portion of or all of the audit report and discussing implementation issues.

(D) An implementation plan that addresses correcting past compliance, improving current compliance, and preventing future non-compliance.[16]

The privilege will not apply to the extent that it is waived expressly by the owner or operator of the facility that prepared or caused to be prepared the environmental audit report.[17] In addition, the privilege is waived where

---

[15] *Id.* § 8-1-303 (b).

[16] *Id.* The privilege does not apply to the following:

(a) Documents, communications, data, reports, or other information that must be collected, developed, maintained, reported, or otherwise made available to a regulatory agency under:
   1. federal or state law or extensions thereof;
   2. A rule or standard adopted by the Commission;
   3. A determination, a permit, or an order made or issued by the Commission or the Director; or
   4. Any other federal, state, or local law, permit, or order.

(b) Information obtained by observation, sampling, or monitoring by any regulatory agency.

(c) Information obtained from a source independent of the environmental audit.

§ 8-1-305 (Exceptions).

[17] *Id.* § 8-1-304(a)(1).

the owner or operator of a facility or person conducting an activity seeks to introduce an environmental audit report as evidence.[18]

Finally, where the owner or operator of a facility authorizes the disclosure of the environmental audit report to any party the privilege will be deemed waived, unless the disclosure is made under the terms of a confidentiality agreement between the owner or operator of a facility and (1) a potential purchaser of the facility or (2) a customer, lending institution, or insurance company with an existing or proposed relationship with the facility.[19]

Disclosure made under the terms of a confidentiality agreement between government officials and the owner or operator of a facility will not waive privilege,[20] nor will disclosure made to an independent contractor retained by the owner or operator of the facility for the purpose of identifying noncompliance with statutory or regulatory requirements and assisting the owner or operator in achieving compliance with reasonable diligence.[21]

The waiver of privilege may be for part or all of the environmental audit report.[22] The waiver of privilege, however, extends only to that part of the environmental audit report expressly waived by the owner or operator of a facility.[23]

In a criminal,[24] civil, or administrative proceeding,[25] a court of record or administrative tribunal, after an in camera review, will require disclosure of material for which the environmental audit privilege is asserted if the court or administrative tribunal determines one of the following:

✓ The privilege is asserted for a fraudulent purpose;

✓ The material is not subject to the privilege;

✓ The material is subject to the privilege and the material shows evidence of noncompliance with (A) federal or state law or extensions

---

[18] *Id.* § 8-1-304(a)(2).

[19] *Id.* § 8-1-304(a)(3)(A).

[20] *Id.* § 8-1-304(3)(B).

[21] *Id.* § 8-1-304(3)(C).

[22] *Id.* § 8-1-304(b).

[23] *Id.* For purposes of determining waiver, "the parties to a legal action may at any time stipulate that the entry of an order that directs that specific information contained in an environmental audit report is or is not subject to the privilege . . ." *See* § 8-1-306.

[24] *Id.* § 8-1-308 (Disclosure in a criminal proceeding).

[25] *Id.* § 8-1-307(a) (Discussing disclosure in civil or administrative proceeding).

of such statutes; or (B) any rule or regulation adopted by the Commission; or (C) a determination, permit, or order issued by the Commission or the Director; and

✓ The person claiming the privilege did not promptly initiate and pursue appropriate efforts to achieve compliance with reasonable diligence.[26]

If the noncompliance constitutes a failure to obtain a required permit, the person is deemed to have made appropriate efforts to achieve compliance if the person filed an application for the required permit not later than ninety days after the date the person became aware of the noncompliance.[27] In the event additional time is required to prepare a permit application, the person must, within ninety days, submit a schedule to the Department of Pollution Control and Ecology that identifies the activities required to complete the application. If that suggested schedule is acceptable to the Department of Pollution Control and Ecology, the filing of the application pursuant to the submitted schedule will constitute reasonable diligence to achieve compliance for a failure to obtain a required permit.

A prosecuting authority who, based on information obtained from a source independent of an environmental audit report, has probable cause to believe an offense has been committed under the environmental laws, rules, or standards adopted by the Commission, or a determination, permit, or order issued by the Commission or Director may obtain an environmental audit report for which the audit privilege is asserted under certain limited circumstances.[28] The prosecuting authority may obtain the report (1) under a search warrant, (2) under a subpoena, or (3) through discovery.

The prosecuting authority must immediately place the report under seal and cannot review or disclose the contents of the report.[29] Not later than thirty days after the date the prosecutor obtained the report, the owner or operator who prepared the report or caused the report to be prepared may file with the appropriate court or administrative tribunal a petition requesting an in camera hearing on whether the report or portions of the

---

[26] *Id.* §§ 8-1-307 & 8-1-308(a).

[27] *Id.* §§ 8-1-307(b) & 8-1-308(b).

[28] *Id.* § 8-1-309(a) (Proceeding to obtain environmental audit report).

[29] *Id.* § 8-1-309(a).

report are (1) privileged or (2) subject to disclosure.[30] The right to claim the privilege is waived if the owner or operator does not file a petition.

A court or administrative tribunal that receives a petition must issue an order scheduling an in camera hearing for a date that is not later than forty-five days after the date the petition was filed under subsection (b). The order must allow the prosecuting authority to do the following:

1. Remove the seal from the environmental audit report.
2. Review the environmental audit report.
3. Place appropriate limits on the distribution and review of the environmental audit report to protect against unnecessary disclosure.[31]

The prosecuting authority may consult with enforcement agencies regarding the contents of the environmental audit report as necessary to prepare for the in camera hearing. However, the contents of the environmental audit report used in preparation for the in camera hearing may not be used in any investigation or in any proceeding against the defendant unless determined by a court or administrative tribunal not to be subject to the audit privilege and must otherwise be kept confidential unless the information is found by the court to be subject to disclosure.

A party asserting the environmental audit privilege has the burden of proving the privilege, including if there is evidence of noncompliance with federal or state law or extensions thereof, proof that appropriate efforts to achieve compliance were promptly initiated and pursued with reasonable diligence.[32] A party seeking disclosure has the burden of proving the privilege is asserted for a fraudulent purpose. The prosecuting authority seeking disclosure has the burden of proving the conditions for disclosure set forth in the privilege statute.

Upon making a determination, the court of record or administrative tribunal may compel disclosure of only those parts of an environmental audit report that are relevant to issues in dispute in the proceeding.[33]

Nothing in the statute will limit, waive, or abrogate the scope of any statutory or common law privilege, including the work product doctrine and the attorney-client privilege.[34] Similarly, nothing in the statutes will limit,

---

[30] *Id.* § 8-1-309(b).

[31] *Id.* § 8-1-309(c).

[32] *Id.* § 8-1-310(a) (Burden of proof).

[33] *Id.* § 8-1-311 (Partial disclosure).

[34] *Id.* § 8-1-312(a).

waive or abrogate the rights of the public as provided for in the Arkansas Freedom of Information Act.[35]

The Arkansas law is not subject to any sunset provision.[36] Moreover, the legislature expressly repealed "all laws and parts of laws in conflict with" the law.

§ 5.3.2 Colorado[37]

Colorado law dictates, under its penalty provisions for violations of the air pollution control laws, that in determining the amount of any civil penalty, the "existence and scope of a regularized and comprehensive environmental compliance program or an environmental audit program" will be one factor considered when setting both civil and criminal penalties, where applicable.[38]

In evaluating Colorado's Title V operating permit program, EPA was asked to consider the role of Colorado's audit privilege statute.[39] Specifically, according to EPA, "Two commenters expressed concern with the EPA proposal to consider Colorado's law (S.B. 94-139) preventing the admission of voluntary environmental audit reports as evidence in any civil, criminal or administrative proceeding as 'wholly external' to Colorado's Program and asserted that these provisions are consistent with congressional intent and EPA policy, and the Federal Government should not interfere in the State's interpretation and exercise of its own prosecutorial discretion." In addition, the EPA reported that "one commenter also stated that, absent the audit privilege, it would be unlikely that voluntarily disclosed information would be identified and further indicated that, although title V may be delegated by EPA, such delegation does not preempt or require the State to defend its laws to EPA."

---

[35] *Id.* § 8-1-312(b).

[36] 1995 Ark. Acts 350.

[37] Colo. Rev. Stat. § 25-7-122 (1993); Colo. Rev. Stat. § 25-15-309 (1993).

[38] Colo. Rev. Stat. § 25-7-122 (1993) (civil penalties). *See also* C.R.S. 25-15-309 (1993). Note that these provisions are inspired by the Justice Department's guidance concerning environmental audits.

[39] 60 Fed. Reg. 4563, Clean Air Act Final Interim Approval of Operating Permits Program; State of Colorado (January 24, 1995) (Final interim approval).

In response to these inquiries, "EPA did not identify this as an approval issue and stated that it is not clear at this time what effect this privilege might have on title V enforcement actions." Instead, EPA said that:

> A national position on approval of environmental programs in states which adopt statutes that confer an evidentiary privilege for environmental audit reports is being established by EPA. Further, EPA disagrees with the commenter's interpretation of congressional intent and EPA policy. Congressional intent was to encourage owners and operators to do self-auditing and correct any problems expeditiously, but this is not the same as providing an evidentiary privilege and enforcement shield. Congress could have provided such a privilege and shield in the Act, but did not. Section 113 of the Act and title V contain no exceptions for withholding self-auditing reports as evidence in any enforcement proceeding. Likewise, 40 CFR part 70 contains no such exceptions. Also, EPA disagrees with the commenter's assumption that, absent the audit privilege provided by Colorado law, it is unlikely that voluntarily disclosed information would otherwise be identified. For example, section 114 of the Act gives EPA the authority to issue information requests and requires disclosure of information regardless of whether it is generated through a self-audit. Colorado has similar authority.[40]

Although EPA agreed in principle "that Colorado has the authority to adopt its own laws regarding environmental matters as long as the area has not been preempted by Congress,"[41] the agency made some thinly veiled threats to the state concerning the potential effect of use of audit privilege legislation on ability of the states to retain federal programs. For example, EPA said:

> . . . title V of the Act and the part 70 regulations give EPA the responsibility to ensure that states implement their operating permit programs in accordance with title V and part 70. Thus, if Colorado's self-audit privilege impedes Colorado's ability to implement and enforce its Program consistent with title V and part 70, EPA may find it necessary to withdraw its approval of the Colorado Program.[42]

---

[40] 60 Fed. Reg. 4564 (1995).

[41] 60 Fed. Reg. 4564 (1995).

[42] 60 Fed. Reg. 4563, 4564–65 (1995).

§ 5.3.3 Idaho[43]

On March 22, 1995, the governor of Idaho signed into law the "Environmental Audit Protection Act,"[44] the purpose of which "is to encourage owners and operators of facilities and other persons conducting activities regulated under applicable environmental laws to conduct voluntary internal environmental audits of their activities, operations, compliance programs and management systems and to assess and improve compliance with applicable environmental laws while protecting confidentiality of communications relating to voluntary internal environmental audits."[45] The legislature's goal was to have "all persons . . . comply with environmental laws and regulations."[46] Idaho determined that, in meeting this goal, "voluntary environmental audits are in the public interest."[47]

The legislature found that "voluntary compliance with environmental laws and regulations will protect the public health, safety and welfare in the most efficient and effective manner."[48] Thus, the legislature enacted the environmental audit privilege legislation in an effort to encourage all persons, corporations, government agencies, and associations to conduct detailed audits or self-evaluations of compliance with environmental statutes and regulations.[49]

To encourage owners and operators of facilities, and persons conducting activities regulated under federal, state, and local environmental laws, regulations, rules, ordinances, and permits, to conduct voluntary internal environmental audits, to develop compliance programs, and to implement management systems, the legislature of the state of Idaho recognizes that an environmental audit privilege is necessary. The legislature of the state of Idaho further recognizes that voluntary disclosure

---

[43] 1995 Idaho Sess. Laws 359.

[44] IDAHO CODE § 9-801.

[45] *Id.* § 9-802(a). In addition, § 9-810 requires the environmental agencies to promulgate rules further defining an environmental audit and timetables for remedial actions to be performed and rules governing the treatment of environmental audit reports and confidential business information within one hundred eighty (180) days of the effective date of the statute. *Id.* § 9-810(1).

[46] *Id.* § 9-802(b).

[47] *Id.* § 9-802(e).

[48] *Id.* § 9-802(c).

[49] *Id.* § 9-802(d).

of information developed during an environmental audit to regulatory agencies should be encouraged and that the development of an environmental program to correct environmental problems and to achieve compliance with regulations should also be encouraged. To that end the legislature is establishing immunities for violations of laws and rules voluntarily disclosed to environmental agencies.[50] The legislature of the state of Idaho is further strengthening protections for confidential business

---

[50] For purposes of this law, the following definitions apply:

1. "Document," as used in this legislation, includes writings, drawings, graphs, charts, photographs, phone records and other data compilations, including electronic, from which information can be obtained or translated.
2. "Environmental agency" shall mean any department or division of the state, local government or health board with authority to enforce any state environmental law.
3. "Environmental audit" means an internal evaluation done pursuant to a plan or protocol that is designed to identify and prevent noncompliance and to improve compliance with statutes, regulations, permits and orders. An environmental audit may be conducted by an owner or operator, by an owner or operator's employees or by an independent contractor. An environmental audit may include:
   (a) One (1) or more facilities;
   (b) Any activity at one (1) or more facilities;
   (c) Impacts on one (1) or more environmental media at a facility or facilities; or
   (d) Management systems related to a facility, an activity or an impact on environmental media.
4. "Environmental audit report" means a set of documents, each labeled "Environmental Audit Report" (or a substantive equivalent label), and prepared as a result of an environmental audit. An environmental audit report may include field notes, records of observations, findings, opinions, suggestions, implementation plans, conclusions, drafts, memoranda, drawings, photographs, computer-generated or electronically recorded information, maps, data, charts, graphs and surveys, provided such supporting information is collected or developed for the primary purpose and in the course of an environmental audit. An environmental audit report may include memoranda and documents analyzing portions or all of the audit report.
5. "Environmental law" means any federal, state or local law, regulation, rule, ordinance or permit terms and conditions designed to protect or enhance the quality of land, water or air for the protection of human health, wildlife, other biota, or the environment.
6. "Person" means any individual, firm, association, partnership, joint stock company, trust, estate, political subdivision, public or private corporation, state or federal governmental department, agency or instrumentality, or any other legal entity which is recognized by the law as the subject of rights and duties.
7. "In camera review" means a hearing or review in a courtroom, hearing room or chambers, to which the general public is not admitted. After such hearing or

information including environmental audits that are submitted to environmental agencies.[51]

No state of Idaho public official, employee, or environmental agency will be required to disclose an environmental audit report prepared by or on behalf of any person, except from the state of Idaho or any political subdivision.[52]

Similarly, the statute does not prohibit a request for information, investigation, or disclosure of information required to be disclosed pursuant to federal and state law, rule, or regulation. Rather, the law mandates that documents, communications, data, reports, and other information that must be collected, developed, and reported pursuant to federal and state law, rule, and regulation must be disclosed in accordance with the applicable law, rule, or regulation.[53]

The environmental audit prohibition against compelled disclosure does not apply to the extent that it is waived expressly by the owner or operator of a facility.[54] A waiver may be full or partial; however, waivers only apply to the portions of the environmental audit report that are specifically waived.

The prohibition against disclosure does not apply if the environmental agency or court after in camera review determines that protection for the audit report is for a fraudulent purpose or the material is not an appropriate subject for an environmental audit.[55]

The party seeking disclosure of the environmental audit report has the burden of proving the disclosure is appropriate,[56] perhaps because the privilege was asserted for a fraudulent purpose. The existence of a written environmental compliance policy or adoption of a plan of action to meet

---

> review, the content of the oral and other evidence and statements of the judge, hearing officer and counsel shall be held in confidence by those participating in or present at the hearing or review, and any transcript of the hearing or review shall be sealed and not considered a public record until or unless its contents are disclosed by a court having jurisdiction over the matter.

Id. § 9-803.

[51] *Id.* § 9-802(2).

[52] *Id.* § 9-804 (prohibition against compelled disclosure).

[53] *Id.* § 9-805 (required disclosures).

[54] *Id.* § 9-806 (exceptions).

[55] *Id.* § 9-806(2). In addition, there is a long list of records that are exempt from disclosure pursuant to the statutes. *See* § 9-340 (records exempt from disclosure).

[56] *Id.* § 9-806(3).

applicable environmental laws constitutes prima facie evidence that an environmental audit report was designed to prevent noncompliance and improve compliance with environmental laws and that the environmental audit is protected from disclosure.

The prohibition against disclosure of the environmental audit report does not extend to certain documents or information, including

1. Documents, communications, data, reports, or other information required to be collected, developed, maintained, reported, or otherwise made available to an environmental agency pursuant to any environmental law or permit;
2. Information obtained by observation, sampling, or monitoring by an environmental agency; or
3. Information obtained from a source independent of the environmental audit.[57]

The environmental audit privilege legislation will not in any way limit, waive, or abrogate the scope or nature of any statutory or common law privilege, including the work product doctrine and the attorney-client privilege.[58]

Any person who makes a voluntary disclosure of an environmental audit report, or relevant portions thereof, that identifies circumstances that may constitute a violation of any state environmental law to the appropriate environmental agency, will be immune from state prosecution, suit, or administrative action for any civil or criminal penalties or incarceration for acts associated with the issues disclosed.[59] Under the law, a disclosure is rebuttably presumed voluntary if three conditions are met: (1) the disclosure is made by the owner or operator in a timely manner, after receipt of the environmental audit report to the environmental regulatory agency having regulatory authority; (2) the disclosure arises out of an environmental audit; and (3) the owner or operator making the disclosure "immediately initiates appropriate efforts to achieve compliance, pursues compliance with due diligence, and expeditiously achieves compliance within a reasonable period after the completion of the environmental audit."[60]

Where audit evidence shows the noncompliance to be the failure to obtain a permit, or other governmental permission, appropriate efforts to

[57] *Id.* § 9-807.
[58] *Id.* § 9-808.
[59] *Id.* § 9-809(1).
[60] *Id.* § 9-809(2).

correct the noncompliance may be demonstrated by the submittal of a permit application or equivalent document within a reasonable time.[61] The person may, but is not required to, enter into a voluntary consent order with the environmental regulatory agency to achieve compliance. Any disclosure to an environmental agency required by law or pursuant to a specific permit condition or order is not considered a voluntary disclosure to the environmental agency.[62]

Immunity against administrative, civil, or criminal penalties or incarceration or other type of enforcement action does not apply if a person or entity has been found by a court to have committed serious violations that constitute a pattern of continuous or repeated violations of environmental laws, regulations, permit conditions, settlement agreements, or consent orders, which were due to separate and distinct events giving rise to the violations within the three-year period prior to date of the disclosure. Such a pattern of continuous repeated violations may also be demonstrated by multiple settlement agreements related to substantially the same alleged violations concerning serious instances of noncompliance with environmental laws that occur within the three-year period immediately prior to the date of the voluntary disclosure.[63]

The statute does not otherwise affect any authority of an environmental agency to require remedial action through a consent order or action in district court or to abate an imminent hazard, associated with the information disclosed in any voluntary disclosure of an environmental violation.[64] In circumstances in which the governor deems an imminent and substantial danger to the public health or the environment exists, the governor may disclose such information contained in the audit, excluding trade secrets, as may be necessary to ensure the protection of the public health or environment.[65]

The Idaho law took effect July 1, 1995, and sunsets on December 31, 1997.

---

[61] *Id.* § 9-809(3).
[62] *Id.* § 9-809(5).
[63] *Id.* § 9-809(6).
[64] *Id.* § 9-809(7).
[65] *Id.* § 9-810(2).

§ 5.3.4 Illinois

The Illinois legislature became the fifth state to pass environmental audit privilege legislation.[66] Senate Bill 1724 was supported unanimously by the Illinois House and passed the Illinois Senate by a vote of 50 to 7 on Dec. 1, 1994. The legislation was forwarded to Governor James Edgar for approval.

On January 24, 1995, the Illinois environmental audit privilege became effective. Its goal was "to encourage owners and operators of facilities and persons conducting other activities regulated under State, federal, regional, or local laws, ordinances, regulations, permits, or orders to conduct voluntary internal environmental audits of their compliance programs and management systems and to assess and improve compliance with those laws." The environmental audit privilege[67] protects the confidentiality of communications relating to voluntary internal environmental audits.[68] The law does not, however, limit, waive, or abrogate existing or future obligations of regulated entities to monitor, record, or report information required under state, federal, regional, or local laws,

---

[66] The first four states that enacted laws granting qualified privilege status to information gleaned from environmental audits were Colorado (S.B. 94-139; COLO. REV. STAT. § 13-25-126.5 *et seq.*), Indiana (P.L. 16-1994; IND. STAT. ANN. § 13-10-3), Kentucky (KY. REV. STAT. ANN. §224.01-040), and Oregon (OR. REV. STAT. § 468.963(1)). These four states represent the "first generation" environmental audit privilege statutes.

[67] For purposes of the environmental audit privilege the following definitions apply:

"Environmental audit" means a voluntary, internal, and comprehensive evaluation of one or more facilities or any activity at one or more facilities regulated under State, federal, regional, or local laws or ordinances, or of management systems related to the facilities or activity, that is designed to identify and prevent noncompliance and to improve compliance with those laws. An environmental audit may be conducted by the owner or operator, by the owner's or operator's employees, or by independent contractors.

"Environmental audit report" means a set of documents, each labeled "Environmental Audit Report: Privileged Document" and prepared as a result of an environmental audit. An environmental audit report shall include a written response to the findings of the audit. An environmental audit report shall also include, as applicable, field notes and records of observations, findings, opinions, suggestions, conclusions, drafts, memoranda, drawings, photographs, computer-generated or electronically recorded information, maps, charts, graphs, and surveys, provided that this supporting information is collected or developed for the primary purpose and in the course of an environmental audit. An environmental audit report, when completed, may have 4 components:

ordinances, regulations, permits, or orders.[69] Nor does the law alter the scope or nature of any statutory or common law privilege, including the work product doctrine and the attorney-client privilege.[70]

Under the law, an "environmental audit report is privileged information and is not admissible as evidence in any legal action in any civil, criminal, or administrative proceeding," except in certain enumerated circumstances.[71] The privilege does not apply to the extent that it is expressly waived by the owner or operator of a facility who prepared or caused to be prepared the environmental audit report.[72]

In addition, in a civil, criminal, or administrative proceeding, a court of record or the board, after an in camera review, will require disclosure of environmental audit, if the court or the board determines one of the following:

✓ The privilege is asserted for a fraudulent purpose.

✓ The material is not subject to the privilege.

✓ The material shows evidence of noncompliance with state, federal or local environmental laws, regulations, ordinances, permits, or orders, and the owner or operator failed to undertake appropriate

---

1. An audit report prepared by the auditor, which may include the scope of the audit, the information gained in the audit, conclusions, and recommendations, together with exhibits and appendices.
2. Memoranda analyzing portions or all of the audit report and discussing potential implementation issues.
3. An implementation plan that addresses correcting past noncompliance, improving current compliance, and preventing future noncompliance.
4. Analytic data generated in the course of conducting the environmental audit.

ILL. STAT. ANN. ch. 415 para 5/52.2(i).

[68] *Id.* § 52.2 (a). The environmental audit privilege does not extend to any of the following:

1. Documents, communications, data, reports, or other information required to be collected, developed, maintained, reported, or otherwise made available to a regulatory agency under this Act or federal, State, or local law, ordinance, regulation, permit, or order.
2. Information obtained by observation, sampling, or monitoring by any regulatory agency.
3. Information obtained from a source independent of the environmental audit.

Id. § 52.2(h).

[69] *Id.* § 52.2(h).

[70] *Id.* § 52.2(j).

[71] *Id.* § 52.5(b).

[72] *Id.* § 52.2(d)(1).

corrective action or eliminate any reported violation within a reasonable time.[73]

An owner or operator asserting the environmental audit privilege has the burden of demonstrating the applicability of the privilege.[74] The state's attorney or attorney general, when seeking disclosure, has the burden of proving the enumerated elements necessary to prove the privilege should not apply.[75]

An owner or operator asserting the environmental audit privilege in response to a request for disclosure must provide to the state's attorney, attorney general, or other person, as the case may be, at the time of filing any objection to the disclosure, all of the following information:

✓ The date of the environmental audit report.

✓ The identity of the entity conducting the audit.

✓ The name and location of the audited facility.

✓ An identification of the portions of the environmental audit report for which the privilege is being asserted.

Whenever a state's attorney or the attorney general has sought and been denied, in any civil, criminal, or administrative proceeding, disclosure of material found by a court or the board to be privileged, the state's attorney or attorney general must not thereafter initiate a written request for the audit material from any owner or operator who was a party to that proceeding.[76]

Within thirty days after the state's attorney or attorney general makes a written request by certified mail for disclosure of an environmental audit report under this subsection, the owner or operator who prepared or caused

---

[73] *Id.* § 52.2(d)(2). In a criminal proceeding, the court or the board, after an in camera review, must require disclosure of material for which the privilege is asserted, if the court or the board determines one of the following:

◆ The privilege is asserted for a fraudulent purpose.

◆ The material is not subject to the privilege.

◆ The material shows evidence of noncompliance with state, federal, or local environmental laws, regulations, ordinances, permits, or orders and the owner or operator failed to undertake appropriate corrective action or eliminate any violation of law identified during the environmental audit within a reasonable time.

[74] *Id.* § 52.2(b)(3).

[75] *Id.* An owner or operator asserting the environmental audit privilege has the burden of demonstrating the applicability of the privilege in the criminal context as well. The state's attorney or the attorney general has the burden of proving that the elements showing the audit should not be privileged are present in criminal context as well as civil.

[76] *Id.* § 52.2(d)(5).

the report to be prepared may file with the appropriate court, or the board, a petition requesting an in camera hearing on whether the environmental audit report or portions of the report are privileged under this Section or subject to disclosure.[77] The circuit court or the board has jurisdiction over a petition filed by an owner or operator under this subsection requesting an in camera hearing on whether the environmental audit report or portions of the report are privileged or subject to disclosure. Failure by the owner or operator to file a petition waives the privilege.

An owner or operator asserting the environmental audit privilege in response to a request for disclosure must include in its request for an in camera hearing all of the following:

- ✓ The year the environmental audit report was prepared.
- ✓ The identity of the entity conducting the audit.
- ✓ The name of the audited facility.
- ✓ The portion of the environmental audit report for which the privilege is being asserted.

Upon the filing of a petition, the court or the board must issue an order scheduling, within forty-five days after the filing of the petition, an in camera hearing to determine whether the environmental audit report or portions of the report are privileged or subject to disclosure.

Upon making a determination, the court or the board must compel the disclosure of only those portions of an environmental audit report that are relevant to issues in dispute in the proceeding.[78]

The Illinois law contains a limited testimonial privilege. If an owner or operator performs or directs the performance of an environmental audit, an officer or employee involved with the environmental audit, or any consultant who is hired for the purpose of performing the environmental audit, may not be examined as to the environmental audit or any environmental audit report as defined by the statute. The limited testimonial privilege will not, however, apply if the state sustains its burden of proving there should be no privilege for the underlying audit document.

Of course, the parties to a conflict involving the environmental audit privilege may, at any time, stipulate to entry of an order directing that specific information contained in an environmental audit report is or is not subject to the privilege.

---

[77] *Id.* § 52.2(e)(1).

[78] *Id.* § 52.2(g).

§ 5.3.5 Indiana[79]

The Indiana legislature adopted a limited self-evaluation privilege for environmental audits. In order to take advantage of this statute, the audit must be labeled "Environmental Audit Report: Privileged Document."[80] The new self-evaluation privilege for environmental audits was designed to supplement existing common law privileges, such as the attorney-client privilege and the work product doctrine.

The privilege is not absolute. The privilege may be waived either by express action or by implication.[81] Information provided to a regulatory agency under an existing permit, order, or statute is not privileged.[82] In addition, the statute grants a court discretion to compel disclosure in a civil or criminal proceeding when the privilege is asserted for a fraudulent purpose. In addition, in a criminal proceeding the audit report may be subject to an in camera review where the report is sought by a search warrant or criminal subpoena.[83]

Additional protections provided by the statute include a mandate that a regulatory agency is precluded from adopting a rule or permit condition that is designed to circumvent the environmental audit privilege. In addition, if the environmental audit reveals the need for a regulatory permit, the entity will be deemed in compliance with law if application for the permit is made within ninety days. Finally, the Indiana statute provides that where an environmental audit is submitted to the state as aa confidential document, the privilege created by statute will not be deemed waived.[84]

§ 5.3.6 Kansas

On April 22, 1995, the Kansas legislature enacted a limited privilege for environmental compliance audits.[85] The act dictated that audit reports,[86]

---

[79] IND. CODE ANN. §§13-10-3-1 to 13-10-12 (West 1995).

[80] *Id.* § 13-10-3-2.

[81] *Id.* § 13-10-3-9.

[82] *Id.* § 13-10-3-11.

[83] *Id.* §§ 13-10-3-4 to 13-10-3-7.

[84] *Id.* § 13-10-3-11.

[85] 1995 Kansas Senate Bill 76, codified as KAN. STAT. ANN. § 60-3332 *et seq.* (1995).

[86] As used in KAN. STAT. ANN. §§ 60-3332 through 60-3339 (1995 Supp.):
(a) "Audit" means a voluntary, internal assessment, evaluation or review, not otherwise required by environmental law, that is performed by the owner or operator,

as that term is defined by statute, will be subject to discovery, but will be inadmissible as evidence in any civil, criminal, or administrative proceedings. For purposes of satisfying the Kansas law, a management system shall be deemed to satisfy the requirements of this act if it contains the following primary characteristics:

(A) a system that covers all parts of the entity's operations regulated under one or more environmental laws;

(B) a system that regularly takes steps to prevent and remedy noncompliance;

(C) a system that has the support of senior management;

(D) the entity implements a system that has policies, entity standards, and procedures that highlight the importance of ensuring compliance with all environmental laws;

(E) the entity's policies, standards, and procedures are communicated effectively to all in the entity whose activities could affect compliance achievement;

---

the owner's or operator's employees, or a qualified auditor and initiated by the owner or operator of a facility for the express and specific purpose of determining whether a facility, operation within a facility or facility management system complies with environmental laws. Once initiated an audit shall be completed within a reasonable period of time. Nothing in this section shall be construed to authorize uninterrupted or continuous auditing.

(b) "Audit report" means a set of documents, each labeled "Audit Report: Privileged Document" and prepared as a result of an audit. An audit report may include the following supporting information, if collected or developed for the primary purpose and in the course of an audit: Field notes and records of observations, opinions, suggestions, conclusions, drafts, memoranda, drawings, photographs, computer-generated or electronically recorded information, maps, charts, graphs and surveys. An audit report, when completed, may have three components:

1. An audit report prepared by the auditor, which may include the scope of the audit, the information gained in the audit, conclusions and recommendations, together with exhibits and appendices;
2. memoranda and documents analyzing all or part of the audit report and discussing potential implementation issues; and
3. an implementation plan that addresses correcting past noncompliance, improving current compliance and preventing future noncompliance.

(c) "Facility" means all contiguous land, structures and other appurtenances and improvements on the land.

(d) "Qualified auditor" means a person or organization with education, training and experience in preparing studies and assessments.

(e) "Environmental law" means any requirement contained in state environmental statutes and in rules and regulations promulgated under such statutes.

(F) specific individuals within both high-level and plant- or operation-level management are assigned responsibility to oversee compliance with such standards and procedures;

(G) the entity undertakes regular review of the status of compliance, including routine evaluation and periodic auditing of day-to-day monitoring efforts, to evaluate, detect, prevent, and remedy noncompliance;

(H) the entity has a reporting system that employees can use to report unlawful conduct within the organization without fear of retribution; and

(I) the entity's standards and procedures to ensure compliance are enforced through appropriate employee performance, evaluation, and disciplinary mechanisms.

The criteria will vary depending on the nature of the entity, including its size, its financial resources and assets, and the environmental risks posed by its operations, and based on a qualitative assessment of the totality of circumstances.

Certain exceptions apply. The privilege will not be recognized if it has been waived by the owner or operator of the audited facility. Similarly, the privilege will not be recognized if it is asserted for a fraudulent purpose. Finally, the privilege cannot be used to hide criminal activities. The statute allows for an in camera review of the audit documents to determine whether the privilege is applicable.

The Kansas statute has a qualified testimonial immunity.[87] Under the statute, there is a rebuttable presumption of immunity from any administrative, civil, or criminal penalties for the violation disclosed if the disclosure is one:

1. made promptly after knowledge of the information disclosed is obtained by the person or entity;
2. made to an agency having regulatory authority with regard to the violation disclosed;
3. arising out of an audit;
4. for which the person or entity making the disclosure initiates action in a diligent manner to resolve the violations identified in the disclosure; and

[87] *Id.* § 60-3338.

5. in which the person or entity making the disclosure cooperates with the appropriate agency in connection with investigation of the issues identified in the disclosure.

A disclosure is not voluntary for purposes of the immunity provision if it is required by state law to be reported to a regulatory authority.

The presumption may be rebutted and penalties may be imposed under state law if it is established that

1. the disclosure was not voluntary;
2. the violation was committed intentionally and willfully by the person or entity making the disclosure;
3. the violation was not fully corrected in a diligent manner; or
4. significant environmental harm or a public health threat was caused by the violation.

In any enforcement action brought against a person or entity regarding a violation for which the person or entity claims to have made a voluntary disclosure within the privilege statute, the burden of proof shifts concerning whether the disclosure is voluntary. The person or entity making the voluntary disclosure claim will have the burden of establishing a prima facie case that the disclosure was voluntary. Once a prima facie case of voluntary disclosure is established, however, the opposing party will have the burden of rebutting the presumption by a preponderance of the evidence.

§ 5.3.7 Kentucky[88]

In order to encourage owners and operators of facilities and persons conducting other regulated environmental activities, or its federal, regional, or local counterparts or extensions, both to conduct voluntary internal environmental audits of their compliance programs and management systems and to assess and improve compliance with statutory and regulatory requirements, "an environmental audit privilege is created to protect the confidentiality of communications relating to voluntary internal environmental audits."[89] With certain enumerated exceptions, environmental audit reports will be privileged and will not be admissible as evidence in any legal action in any civil, criminal, or administrative

---

[88] KY. REV. STAT. ANN. § 224.01-040 (Michie 1994).

[89] *Id.* § 224.01-040(2) (1994). For purposes of the statute, the following definitions were established:

proceeding.[90] The privilege does not apply to the extent that "it is waived expressly or waived by implication by the owner or operator of a facility or persons conducting an activity that prepared or caused to be prepared the environmental audit report."

Similarly, if the owner or operator of a facility or person conducting an activity seeks to introduce an environmental audit report as evidence, then the privilege for the entire report will be waived. Kansas, thus, allows no partial waivers. Seeking to introduce any part of the report will constitute waiver of the privilege for the *entire* report.[91]

In a civil or administrative proceeding, a court of record, after a private review consistent with the Kentucky Rules of Civil Procedure, must require disclosure of privileged environmental audit material, if the court determines that any one of three factors warrants disclosure:

1. The privilege is asserted for a fraudulent purpose;
2. The material is not subject to the privilege; or

(a) "Environmental audit" means a voluntary, internal, and comprehensive evaluation of one (1) or more facilities or an activity at one (1) or more facilities regulated under this chapter, or federal, regional, or local counterparts or extensions thereof, or of management systems related to that facility or activity, that is designed to identify and prevent noncompliance and to improve compliance with statutory or regulatory requirements. An environmental audit may be conducted by the owner or operator, by the owner's or operator's employees, or by independent contractors; and

(b) "Environmental audit report" means a set of documents, each labeled "environmental audit report: privileged document" and prepared as a result of an environmental audit. An environmental audit report may include field notes and records of observations, findings, suggestions, conclusions, drafts, memoranda, drawings, photographs, computer-generated or electronically recorded information, maps, charts, graphs, and surveys, provided the supporting information is collected or developed for the primary purpose and in the course of an environmental audit. An environmental audit report, when completed, shall have three (3) components:

1. An audit report prepared by an auditor, which shall include the scope and date of the audit and the information gained in the audit together with exhibits and appendices, and may include conclusions and recommendations;
2. Memoranda and documents analyzing part or all of the audit report and discussing implementation issues; and
3. An audit implementation plan that addresses correcting past noncompliance, improving current compliance, and preventing future noncompliance.

[90] *Id.* § 224.01-040(3).

[91] *Id.* § 224.01-040(b).

3. Even if subject to the privilege, the material shows evidence of noncompliance with environmental law, or with the federal, regional, or local counterparts or extensions thereof, and appropriate efforts to achieve compliance were not promptly initiated and pursued with reasonable diligence.

In a criminal proceeding, a court of record, after a private review, must also require disclosure of privileged environmental audit material if any of the above three factors are found. The criminal context, however, adds a fourth factor:

> The material contains evidence relevant to commission of certain enumerated offenses and the prosecuting authority has a need for the information.

An environmental audit report found subject to disclosure for this reason, will not be deemed privileged in any criminal proceeding.

The party asserting the environmental audit privilege has the burden of proving the privilege, including, if there is evidence of noncompliance with this chapter or the federal, regional, or local counterparts or extensions thereof, proof that appropriate efforts to achieve compliance were promptly initiated and pursued with reasonable diligence.[92] A party seeking disclosure of the otherwise privileged audit, including a prosecuting authority seeking disclosure, has the burden of proving that the privilege is asserted for a fraudulent purpose.

The prosecuting authority, having probable cause to believe certain enumerated environmental crimes have been committed, based upon information obtained from a source independent of an environmental audit report, may obtain an otherwise privileged environmental audit report pursuant to a search warrant, criminal subpoena, or discovery as allowed by the Kentucky Rules of Criminal Procedure or any applicable statute. The prosecuting authority must immediately place the report under seal and must not review or disclose its contents.

Within twenty days of the prosecuting authority's obtaining an environmental audit report, the owner or operator who prepared or caused to be prepared the report, or the prosecuting authority, may file with the

[92] *Id.* § 224.01-040.

appropriate court a petition requesting a private hearing on whether the environmental audit report or portions thereof are privileged under this section or subject to disclosure. Failure by the owner or operator to file such a petition will waive the privilege.

Upon filing a petition, the court must issue an order scheduling a private hearing, within thirty days of the filing of the petition, to determine whether the environmental audit report or parts of the report are privileged or subject to disclosure. The order must also allow the prosecuting authority immediately to remove the seal from the report and to review the entire report and must place appropriate limitations on distribution and review of the report to protect against unnecessary disclosure.

The prosecuting authority may consult with enforcement agencies regarding the contents of the report as necessary to prepare for the private hearing. The information used in preparation for the private hearing must not be used in any investigation or in any proceeding against the defendant, and must otherwise be kept confidential, unless and until the information is found by the court to be subject to disclosure. Upon making a determination, the court may compel the disclosure only of those portions of an environmental audit report relevant to issues in dispute in the proceeding.

Parties to a dispute may, of course, at any time stipulate to the entry of an order directing that specific information contained in an environmental audit report is or is not subject to the statutory environmental audit privilege.

The environmental audit privilege established in Kentucky does not extend to

(a) Documents, communications, data, reports, or other information required to be collected, developed, maintained, reported, or made available to a regulatory agency pursuant to state environmental law or administrative regulations promulgated pursuant thereto, or other federal, state, or local law, ordinance, regulation, permit, or order, and any information developed relating to any release subject to state environmental law;

(b) Information obtained by observation, sampling, or monitoring by any regulatory agency; or

(c) Information obtained from a source independent of the environmental audit.

Nothing in the Kentucky law limits, waives, or otherwise abrogates the scope or nature of any statutory or common law privilege, including the work product doctrine and the attorney-client privilege.[93] Nor does the law limit, waive, or abrogate any reporting requirement state law or permit conditions.

§ 5.3.8 Michigan

On March 18, 1996, the governor of Michigan approved an act to amend the Public Acts of 1994. The amendment, entitled Part 148 Environmental Audit Privilege and Immunity, provides a privilege to an environmental audit report,[94] defined by the act as "a document or set of documents, each labeled at the time it is created 'environmental audit report: privileged document' and created as a result of an environmental audit.'"[95]

The privilege not only protects the actual audit report from disclosure, but also protects auditors and recipient of the audit report from testifying about information in the report that is privileged.[96]

The act clearly excludes certain documents and reports from the privilege, although they might be included in an audit report.[97] This exclusion applies to information required to be reported by law, information obtained by a regulatory agency, pretreatment monitoring results, information obtained from independant sources, and machinery and equipment maintenance records.[98]

The act permits express waiver of the privilege.[99] Disclosure of the report, however, by the person who received the report to that person's employee, legal representative, or agent does not constitute a waiver of the privilege.[100] Waiver by disclosure also does not occur when made under the terms of confidentiality.[101]

---

[93] *Id.*
94 1996 Mich. Pub. Acts 132, § 14802(2).
[95] *Id.* § 14801(b).
[96] *Id.* § 14802(4).
[97] *Id.* § 14802(3).
[98] *Id.* § 14803(3)(a)-(e).
[99] *Id.* § 14803(1).
[100] *Id.* § 14803(2)(a), (b), (c).
[101] *Id.* § 14803(3)(a), (b).

As with other state audit privilege legislation, the privilege may be asserted when disclosure of the report is requested.[102] Objection to disclosure leads to an in camera review and hearing.[103] The court then decides the status of the audit report based on the asserted purpose for the privilege, whether the materials contained in the report qualify for the privilege, and, even if subject to the privilege, whether evidence of noncompliance with the laws appears.[104] The court's finding may be appealed.[105]

The act allows an environmental audit report to be seized pursuant to a search warrant; however, the report is immediately placed under seal with the court.[106] Parties eligible to assert the privilege receive notice of the filing and have thirty days to assert the privilege and file an objection to disclosure.[107] Failure to make an objection constitutes waiver of the privilege.[108] The court then conducts an in camera review and either finds the privilege valid or permits disclosure.[109]

In addition to providing a privilege, the act also permits a person to be immune from penalties if the person voluntarily discloses audit findings of noncompliance and performs all requirements listed in the act that make the person's actions voluntary.[110]

Finally, the act includes a provision that calls for an evaluation of the environmental audit privilege and immunity amendment within five years of its effective date.[111]

§ 5.3.9 Minnesota

Minnesota's environmental audit privilege statute defines an "Environmental audit" as a "systematic, documented, and objective review by a regulated entity of one or more facility operations and practices related

[102] *Id.* § 14804(1).
[103] *Id.*
[104] *Id.* § 14804(4) (a), (b); (5).
[105] *Id.* § 14804(6).
[106] *Id.* § 14805(1).
[107] *Id.*
[108] *Id.*
[109] *Id.* § 14805(4), (5), (6).
[110] *Id.* § 14809(1)-(6).
[111] *Id.* § 14810(2).

to compliance with one or more environmental requirements and, if deficiencies are found, a plan for corrective action."[112] In order to take advantage of the Minnesota statute, the final audit document must be designated as an "audit report" and must include the date of the final written report of findings for the audit.[113]

The state must defer for at least ninety days any enforcement action against the owner or operator of a facility if a report has been submitted to the commissioner. If the report includes a performance schedule, and the performance schedule is approved, the state must defer enforcement for the term of the approved performance schedule unless the owner or operator of the facility fails to meet an interim performance date contained in the schedule.

If, within ninety days after the required report is received by the commissioner or within the time specified in an approved performance schedule, the owner or operator of a facility corrects the violations identified in the audit or self-evaluation and certifies to the commissioner that the violations have been corrected, the state may not impose any administrative, civil, or criminal penalties against the owner or operator of the facility for the reported violations. There are, of course, exceptions to this general rule. The state may at any time bring

1. a criminal enforcement action against any person who knowingly commits a violation;
2. a civil or administrative enforcement action, which may include a penalty against the owner or operator of a facility if:
   (i) less than one year has elapsed since the final resolution of a notice of violation, an administrative penalty order, or a civil or criminal lawsuit that resulted in an enforcement action being taken against the owner or operator of a facility for a violation of a requirement that was also shown as having been violated in the report; or
   (ii) a violation caused serious harm to public health or the environment; or

[112] 1996 Minn. Laws 359.

[113] *Id.*

3. the enforcement action against the owner or operator of a facility to enjoin an imminent threat to public health or the environment.[114]

§ 5.3.10 Mississippi

Mississippi enacted a limited self-evaluation privilege on April 7, 1995.[115] The statute was designed to encourage owners and occupiers of property to perform environmental audits on their private facilities. The privilege extends to civil, criminal, and administrative proceedings and will protect anyone who conducts the audit or orders others to prepare an audit. The privilege can be expressly waived. Similarly, the privilege can be lost if the regulated entity fails to come into compliance within a reasonable amount of time after discovering the violation.

The privilege does not apply to any document prepared pursuant to state, federal, or local environmental laws, regulations, or permit requirements. Similarly, the privilege does not apply to any documents that may have existed before the audit report was prepared.

The Mississippi statute provides for an in camera review of documents to determine if the privilege applies or not. The party asserting the privilege has the burden of proving its applicability. The party seeking discovery of the document, in contrast, has the burden of proving no privilege in fact exists.

§ 5.3.11 New Hampshire

On March 18, 1996, the governor of New Hampshire, in the interest of protecting the health and welfare of the public and the environment, signed into law legislation that encourages businesses to conduct environmental audits by providing an audit privilege.[116] The environmental audit privilege creates incentives for voluntary compliance with environmental laws, as it provides a limitation on the penalties that may be imposed on a business that engages in self-audits.[117]

---

[114] *Id.*

[115] MISS. CODE ANN. § 49-2-51 *et seq.*

[116] 1996 N.H. Laws 4.

[117] *Id.* at 4:1(II).

The New Hampshire environmental audit privilege protects the environmental audit report, which is composed of specific documents that are prepared in association with the audit, dated, and labeled "environmental audit report: privileged document."[118] Among the protected documents in the environmental audit report is a written plan that identifies the auditor, the scope of the audit, and the anticipated dates of the audit.[119] Documents generated by the auditor in the course of the audit, including those that summarize the audit findings, draw conclusions, and provide recommendations, also are contained in the privileged report.[120]

The audit privilege further extends to documents created by the auditee in response to the aforementioned materials provided by the auditor.[121] The implementation plan prepared by the auditor, which identifies areas of non-compliance discovered in the course of the audit and recommends corrective action and preventive measures for further noncompliance, is also part of the audit report and therefore is protected by the privilege.[122]

As extensive as this privilege may appear, there are multiple exceptions to its use. The privilege does not apply when

a) it is waived, either expressly or impliedly, by the auditee.

b) it is waived by the auditee, who wishes to introduce part or all of the report as evidence.

c) the auditee fails to promptly initiate and diligently pursue actions to achieve compliance when the audit report evidences noncompliance.

d) the audit report reveals a threat of imminent and substantial harm to the public or the environment.

e) the court determines, after an in camera review, that the auditee had a fraudulent intent in commissioning the audit, or that the auditee was aware of government investigation, or that a criminal offense has occurred for which the state has a compelling need for the otherwise unavailable information.[123]

---

118 *Id.* at Chapter 147-E:1(V).
119 *Id.* at Chapter 147-E:2(I).
120 *Id.* at Chapter 147-E:2(II).
121 *Id.* at Chapter 147-E:2(II) - (III).
122 *Id.* at Chapter 147-E:2(IV).
123 *Id.* at Chapter 147-E:4(I) - (IV).

The law also excludes documents from the privilege if they are required to be available to the regulatory agency; if the regulatory agency obtains the information through its own observation, sampling, or monitoring; or if the documents are developed in the regular course of business.[124] The course-of-business exclusion includes information that may have been reviewed by the auditor, formed a basis for the audit report, or even used in the audit report.[125]

In the event that a party asserts the privilege and one wishes to contest this assertion, the court will conduct an in camera review of the environmental audit report.[126] The party opposing disclosure may file objections, a hearing may be held, and the court will issue an order regarding the status of the documents.

The New Hampshire law also provides for a penalty waiver when certain steps are followed after an audit reveals a violation.[127] There are also explicit exceptions to qualifying for a penalty waiver.[128]

§ 5.3.12 Oregon

In 1993, the Oregon legislature also enacted a statute designed to offer a measure of comfort to executives weighing the benefits of conducting environmental audits of their businesses against the danger that audit results could be used to build a case against them, either by government agencies or private litigants.[129] Oregon's environmental audit privilege law supplements current common law privileges and is the first attempt to protect internal comprehensive evaluations of environmental performance by either business or local governments against disclosure in any state or private enforcement action. The Oregon statute provides that internal environmental audits prepared by businesses or local governments to identify areas of environmental noncompliance or to improve compliance efforts are privileged and may not be introduced into evidence in any civil, criminal, or administrative proceeding.

---

[124] *Id.* at Chapter 147-E:5(I) - (III).

[125] *Id.* at Chapter 147-E:5(III)(a)(b)(c).

[126] *Id.* at Chapter 147-E:6.

[127] *Id.* at Chapter 147-E:9(I).

[128] *Id.* at Chapter 147-E:9(II).

[129] 1993 Ore. Laws 422.

Oregon's audit privilege is expressly designed to "encourage owners and operators of facilities and persons . . . both to conduct voluntary internal environmental audits of their compliance programs and management systems and to assess and improve compliance . . ." overall. The Environmental Audit Reports are "privileged and shall not be admissible as evidence in any legal action in any civil, criminal or administrative proceeding. . . ." An Environmental Audit is defined as a "voluntary, internal and comprehensive evaluation of one or more facilities or an activity at one or more facilities . . . that is designed to identify and prevent noncompliance or to improve compliance . . ." with environmental laws and regulations. The audit privilege applies only to a set of documents labeled "Environmental Audit Report: Privileged Document." The Environmental Audit Report may include a broad array of documents such as field notes, records of observations, findings, opinions, maps, charts, graphs, surveys, and computer generated or electronically recorded information, provided the audit is "collected or developed" for the primary purpose and in the course of the audit. When completed, the Environmental Audit Report may have three components:

1. An audit report prepared by the auditor, including the scope of the audit, the information gained in the audit, conclusions, and recommendations, along with exhibits and appendices;
2. Memoranda and documents analyzing portions or all of the audit report and implementation issues; and
3. An implementation plan that addresses correcting past noncompliance, improving compliance, and preventing future noncompliance.[130]

The self-evaluation privilege established by the statute is by no means absolute. Many exceptions could negate the privilege. For example, the privilege applies only to a set of documents that is both labeled as an environmental audit report and tagged as privileged. The privilege also does not cover documents, data, or reports that are required by law; information an agency develops by its own observations or sampling; or information obtained from a source separate from the environmental audit.

---

[130] But note that the environmental audit privilege itself does not extend to
1. Documents, communications, data, reports, or other information required to be collected, developed, maintained, reported, or otherwise provided to regulatory agencies.

There are a number of other circumstances that can trigger loss of the privilege. The environmental audit privilege may be waived, either expressly or by implication, by a facility's owner or operator or the persons preparing the report. In addition, a court may require disclosure if it determines the privilege was fraudulently asserted, that it does not apply, or that the material, though privileged, shows noncompliance with environmental laws or lack of due diligence to comply. The privilege is, likewise, unavailable if the material contains information related to the commission of a crime and a prosecutor shows a "compelling need" for it and the inability to get it by other means.[131]

The party asserting the environmental audit privilege has the burden of proving the privilege, including proving that appropriate efforts to achieve compliance were promptly initiated and pursued with reasonable diligence. Parties seeking disclosure have the burden to prove that the privilege is asserted for fraudulent purposes. A district attorney or state attorney general seeking disclosure on grounds of compelling need has the burden of proving the conditions for disclosure.[132]

§ 5.3.13 South Carolina

On June 4, 1996, the Governor of South Carolina approved the South Carolina Environmental Audit Privilege and Voluntary Disclosure Act of 1996.[133] In an effort to protect the environment, the legislation encourages owners and operators of facilities to conduct voluntary internal environmental audits by providing an audit privilege to protect

---

2. Information obtained by observation, sampling, or monitoring by an agency.
3. Information obtained from a source independent of the environmental audit.

[131] *See generally* James D. Lawlor, Esq., Executive Legal Alert, AIR & WATER POLLUTION REPORT (Feb. 28, 1994); W. Hugh O'Riordan, *Oregon's Environmental Audit Privilege: An Incomplete First Step Toward Encouraging Voluntary Compliance,* WASHINGTON LEGAL FOUNDATION LEGAL OPINION LETTER (Feb. 11, 1994).

[132] A district attorney or the attorney general having independent probable cause to believe an offense has been committed may obtain a privileged Environmental Audit Report through a search warrant, subpoena, or through discovery. Once obtained, the Environmental Audit Report must be placed under seal and shall not be reviewed or disclosed until after thirty days have passed. Affected parties must file a petition with the appropriate court requesting an in camera hearing or lose the privilege. Once a petition is filed, the court must schedule an in camera hearing within forty-five days.

[133] 1996 S.C. Acts 384.

communications related to the audit and by providing a limited protection from penalties in instances of voluntary disclosure.[134]

Although the majority of the legislation mirrors audit privilege legislation enacted in other states regarding assertion of the privilege, court determination of its availablity, and voluntary disclosure,[135] South Carolina also included two distinct provisions. The first enables the Workers' Compensation Commission to obtain and admit evidence from an audit report regardless of the privilege the chapter provides.[136] The second clause allows information in the audit report to be available to solicitors and the attorney general for criminal proceedings.[137] Disclosure under either of these clauses is limited to use in the specific proceeding.[138]

§ 5.3.14 South Dakota

South Dakota Governor William Janklow recently signed an environmental audit privilege bill (SB 24) into law. The law creates a presumption against the imposition of any civil or criminal penalties for violations discovered during an audit and reported to the Department of Environmental & Natural resources within sixty days of the audit. The law precludes the state from routinely requesting audits, but allows use of discovery, where appropriate.[139]

§ 5.3.15 Texas[140]

The Texas legislature adopted its own Environmental Audits statute in May 1995, proposing to encourage voluntary compliance with environmental and occupational health and safety laws.[141] Audit reports are

---

[134] *Id.* § 48-57-10 (A).

[135] *Id.* Environmental Audit Report is defined in 48-57-20(3); the report is declared privileged in 48-57-30(A), 48-57-50, 48-57-60; the privileged is waived in 48-57-40; and immunity for voluntary disclosure is provided in 48-57-100.

[136] *Id.* § 48-57-30(C).

[137] *Id.* § 48-57-30(D).

[138] *Id.* § 48-57-30(C)(D).

[139] *See generally Several Stat Audit Privilege Bills Advance, but Setbacks Affect Bills in Florida, Vermont,* 5 DUE DILIGENCE GUIDE (BNA) at 18 (March 1996).

[140] TEXAS REV. CIV. STAT. ANN. art. 4447cc, §§ 1 to 13 (West 1996).

[141] *Id.* § 1.

considered privileged, and are not admissible as evidence, or subject to discovery, except as otherwise provided by the act.[142] In order to facilitate the identification of an audit, each document in the audit should be labeled "Compliance Report: Privileged Document," or in a similar fashion.[143] However, failure to label documents in such a manner does not create a waiver of the privilege or a presumption that the privilege does not apply.[144] In addition, any documentation required to be collected by a regulatory agency, information obtained by the agency in person, or any information obtained by a source not involved in the preparation of the audit is considered nonprivileged.[145]

The audit privilege may be expressly waived by the owner or operator preparing the report or causing such report to be prepared. However, the waiver may be estopped if the disclosure of an audit is made pursuant to the terms of a confidentiality agreement between the preparer of the audit and a third party or if the disclosure is made to an employee or agent of the owner or operator preparing the report.[146] Disclosure may be required by the court or a competent administrative hearings official with jurisdiction over such matters through an in camera review, if the privilege asserted is suspected to be fraudulent,[147] the audit shows noncompliance with environmental laws,[148] or there is reasonable cause to suspect a criminal offense has been committed under such laws.[149] The party seeking disclosure has the burden of proving fraudulent purpose or evidence of noncompliance with the laws.[150] However, under the suspicion of a criminal act, the attorney representing the state may obtain a copy of the audit through a search warrant or criminal subpoena, and the burden rests with the owner or operator who prepared the report to request an in camera review with the proper court to determine the audit's privileged status.[151]

---

[142] *Id.* § 5.
[143] *Id.* § 4.
[144] *Id.*
[145] *Id.* § 8.
[146] *Id.* § 6.
[147] *Id.* § 7.
[148] *Id.*
[149] *Id.* § 9.
[150] *Id.* § 7.
[151] *Id.* § 9.

Failure to file such a petition by the owner or operator having the report prepared waives the audit's privileged status.[152]

Immunity for voluntary disclosure of a violation of an environmental law is given as to administrative, civil, or criminal penalties.[153] A disclosure is not considered voluntary, however, if the disclosure is one required by a regulatory agency as a result of a specific condition of an enforcement order or decree.[154] In addition, the immunity privilege will not apply if the subject matter of the disclosure was committed intentionally, knowingly, or recklessly by the person claiming the privilege, or an employee or agent of that person.[155] Finally, the facility seeking such a privilege must, prior to seeking the privilege, give notice to the appropriate regulatory agency of the fact that it is planning to commence an audit of the facility, specifying the portion of the facility to be audited and the time at which the audit will commence and end.[156]

---

152 *Id.*

15 *Id.* § 10. Under such, a voluntary disclosure will only be considered voluntary if

1. [T]he disclosure was made promptly after knowledge of the information disclosed is obtained by the person;
2. [T]he disclosure was made in writing by certified mail to an agency that has regulatory authority with regard to the violation disclosed;
3. [A]n investigation of the violation was not initiated or the violation was not independently detected by an agency with enforcement jurisdiction before the disclosure was made using certified mail;
4. [T]he disclosure arises out of a voluntary environmental or health and safety audit;
5. [T]he person who makes the disclosure initiates an appropriate effort to achieve compliance, pursues that effort with due diligence, and corrects the noncompliance within a reasonable time;
6. [T]he person making the disclosure cooperates with the appropriate agency in connection with an investigation of the issues identified in the disclosure; and
7. [T]he violation did not result in injury to one or more persons at the site or substantial off-site harm to persons, property, or the environment.

154 *Id.*

155 *Id.*

156 *Id.*

§ 5.3.16 Utah[157]

The governor of Utah signed the Environmental Self Evaluation Act into law on March 20, 1995.[158] Thus, Utah too joined the growing minority of states that are determined "to enhance the environment through voluntary compliance with environmental laws; and provide incentives to voluntarily identify and remedy environmental compliance problems"[159] by enacting environmental audit privilege legislation.[160]

Information divulged in violation of the environmental audit privilege law may not be admitted into evidence in an administrative or judicial proceeding.[161] Persons divulging privileged environmental audit information are liable for all damages that are proximately caused by the disclosure.[162] In addition, persons divulging such information may be considered guilty of a class B misdemeanor, subject to a civil penalty not to exceed $10,000 and subject to penalties for contempt of court.[163]

---

[157] Environmental Self Evaluation Act, 1995 Ut. SB 84.

[158] The effective date of that Act is March 31, 1995. UTAH CODE ANN. § 19-7-108. The Act applies to all civil and criminal judicial and administrative proceedings commenced after that date. § 19-7-108.

[159] § 19-7-102.

[160] For purposes of the law, the following definitions were adopted

> "Administrative proceeding" means an adjudicatory proceeding conducted by the department or other government entity with authority to enforce any environmental law, including any notice of violation proceeding, any department proceeding listed in Section 19-1-305, or any proceeding conducted pursuant to Title 63, Chapter 46B, Administrative Procedures Act.
>
> "Environmental audit report" means any document, information, report, finding, communication, note, drawing, graph, chart, photograph, survey, suggestion, or opinion, whether in preliminary, draft, or final form, prepared as the result of or in response to an environmental self-evaluation.
>
> "Environmental law" means any requirement contained in this title, or in rules made under this title, or in any rules, orders, permits, licenses, or closure plans issued or approved by the department, or in any other provision or ordinance addressing protection of the environment.
>
> "Environmental self-evaluation" means a self-initiated assessment, audit, or review, not otherwise expressly required by an environmental law, that is performed to determine whether a person is in compliance with environmental laws. A person may perform an environmental self-evaluation through the use of employees or the use of outside consultants.

[161] § 19-7-104(a)(1).

[162] § 19-7-104(a)(2). Government employees are, however, subject to Title 63, Chapter 30, Utah Governmental Immunity Act. *Id.*

[163] This provision accepts attorneys, court officers, and agents in pursuant of their duties.

An environmental audit report obtained pursuant to the in camera review process is considered a protected record. Department employees and attorneys may not disclose the report except in strict accordance with the provisions of that law.[164]

The person seeking disclosure of an environmental audit report in an administrative proceeding must request an in camera review of the audit by a court of record. The person seeking disclosure of the report has the burden of proving that the report is not privileged or excepted pursuant to the Utah Rule of Evidence, Rule 508. If the court determines that all or part of the environmental audit report is not privileged, then the court must order the disclosure of the nonprivileged portions of the environmental audit report. The privileged portions of the environmental audit report may not be disclosed.

The person asserting the environmental self-evaluation privilege has the burden of establishing a prima facie case of privilege.[165]

The Utah law has a testimonial privilege.[166] A person, including the person's officers and employees, who performs or assists in performing an environmental self-evaluation as defined by the Utah law cannot be examined without the consent of the person for whom the environmental self-evaluation was conducted. This rule also applies to consultants hired for the purposes of conducting the audit. The testimonial privilege may, however, be waived if the person or consultant is ordered to testify in court. Similarly, the testimonial privilege will not apply if the self-evaluations or audit report is subject to discovery or is admissible under the Utah Rules of Evidence, Rule 508.[167]

§ 5.3.17 Virginia

The Virginia law signed into law by Governor George Allen also contains a testimonial privilege.[168] H.R. 1845, sponsored by Rep.

---

[164] § 19-7-104. Section 4.

[165] § 19-7-106.

[166] *See* § 19-7-107.

[167] *Id.*

[168] *See* 63 U.S.L.W. 2622 (April 11, 1995); *See also* Special Report, *States Taking Legislative Initiative in Encouraging Corporate Audits,* BNA State Environmental Daily (March 16, 1995).

Whittington Clement (D-Danville) protects companies by stating that "no person involved in the preparation of or in possession of a document shall be compelled to disclose such document or information about its contents or the details of its preparation."[169]

The environmental audit privilege in Virginia does not extend to criminal proceedings, or to cases in which such violations demonstrate a "clear, imminent and substantial danger to the public health or environment" or to self-assessments conducted in bad faith. The agency bears the burden of proving entitlement to this abrogation of the general rule of privilege. A party seeking disclosure must prove the need or applicability of an exception to have the document admitted as evidence in a legal proceeding.

The law contains a limited penalty immunity provision that is similar to the Colorado provision. Companies that discover and admit environmental violations or irregularities during a self-assessment will be immune from administrative or civil penalty for such violations that might be required under statute, regulation, or administrative order.

§ 5.3.18 Wyoming[170]

The Wyoming legislature, likewise, determined that "protection of the environment rests principally on voluntary public compliance with environmental laws." For this reason, the people of Wyoming joined the growing minority of states that determined that "limited expansion of existing privileges will encourage voluntary public compliance and will improve environmental quality," because "the voluntary provisions of this act will not inhibit mandatory exercise of the regulatory authority by those entrusted with protecting our environment."

In enacting the Wyoming version of environmental audit privilege legislation, the legislature decided to revisit the question annually. Each year after the effective date of the act, the department of environmental quality must prepare a report to the joint minerals, business, and economic development interim committee as to the effectiveness of the voluntary

[169] Special Report *States Taking Legislative Initiative in Encouraging Corporate Audits*, BNA State Environmental Daily (March 16, 1995).

[170] 1995 Wyo. Sess. Laws 58.

environmental audit reporting process authorized under this act and determine if the process meets purposes and findings specified in this act. The committee must report its findings and recommendations to the fifty-fourth legislature at the 1998 budget session and every two years thereafter. The act took effect upon approval by the governor and has no statutory sunset.

Owners and operators of facilities and persons whose activities are regulated under the state environmental laws may conduct a voluntary internal environmental audit of compliance programs and management systems to assess and improve compliance with this act. An environmental audit privilege is created to protect the confidentiality of communications relating to these audits.[171] The creation of this statutory privilege will not, by

---

[171] WYO. STAT. § 35-11-1105(b). For purposes of the statute, the following definitions apply:

> (i) "Environmental audit" means a voluntary, internal and comprehensive evaluation of one (1) or more facilities or an activity at one (1) or more facilities regulated under this act, or of management systems related to the facility or activity, that is designed to identify and prevent noncompliance and to improve compliance with this act. An environmental audit may be conducted by the owner or operator, by the owner's or operator's employees or by independent contractors. Once initiated the voluntary environmental audit shall be completed within one hundred eighty (180) days. Nothing in this section shall be construed to authorize uninterrupted voluntary environmental audits;
>
> (ii) "Environmental audit report" means a set of documents, each labeled "Environmental Audit Report: Privileged Document," prepared as a result of an environmental audit and may include field notes and records of observations, findings, opinions, suggestions, conclusions, drafts, memoranda, drawings, photographs, computer-generated or electronically recorded information, maps, charts, graphs and surveys if supporting information is generated or developed for the primary purpose and in the course of an environmental audit. An environmental audit report, when completed, shall have three (3) components:
>
> > (A) An audit report prepared by the auditor, including the scope, commencement and completion dates of the audit, the information gained in the audit, conclusions and recommendations, together with exhibits and appendices;
> >
> > (B) Memoranda and documents analyzing the audit report and discussing implementation issues; and
> >
> > (C) An audit implementation plan that corrects past noncompliance, improves current compliance and prevents future noncompliance.
>
> (iii) "In camera review" means a hearing or review in a courtroom, hearing room or chambers to which the general public is not admitted. After such hearing or review, the content of oral and other evidence and statements of the judge and counsel shall be held in confidence by those participating in or present at the hearing or review, and

statute, "limit, waive or abrogate the scope or nature of any statutory or common law privilege, including the work product doctrine and the attorney-client privilege."

The environmental audit report[172] is privileged and will not be admissible as evidence in any civil, criminal, or administrative proceeding, except if the owner or operator of a facility may waive this privilege in whole or in part. If an owner or operator of a facility or person conducting an activity seeks to introduce any part of an environmental audit report as evidence in any proceeding, including reporting of environmental violations,[173] the privilege is waived as to those sections of the report dealing with that media sought to be introduced into evidence.

In a civil or administrative proceeding, the court or hearing officer after in camera review consistent with the Wyoming Rules of Civil Procedure, may require disclosure of all or part of the report if it determines:

✓ The privilege is asserted for a fraudulent purpose;

✓ The material is not subject to the privilege;

✓ The material shows evidence of noncompliance with this act or any federal environmental law or regulation and appropriate efforts to achieve compliance were not initiated as promptly as circumstances permit and pursued with reasonable diligence; or

✓ The information contained in the environmental audit report demonstrates a substantial threat to the public health or environment

---

any transcript of the hearing or review shall be sealed and not considered a public record until its contents are disclosed, pursuant to this section, by a court having jurisdiction over the matter.

§ 35-11-1105(a).

[172] The environmental audit privilege does not extend to:

(i) Documents, communications, data, reports or other information required to be collected, developed, maintained, reported or otherwise made available to a regulatory agency pursuant to any regulatory requirement of this act or any other federal or state law or regulation;

(ii) Information obtained by observation, sampling or monitoring by any regulatory agency;

(iii) Information obtained from a source independent of the environmental audit; or

(iv) Documents existing prior to the commencement of the environmental audit; or

(v) Documents prepared subsequent to and independent of the completion of the environmental audit.

[173] The reporting violations specified in the statute are those enumerated under WYO. STAT. § 35-11-1106(a).

or damage to real property or tangible personal property in areas outside the facility property.

In a criminal proceeding, the court may require disclosure, after in camera review, of all or part of the report if it determines any of the four enumerated conditions justifying disregarding the audit privilege in the civil context exists or the material contains evidence relevant to the commission of a criminal offense under state or any federal environmental law. If the district attorney or the attorney general has a need for the information, the information is not otherwise available, and the district attorney or attorney general is unable to obtain the substantial equivalent of the information by any means without incurring unreasonable cost and delay, then the privilege may likewise be disregarded.

The party asserting the environmental audit privilege has the burden of proving the privilege, including proof that appropriate efforts to achieve compliance with state or federal environmental law or regulation were promptly initiated and pursued with reasonable diligence. A party seeking disclosure of an audit has the burden of proving that the privilege is asserted for a fraudulent purpose. A district attorney or the attorney general seeking disclosure for purposes of proving criminal conduct has the burden of proving the specified conditions for disclosure.

A district attorney or the attorney general, having probable cause to believe an environmental crime has been committed under Wyo. Stat. § 35-11-901 based upon information obtained from a source independent of an environmental audit report, may obtain a report for which a privilege is asserted under this section pursuant to a search warrant, criminal subpoena or discovery. The district attorney or the attorney general must immediately place the report under seal and must not review or disclose its contents.

Within twenty days after the district attorney or the attorney general has obtained an environmental audit report, the owner or operator of the facility in which the report was prepared may file with the appropriate court a petition requesting an in camera hearing on the privilege or confidentiality of the environmental audit report. Failure by the owner or operator to file the petition will waive the privilege.

Upon filing a petition, the court must within thirty days after filing the petition issue an order scheduling an in camera hearing to determine if the environmental audit report is privileged under this section or subject to disclosure. The order must allow the district attorney or the attorney general to review the report and must place appropriate limitations on distribution and review of the report to protect against unnecessary disclosure. The district attorney or the attorney general may consult with appropriate law enforcement agencies regarding the contents of the report as necessary to prepare for the in camera hearing.

Information used in preparation for the in camera hearing cannot be used in any investigation or in any proceeding against the defendant and must otherwise be kept confidential, unless the information is found by the court to be subject to disclosure. Upon making a determination pursuant to an in camera review, the court may compel disclosure of those portions of an environmental audit report relevant to issues in dispute in the proceeding.

The parties may at any time stipulate to entry of an order directing whether specific information contained in an environmental audit report is subject to the environmental audit privilege.

If an owner or operator of a facility regulated under this act voluntarily reports to the department a violation disclosed by the audit conducted under Wyo. Stat. § 35-11-1105 within sixty days of the completion date of the audit, the department must not seek civil penalties or injunctive relief for the violation reported,[174] unless one of four criteria are found:

(i) The facility is under investigation for any violation of this act at the time the violation is reported;

(ii) The owner or operator does not take action to eliminate the violation within the time frame specified in an order affirmed by the council or otherwise made final pursuant to Wyo. Stat. § 35-11-701(c)(ii);

(iii) The violation is the result of gross negligence or recklessness; or

(iv) The department has assumed primacy over a federally delegated environmental law and a waiver of penalty authority would result in a state program less stringent than the federal program or the waiver would violate any federal rule or regulation required to maintain state primacy.[175]

---

[174] § 35-11-1106(a) (Limitation on civil penalties; voluntary reports of violations).

[175] If a federally delegated program requires the imposition of a penalty for a violation, the voluntary disclosure of the violation will, to the extent allowed under federal law or regulation, be considered a mitigating factor in determining the penalty amount.

The penalty immunity provisions of the law do not prevent a cause of action for injunctive relief under Wyo. Stat. § 35-11-115.

Reporting a violation is mandatory if required by state law, any departmental rule or regulation, federal law or regulation, local ordinance or resolution, any order of the council, or by any court. Thus, mandatory reporting is not considered "voluntary" for purposes of the environmental audit privilege law.

The elimination of administrative or civil penalties under the environmental audit privilege law does not apply if a person or entity has been found by a court to have committed serious violations that constitute a pattern of continuous or repeated violations of environmental laws, rules, regulations, permit conditions, settlement agreements, or orders of consent and that were due to separate and distinct events giving rise to the violations, within the three-year period prior to the date of the disclosure. A pattern of continuous or repeated violations may also be demonstrated by multiple settlement agreements related to substantially the same alleged violations concerning serious instances of noncompliance with environmental laws that occurred within the three-year period immediately prior to the date of the voluntary disclosure.

§ 5.4 Pending Federal Legislation

Representatives Hyde (R-Ill.), Hefley (R-Colo.), and DeLay (R-Tex.) introduced environmental audit privilege legislation into the House of Representatives on February 24, 1995.

Senator Mark Hatfield (R-Ore.) is expected to reintroduce a version of S.2371 in Congress along with Senator Brown (R-Colo.) as a cosponsor.

§ 5.5 Other State Audit Legislation

Some states have enacted legislation that encourages environmental audits but does not expressly provide an audit privilege. Below are brief descriptions of some of these statutes.

§ 5.5.1 California

California provides certain financial incentives for those who conduct audits. For example, California law makes certain loan programs available to owners of USTs that need to upgrade their tanks.[176] In order to make application for such financing an applicant must complete a loan application[177] which includes an environmental audit.

Other means the California legislature has used to encourage environmental audits include endorsing the use of environmental quality assessments,[178] which are also called environmental audits, and recommending "that state and local governments initiate programs to encourage the broader use of the assessment process, and while the state should take steps to actively encourage the private sector use of environmental assessments, it is not the intention of the Legislature to require that any person or business utilize environmental assessors."[179] The legislature explained:

> Environmental assessments can encourage voluntary compliance with both the letter and the spirit of the law as well as encourage cost-effective process improvements. By reducing potential liability, assessments can reduce the long-term costs of hazardous substance management. In addition, the use of assessments can help to build public confidence that hazardous substances are being managed in an increasingly safe manner. The use of independent environmental assessments is an important emerging feature of specific state hazardous substance management programs.[180]

---

[176] CAL. GOV'T CODE § 15399.13 (1993) (Operative until January 1, 1998).

[177] *Id.* The statutes requires three of the following must be provided:

(a) Evidence of eligibility.

(b) An environmental audit, as specified in Section 5268 of Title 10 of the California Code of Regulations.

(c) Financial and legal documents necessary to demonstrate the applicant's ability to repay and provide collateral for the loan. The department shall develop a standard list of documents required of all applicants, and may also request from individual applicants additional financial and legal documents not provided on this list.

(d) An explanation of the reasons the project tank is not in compliance with applicable local, state, or federal standards, and evidence that tanks not included in the list of project tanks are currently in compliance with applicable local, state, or federal standards.

(e) A detailed cost estimate of the tasks which are required to be completed in order for the project tanks to comply with applicable local, state, or federal standards.

(f) Any other information which the department determines to be necessary to include in an application form.

[178] CAL. HEALTH & SAFETY CODE § 25570(b)(1)(1993).

[179] CAL. HEALTH & SAFETY CODE § 25570 (1993).

[180] *Id.* (2).

The legislature noted that "many of California's major businesses have internal environmental assessment programs. Larger firms often maintain in-house staffs."[181] Thus, the legislature directed the state agencies to "provide a list of registered independent third-party assessors for use by small- and medium-sized firms seeking technical assistance to achieve and maintain regulatory compliance."[182]

The California legislature then created a modified self-evaluation privilege that states that:

> . . . . Notwithstanding any other provision of law, the findings of any environmental audit or assessment conducted by or for a small business pursuant to this section shall remain the property of the business.[183]

On February 16, 1996, California issued a new proposed policy on environmental audits that was designed to expand and clarify the California law with the new U.S. EPA guidance.[184] The California EPA issued its final policy on Incentives for Self Evaluation (Policy) on July 8, 1996. Designed to encourage due diligence and self-policing, self-disclosure, and self-correction, the policy provides a variety of incentives for regulated entities that are "good actors" and meet specific conditions. Incentives include no gravity-based penalties, reduction of gravity-based penalties by 75%, no criminal recommendations, and no routine request for audits.

The California EPA Policy, however, goes beyond the U.S. EPA audit policy by offering a fee-for-service certification program. This program would allow companies, for a fee, to have California EPA conduct a review of the corporation's audit or due diligence program and certify it as meeting the requirements of the policy. This addition is designed to reduce uncertainty many corporations feel concerning use of internal audits for prosecution.

---

181 *Id.* (3).

182 *Id.* (4).

183 *Id.* The definition of "small business," for the purposes of this section, was the same as that of the federal Small Business Administration.

184 *See generally California Unveils Draft Audit Policy That Expands On, Clarifies U.S. EPA Approach,* 5 Due Diligence Guide (BNA), at 19-20 (March 1996).

§ 5.5.2 New Jersey[185]

New Jersey law protects a foreclosing lender from liability under its state equivalent of CERCLA by allowing, but not requiring, the foreclosing lender to perform an environmental audit, which must then be forwarded to the Department of Environmental Protection. Upon receipt of the audit, the department must, within ninety days, transmit its finding to the holder. The department can charge "reasonable fees and adopt any additional regulations necessary to provide guidelines for the submission and review of such audits."

§ 5.5.3 New Mexico[186]

New Mexico law defines "environmental audit" as "a systematic assessment, analysis and evaluation by a regulated entity[187] of its compliance with environmental laws and regulations" administered by the environmental improvement board[188] and the environmental improvement division of the health and environment department,[189] "applicable to its operation."[190]

The law requires that no regulations may be adopted pursuant to the Environmental Compliance Act until after a public hearing is held.[191] For purposes of this law, "regulation" is defined as including

> any amendment or repeal thereof. Hearings on regulations shall be held pertaining to that environmental area which is substantially affected by the regulation. In making a regulation, the board shall give the weight it deems appropriate to all relevant facts and circumstances presented at the hearing, including but not limited to:

---

[185] 1993 N.J. Laws 112.

[186] N.M. Stat. Ann. § 74-7-3 (1993).

[187] "Regulated entity" means any person, partnership, corporation, firm, association, governmental or other entity organized and engaging in any business or activity in the state which deals with or has an impact on the environment of this state or which must by law comply with federal or state environmental protection regulations. *Id.* § 74-7-3(E).

[188] *Id.*

[189] *Id.* § 74-7-3(C).

[190] *Id.* § 74-7-3(D).

[191] N.M. Stat. Ann. § 74-7-5 (1993).

1. the protection of the health and welfare of both the general public and the individual worker and the maintenance of the delicate ecological balance;
2. the necessity for and technical practicability and economic reasonableness of taking action with respect to environmental auditing programs;
3. the need to protect private proprietary processes;
4. the level of management support within the specific regulated entity for the environmental auditing program;
5. a regulated entity's established procedures to ensure compliance and correction of any environmental standards that are violated; and
6. compliance with the requirements of the following federal laws and their associated standards, regulations and state implementing directives:
   (a) the National Environmental Policy Act of 1969;
   (b) the Federal Water Pollution Control Act;
   (c) the Safe Drinking Water Act;
   (d) the Resource Conservation and Recovery Act of 1976;
   (e) the Used Oil Recycling Act of 1980;
   (f) the Clean Air Act;
   (g) the Toxic Substances Control Act;
   (h) the Occupational Safety and Health Act of 1970;
   (I) the Noise Control Act of 1972;
   (j) the Hazardous Materials Transportation Act; and
   (k) the Comprehensive Environmental Response, Compensation and Liability Act of 1980.[192]

---

[192] Notice of the hearing shall be given at least thirty days prior to the hearing date and shall state the subject, time and place of the hearing and the manner in which interested persons may present their views. The notice shall state where interested persons may secure copies of any proposed regulation. The notice shall be published in a newspaper of general circulation in the area affected. Reasonable effort shall be made to give notice to all persons who have made a written request to the board for advance notice of hearings. Id. § 74-7-5(B).

At the hearing, the board shall allow all interested persons reasonable opportunity to submit data, views or arguments, orally or in writing, pertaining to the feasibility of conducting environmental audits. *Id.* § 74-7-5(C)

No regulation or amendment or repeal thereof adopted by the board shall become effective until thirty days after its filing pursuant to the State Rules Act [14-3-24, 14-3-25, 14-4-1 to 14-4-9 NMSA 1978]. *Id.* § 74-7-5(D)

Any person who is affected by a regulation adopted by the board may appeal to the court of appeals for further relief. All appeals shall be upon the transcript made at the hearing

§ 5.5.4 Pennsylvania

Section 507(a) of the CAA sets forth seven requirements that states must meet to have an approvable small business assistance plan (SBAP).[193] Two of these requirements require that the state consider environmental audit protocols in the establishment of the SBAP programs. Pennsylvania agreed to meet the second of these seven requirements through the outreach and audit programs. According to the state, pamphlets will be distributed containing information regarding accidental release prevention and pollution prevention. In addition, pollution prevention and accidental release information will be provided during onsite audits, which may be requested by the small businesses.[194]

In addition, the fifth requirement (to develop adequate mechanisms for informing small business stationary sources of their obligations under the CAA, including mechanisms for referring such sources to qualified auditors or, at the option of the state, for providing audits of the operations of such sources to determine compliance with the CAA) was met by the state of Pennsylvania by planning to maintain a toll-free telephone line to allow easy access to information regarding federal and/or state requirements.[195] In meeting this requirement, the state will also inform affected small businesses, in a timely manner by certain proactive mechanisms, and will provide material, through the outreach portion of the program on environmental audits, to assist small businesses in meeting the requirements of CAA. Under the plan in Pennsylvania, the "environmental audit will determine applicable requirements, compliance status, control options and pollution prevention alternatives."[196]

---

and shall be taken to the court of appeals within thirty days after filing of the regulation pursuant to the State Rules Act. *Id.* § 74-7-5(E).

[193] Sections 7.7 through 7.9 of the 1992 Pennsylvania Air Pollution Control Act, authorize the establishment of a Small Business Assistance Program which meets the requirements of section 507 of the CAA. In developing the Program submittal, the Commonwealth delegated the majority of its functions to the Department of Environmental Resources (DER).

[194] 60 Fed. Reg. 1738 (Jan. 5, 1995) (Approval and Promulgation of Air Quality Implementation Plans; Commonwealth of Pennsylvania Small Business Assistance Program) (Final rule).

[195] 60 Fed. Reg. 1738, 1739.

[196] *Id.* at 1740.

§ 5.5.5 Rhode Island[197]

Rhode Island law states that in order to incur owner or operator liability under the state Superfund act the person must "participate . . . in the management." The statute defines "participating" or "participate" "in the management of a dwelling, dwelling unit or child care facility" as the actual participation by a holder in the management or operational affairs of a dwelling, dwelling unit, or child care facility, without limitation where a holder exercises control at a level comparable to that of a manager of the enterprise with responsibility for day-to-day decision making with respect to all or substantially all of the operational (as opposed to financial or administrative) aspects of the dwelling, dwelling unit, or child care facility.

According to the legislature, conducting, or requiring the borrower to conduct, an environmental assessment or audit of the dwelling, dwelling unit, or child care facility will not, in and of itself, constitute participation in the management of a dwelling, dwelling unit, or child care facility.

[197] R. I. GEN. LAWS § 130-A:1 (1993).

Chapter 6

Environmental Self-Regulatory Organizations and the Environmental Report Card

§ 6.1 Overview of the Environmental Report Card Concept

The concept of companies voluntarily producing an "environmental report card" that they disclose to the public has become increasingly popular.[1] Some of these programs require that management complete the report card. Others require that outside auditors do the environmental evaluation.

[1] For a discussion of consumer reliance on voluntary codes of environmental and other social corporate conduct, *see* James S. Hoyte, *Comments on Safoff's Environmentalism As Populism, in* W. MICHAEL HOFFMAN ET AL., BUSINESS ETHICS AND THE ENVIRONMENT: THE PUBLIC POLICY DEBATE (1990).

Regardless of the mechanisms used to produce the report card, these programs have had an important impact on increasing the level of environmental awareness and compliance on the part of corporate America.[2] Such programs encourage environmental vigilance beyond that which is required by law or regulation.[3] These programs have become a very important market-based incentive for enhanced environmental care.[4]

---

[2] Not everyone thinks the corporate world needs another set of reporting guidelines. Many groups already offer advice on reporting environmental information, including Business Council for Sustainable Development, Coalition for Environmentally Responsible Economies, Council on Economic Priorities, World Industry Council for the Environment, Investor Responsibility Research Center, and Global Environmental Management Initiative.

[3] *See generally* MICHAEL SILVERSTEIN, THE ENVIRONMENTAL ECONOMIC REVOLUTION: HOW BUSINESS WILL THRIVE AND THE EARTH SURVIVE IN YEARS TO COME (1993); WALTER CODDINGTON, ENVIRONMENTAL MARKETING: POSITIVE STRATEGIES FOR REACHING THE GREEN CONSUMER (1993); ROBERT REHAK, GREENER MARKETING AND ADVERTISING: CHARTING A RESPONSIBLE COURSE (1993); THEODORE PANAYOTOU, GREEN MARKET: THE ECONOMICS OF SUSTAINABLE DEVELOPMENT (1993); FRANK B. FREIDMAN, PRACTICAL GUIDE TO ENVIRONMENTAL MANAGEMENT (1991); BRUCE SMART, BEYOND COMPLIANCE: A NEW INDUSTRY VIEW OF THE ENVIRONMENT (1992); DONELLA H. MEADOWS ET AL., BEYOND THE LIMITS: CONFRONTING GLOBAL COLLAPSE, ENVISIONING A SUSTAINABLE FUTURE (1992); STEPHAN SCHMIDHEINY, CHANGING COURSE: A GLOBAL PERSPECTIVE ON DEVELOPMENT AND THE ENVIRONMENT (1992); MANAGING THE GLOBAL ENVIRONMENTAL CHALLENGE (1992); FRANCES CAIRCROSS, COSTING THE EARTH: ECOLOGY AND THE HUMAN SPIRIT (1992); ENVIRONMENTAL DIVIDENDS: CUTTING MORE CHEMICAL WASTES (1992); AL GORE, JR., EARTH IN THE BALANCE (1991); ROBERT COSTANZA, ECOLOGICAL ECONOMIES: THE SCIENCE AND MANAGEMENT OF SUSTAINABILITY (1991); PATRICK CARSON ET AL., GREEN IS GOLD: BUSINESS TALKING TO BUSINESS ABOUT THE ENVIRONMENTAL REVOLUTION (1991); JOEL S. HIRSHHORN, PROSPERITY WITHOUT POLLUTION: THE PREVENTION STRATEGY FOR INDUSTRY AND CONSUMERS (1991); JOHN DAVIS, GREENING BUSINESS: MANAGING FOR SUSTAINABLE DEVELOPMENT (1991); BRUCE PIASECKI, IN SEARCH OF ENVIRONMENTAL EXCELLENCE (1990); CHERYL SIMON SILVER, ONE EARTH ONE FUTURE: OUR CHANGING GLOBAL ENVIRONMENT (1990). *See also* CONRAD B. MACKERRON, BUSINESS IN THE RAIN FORESTS: CORPORATIONS, DEFORESTATION AND SUSTAINABILITY (1993); DOUGLAS COGAN, THE GREENHOUSE GAMBIT, BUSINESS AND INVESTMENT RESPONSES TO CLIMATE CHANGE (1992).

[4] *See* Nazli Choucri, MULTINATIONAL CORPORATIONS AND THE GLOBAL ENVIRONMENT, *in* CHOUCRI NAZLI ET AL., GLOBAL ACCORD 208 (MIT 1993): "We begin with three facts: (1) Since multinational corporations conduct most of the world's economic activity, they are the major environmental actors as producers, managers, and distributors of goods and services; (2) by necessity, these firms are generic polluters as they engage in a wide range of hazardous and pollution-intensive activities; and (3) corporations are also central to the 'solution.'"

Among industry's voluntary codes of conduct[5] are

✓ the Business Charter for Sustainable Development, prepared by the International Chamber of Commerce in 1990;[6]

✓ the Coalition for Environmentally Responsible Economies (CERES) principles;[7]

✓ the Global Environmental Management Initiative (GEMI);[8]

✓ the Public Environmental Reporting Initiative (PERI) guidelines;

✓ the Business for Social Responsibility (BSR);[9]

✓ the Chemical Manufacturers Association (CMA) Responsible Care program;[10]

✓ the International Standards Organization (ISO) standards;

✓ Green Seal's Environmental Partners Program;[11]

---

[5] These industry run programs should be distinguished from the EPA, DOI, and DOE voluntary programs in which industry agrees to assume voluntary regulatory oversight in specified areas.

[6] This organization requires that members have greater openness and the periodic provision of "appropriate information" to stakeholders.

7 CERES requires signatory companies to complete a standardized annual report on their environmental performance. *See generally* ERNEST CALLENBACH ET AL., ECOMANAGEMENT: THE ELMWOOD GUIDE TO ECOLOGICAL AUDITING AND SUSTAINABLE BUSINESS (1993).

8 GEMI was created in 1990 to foster worldwide environmental excellence. Initial member corporations included: Procter & Gamble, Du Pont, AT&T, Monsanto, Digital Equipment, Merck, Boeing, and Dow Chemical. At its 1993 annual meeting, GEMI members presented environmental report cards on their respective corporate performance. The report cards were designed to identify areas in which companies were in need of improvement, individually, by industry sector, and collectively. *See* SUZANNE T. THOMAS, *Pollution Prevention and Waste Minimization* (BNA 1995) at 61, citing E. L. Grayson and E. G. Reyna, *Pollution Prevention, The Prescription for Business*, HAZARDOUS MATERIALS CONTROL 12 (May 1993). *See also* BRUCE W. PIASECKI, CORPORATE ENVIRONMENTAL STRATEGY: THE AVALANCHE OF CHANGE SINCE BHOPAL (Wiley 1995).

[9] For a detailed discussion of this program see JOEL MAKOWER, BEYOND THE BOTTOM LINE: PUTTING SOCIAL RESPONSIBILITY FOR YOUR BUSINESS AND THE WORLD (1994).

[10] Subscription to Responsible Care principles is now a condition of membership in CMA.

[11] Green Seal is an independent nonprofit organization promoting the manufacture and sale of environmentally preferable consumer products. The purpose of the Environmental Partners Program is to "promote environmentally responsible business practices, including the purchase, sale, and use of products and services which are less damaging to the environment." Environmental Partners must agree to purchase Green Seal certified products and other products less damaging to the environment. This program

✓ the American Petroleum Institute's Strategies for Today's Environmental Partnership (STEP);

✓ the European Chemical Industry Council's (CEFIC) sectoral environmental reporting guidelines;[12]

✓ the Confederation of British Industry (CBI).[13]

EPA also has undertaken numerous voluntary programs that integrate industry voluntarism with regulatory oversight. These programs include, *inter alia:*

✓ EPA's 33/50 Program and

✓ EPA's Environmental Leadership Program.[14]

Within the current climate of voluntary initiatives[15] and proliferation of self-regulatory organizations focusing on environmental compliance matters,[16] a number of companies have begun to produce free-standing environmental reports or to feature environmental information in their annual reports. This practice is increasing in countries such as Germany[17] and Canada. In many countries, however, environmental disclosure is not at all common. For example, in Singapore less than one-fifth of the companies whose annual reports were reviewed contained environmental reporting.

---

requires companies to recycle office materials, to assist Green Seal in setting environmental product standards, to appoint an Environmental Partners Liaison, and to contribute to an annual report that is released to the public.

[12] This group established sectoral environmental reporting guidelines that are to be followed by the majority of chemical companies in producing reports.

[13] This group also produced guidelines for public disclosure of environmental information.

[14] See § 6.9 and accompanying text for a more detailed discussion of EPA's voluntary cleanup programs.

[15] *See* Suzanne T. Thomas, *Pollution Prevention and Waste Minimization,* at 23 (BNA 1995) (In the future there will be attempts to reduce regulation by encouraging the regulated community to voluntarily prevent pollution at source).

[16] *See* Bruce W. Piasecki, Corporate Environmental Strategy: The Avalanche of Change Since Bhopal (Wiley 1995); Joel Makower, The E Factor (Tilden Press 1993); Frank Freidman, The Practical Guide to Environmental Management (ELI 1992); Michael Baram, Managing Chemical Risks (Tufts Press 1992).

[17] Although, as in the United States, the practice is more common among the larger corporations.

Similarly, in Japan, not a single member company of the Keidanren—the country's largest and most powerful economic lobby—issued an environmental report.

Given the proliferation of different reporting guidelines and formats, the information reported by companies is still largely unstandardized and, thus, incomparable between firms. To address this, a common core of standard reporting elements has been called for, "which would go some way towards improving the comparability and consistency of information, while leaving companies the flexibility to innovate and put their own hallmark on their environmental reports."[18]

One of the critical questions in selecting a written environmental policy is which green programs are most appropriate for the registrant? One of those listed above? Or is a program designed solely for the individual registrant going to be most effective?

§ 6.2 The Coalition for Environmentally Responsible Economies (CERES)

On September 7, 1989, CERES proposed a voluntary corporate code of conduct then titled the Valdez Principles.[19] The Valdez Principles asked corporations to go beyond mere compliance with minimum environmental standards set by federal, state, and local governments and aggressively protect the environment.[20] The audit and disclosure provisions of the Valdez Principles are particularly controversial because they potentially expose signatory corporations to increased litigation, increased costs, and disclosure liability beyond that required by federal and state law.[21]

Disclosure is a key tenet of the CERES principles. Joan Bavaria, CEO and cochair of CERES, states that "[w]hile CERES is a program of continuous future improvement toward that elusive goal of environmental

---

[18] *See* Special Report, *Corporate Environmental Reporting: Embraced or Resisted?*, INT'L ENV'T DAILY (BNA) (Apr. 20, 1994).

[19] Coalition for Environmentally Responsible Economies [hereinafter CERES], Valdez Principles (Sept. 7, 1989) [hereinafter Valdez Principles].

[20] *See id.*

[21] *See, e.g.*, Barnaby J. Feder, *Who Will Subscribe to the Valdez Principles?*, N.Y. TIMES, Sept. 10, 1989, § 3, at 6, col. 1.

management perfection, it is not a certification of corporate environmental policies." She emphasizes that "CERES affiliates are sticking their necks out" by making a public pledge to improve steadily over time and report annually on their progress. "CERES companies want to be leaders, and as leaders, they will be visible, vulnerable, and judged." One endorsing company calls it "living in a glass house."[22]

The Valdez Principles are premised on an intent by corporations to make consistent, measurable progress in taking responsibility for the environment.[23] Since the standard of responsibility set under the Principles is to attain continuous progress rather than consistent compliance, there is no inherent limitation on corporate environmental responsibility for companies adopting the Principles.

The Valdez Principles were developed by a proactive investor group.[24]

---

[22] *See Are Voluntary Environmental Charters Effective?*, 9 BUSINESS & THE ENVIRONMENT (September 1994) (quoting Bavaria in the Summer 1994 issue of the CERES newsletter "On Principle").

[23] *See generally* Valerie Ann Zondorak, *A New Face In Corporate Environmental Responsibility: The Valdez Principles,* 18 B.C. ENVTL. AFF. L. REV. 457 (1991).

[24] Numerous no-action letters have discussed shareholder proposals concerning the Valdez principles. *See, e.g.*, The Procter & Gamble Company, 1994 SEC No-Act. LEXIS 609 (July 15, 1994) (Securities Exchange Act of 1934, Rule 14a-8(c)(10)); Sigma-Aldrich Corporation, 1994 SEC No-Act. LEXIS 300 (March 4, 1994) (Securities Exchange Act of 1934, Rule 14a-8); Paul M. Neuhauser, RECONSIDERATION, 1994 SEC No-Act. LEXIS 219 (February 16, 1994) (Securities Exchange Act of 1934, Rule 14a-8); Amoco Corporation, 1994 SEC No-Act. LEXIS 124 (January 31, 1994) (Securities Exchange Act of 1934, Rule 14a-8); International Business Machines Corporation, 1994 SEC No-Act. LEXIS 141 (January 31, 1994) (Securities Exchange Act of 1934, Rule 14a-8); RJR Nabisco, Inc., 1994 SEC No-Act. LEXIS 97 (January 21, 1994) (Securities Exchange Act of 1934, Rule 14a-8); UJB Financial Corp., 1994 SEC No-Act. LEXIS 14 (January 6, 1994) (Securities Exchange Act of 1934, Rule 14a-8); International Paper Company, 1993 SEC No-Act. LEXIS 1188 (December 22, 1993) (Securities Exchange Act of 1934, Rule 14a-8); Occidental Petroleum Corporation, 1993 SEC No-Act. LEXIS 338 (February 25, 1993) (Securities Exchange Act of 1934, Rule 14a-8); Bristol Myers Squibb Corporation, 1993 SEC No-Act. LEXIS 139 (February 3, 1993) (Securities Exchange Act of 1934, Rule 14a-8); ITT Corporation, 1993 SEC No-Act. LEXIS 149 (February 3, 1993) (Securities Exchange Act of 1934, Rule 14a-8); Pacific Telesis Group, 1993 SEC No-Act. LEXIS 176 (February 1, 1993)(Securities Exchange Act of 1934, Rule 14a-8(c)(11)); Eastman Kodak Company, 1993 SEC No-Act. LEXIS 82 (January 27, 1993) (Securities Exchange Act of 1934, Rule 14A-8); Eastman Kodak Company, 1993 SEC No-Act. LEXIS 97 (January 27, 1993) (Securities Exchange Act of 1934, Rule 14a-8); Eastman Kodak Company, 1993 SEC

No-Act. LEXIS 100 (January 27, 1993) (Securities Exchange Act of 1934, Rule 14a-8); Eastman Kodak Company, 1993 SEC No-Act. LEXIS 102 (January 27, 1993) (Securities Exchange Act of 1934, Rule 14a-8; R.R. Donnelley & Sons Company, 1993 SEC No-Act. LEXIS 78 (January 26, 1993)(Securities Exchange Act of 1934, Rule 14a-8(c)(7)); McDonald's Corporation, 1993 SEC No-Act. LEXIS 80 (January 26, 1993)(Securities Exchange Act 1934, Rule 14a-8(c)(3)); Louisiana-Pacific Corporation, 1993 SEC No-Act. LEXIS 85 (January 26, 1993) (Securities Exchange Act of 1934,Rule 14a-8); Homestake Mining Company, 1993 SEC No-Act. LEXIS 91 (January 26, 1993) (Securities Exchange Act of 1934, Rule 14a-8); Eastman Kodak Company, 1993 SEC No-Act. LEXIS 104 (January 26, 1993) (Securities Exchange Act of 1934, Rule 14a-8); Texas Instruments Incorporated, 1993 SEC No-Act. LEXIS 112 (January 26, 1993) (Securities Exchange Act of 1934, Rule 14a-8); American Express Company, 1993 SEC No-Act. LEXIS 44 (January 13, 1993) (Securities Exchange Act of 1934, Rule 14a-8); Exxon Corporation, 1993 SEC No-Act. LEXIS 34 (January 11, 1993) (Securities Exchange Act of 1934, Rule 14a-8); Westinghouse Electric Corporation, 1993 SEC No-Act. LEXIS 2 (January 4, 1993) (Securities Exchange Act of 1934, Rule 14a-8); Archer Daniels Midland Company, 1992 SEC No-Act. LEXIS 848 (August 11, 1992) (Securities Act of 1934, Rule 14a-8); McDonald's Corporation, 1992 SEC No-Act. LEXIS 489 (April 3, 1992) (Securities Exchange Act of 1934, Rule 14a-8); Union Carbide Corporation, 1992 SEC No-Act. LEXIS 469 (March 27, 1992) (Securities Exchange Act of 1934, Rule 14a-8); PepsiCo, Inc., 1992 SEC No-Act. LEXIS 479 (March 25, 1992) (Securities Exchange Act of 1934, Rule 14a-8); Newmont Mining Corporation, 1992 SEC No-Act. LEXIS 409 (March 23, 1992) (Securities Exchange Act of 1934, Rule 14a-8); Texaco, Inc., 1992 SEC No-Act. LEXIS 356 (March 13, 1992) (Securities Exchange Act of 1934, Rule 14a-8); Chevron Corporation, 1992 SEC No-Act. LEXIS 338 (March 11, 1992) (Securities Exchange Act of 1934—Rule 14a-8); Burlington Resources, Inc., 1992 SEC No-Act. LEXIS 322 (March 9, 1992) (Securities Exchange Act of 1934, Rule 14a-8); Weyerhaeuser Company, 1992 SEC No-Act. LEXIS 325 (March 6, 1992) (Securities Exchange Act of 1934, Rule 14a-8); American Electric Power Company, 1992 SEC No-Act. LEXIS 309 (March 3, 1992) (Securities Exchange Act of 1934, Rule 14a-8); United Technologies Corporation, 1992 SEC No-Act. LEXIS 265 (February 25, 1992) (Securities Exchange Act of 1934, Rule 14a-8); Eastman Kodak Company, 1992 SEC No-Act. LEXIS 213 (February 19, 1992) (Securities Exchange Act of 1934, Rule 14a-8); Eastman Kodak Co., 1992 SEC No-Act. LEXIS 172 (February 10, 1992) (Securities Exchange Act of 1934, Rule 14a-8); B.M. Squibb Co., 1992 SEC No-Act. LEXIS 138 (February 4, 1992)(Securities Exchange Act of 1934—Rule 14a); 1992 SEC No-Act. LEXIS 72, Securities Exchange Act of 1934—Rule 14a-8, January 17, 1992, American Telephone and Telegraph Company; 1992 SEC No-Act. LEXIS 33, Securities Exchange Act of 1934—Rule 14a-8, January 14, 1992, Eastman Kodak Company; 1992 SEC No-Act. LEXIS 39, Securities Exchange Act of 1934— Rule 14a-8, January 14, 1992, E. I. Du Pont De Nemours and Company; 1991 SEC No-Act. LEXIS 1361, Securities Exchange Act of 1934—Rule 14a-8, December 12, 1991, Emerson Electric Co.; 1991 SEC No-Act. LEXIS 884, Securities Exchange Act of 1934—Rule 14a-8, July 12, 1991, Procter & Gamble Company; 1991 SEC No-Act. LEXIS 578, Securities Exchange Act of 1934—Rule 14a-8, April 4, 1991, Champion International Corporation; 1991 SEC No-Act. LEXIS 583, Securities Exchange Act of 1934—Rule 14a-8, April 4, 1991, McDonald's Corporation; 1991 SEC No-Act. LEXIS 496, Securities Exchange Act of 1934—Rule 14a-8, March 28, 1991, Houston Industries Incorporated; 1991 SEC No-Act. LEXIS 500, Securities Exchange Act of 1934—Rule 14a-8, March 28, 1991, Texaco, Inc.; 1991 SEC

No-Act. LEXIS 477, Securities Exchange Act of 1934—Rule 14a-8, March 15, 1991, McDonald's Corporation; 1991 SEC No-Act. LEXIS 490, Securities Exchange Act of 1934—Rule 14a-8, March 13, 1991, Newmont Mining Corporation; 1991 SEC No-Act. LEXIS 424, Securities Exchange Act of 1934—Rule 14a-8, March 8, 1991, Amoco Corporation; 1991 SEC No-Act. LEXIS 432, Securities Exchange Act of 1934—Rule 14a-8, March 8, 1991, E. I. Du Pont De Nemours and Company; 1991 SEC No-Act. LEXIS 377, Securities Exchange Act of 1934—Rule 14a-8, March 6, 1991, The Dow Chemical Company; 1991 SEC No-Act. LEXIS 404, Securities Exchange Act of 1934—Rule 14a-8, March 6, 1991, Tribune Company; 1991 SEC No-Act. LEXIS 415, Securities Exchange Act of 1934—Rule 14a-8, March 6, 1991, Champion International Corporation; 1991 SEC No-Act. LEXIS 430, Securities Exchange Act of 1934—Rule 14a-8, March 6, 1991, Texaco, Inc.; 1991 SEC No-Act. LEXIS 400, Securities Exchange Act of 1934—Rule 14a-8, March 4, 1991, Champion International Corporation; 1991 SEC No-Act. LEXIS 384, Securities Exchange Act of 1934—Rule 14a-8, March 1, 1991, Amoco Corporation; 1991 SEC No-Act. LEXIS 397, Securities Exchange Act of 1934—Rule 14a-8, March 1, 1991, Exxon Corporation; 1991 SEC No-Act. LEXIS 347, Securities Exchange Act of 1934—Rule 14a-8(c)(7), February 27, 1991, E. I. Du Pont De Nemours and Company; 1991 SEC No-Act. LEXIS 382, Securities Exchange Act of 1934—Rule 14a-8, February 27, 1991, Eastman Kodak Company; 1991 SEC No-Act. LEXIS 393, Securities Exchange Act of 1934—Rule 14a-8, February 27, 1991, Eastman Kodak Company; 1991 SEC No-Act. LEXIS 412, Securities Exchange Act of 1934—Rule 14a-8, February 27, 1991, Chevron Corporation; 1991 SEC No-Act. LEXIS 202, Securities Exchange Act of 1934—Rule 14a-8(a)(3)(i), February 13, 1991, Chevron Corporation; 1991 SEC No-Act. LEXIS 212, Securities Exchange Act of 1934—Rule 14a-8(a)(3)(i), February 13, 1991, Chevron Corporation; 1991 SEC No-Act. LEXIS 213, Securities Exchange Act of 1934—Rule 14a-8(a)(3)(i), February 13, 1991, Chevron Corporation; 1991 SEC No-Act. LEXIS 250, Securities Exchange Act of 1934—Rule 14a-8(a)(3), February 13, 1991, ARCO; 1991 SEC No-Act. LEXIS 142, Securities Exchange Act of 1934—Rule 14a-8, January 25, 1991, General Electric Company; 1991 SEC No-Act. LEXIS 146, Securities Exchange Act of 1934—Rule 14a-8, January 25, 1991, Union Carbide Corporation; 1991 SEC No-Act. LEXIS 107, Securities Exchange Act of 1934—Rule 14a-8, January 22, 1991, The New York Times Co.; 1991 SEC No-Act. LEXIS 30, Securities Exchange Act of 1934—Rule 14a-8, January 4, 1991, Washington Post Co.; 1991 SEC No-Act. LEXIS 24, Securities Exchange Act of 1934—Rule 14a-8, January 3, 1991, American Electric Power; 1991 SEC No-Act. LEXIS 26, Securities Exchange Act of 1934—Rule 14a-8, January 3, 1991, IBM The Valdez Principles; 1990 SEC No-Act. LEXIS 1358, Securities Exchange Act of 1934—Rule 14a-8, December 26, 1990, The Mead Corp.; 1990 SEC No-Act. LEXIS 1316, Securities Exchange Act of 1934—Rule 14a-8, December 18, 1990, American Electric Power Co.; 1990 SEC No-Act. LEXIS 1167, Securities Exchange Act of 1934—Rule 14a-8, October 5, 1990, Centerior Energy Corp.; 1990 SEC No-Act. LEXIS 670, Securities Exchange Act of 1934—Rule 14a-8, April 18, 1990, Mobil Corporation; 1990 SEC No-Act. LEXIS 494, Securities Exchange Act of 1934—Rule 14a-8, March 19, 1990, Eastman Kodak Company; 1990 SEC No-Act. LEXIS 454, Securities Exchange Act of 1934—Rule 14a-8, March 13, 1990, Kimberly-Clark Corp.; 1990 SEC No-Act. LEXIS 476, Securities Exchange Act of 1934—Rule 14a-8, March 13, 1990, Aetna Life & Casualty Co.; 1990 SEC No-Act. LEXIS 452, Securities Exchange Act of 1934—Rule 14a-8, March 12, 1990, Sun Co., Inc.; 1990 SEC No-Act. LEXIS 477, Securities Exchange Act of 1934—Rule 14a-8, March 12, 1990, Sun Co., Inc.; 1990 SEC No-Act. LEXIS 387, Securities Exchange Act of 1934—Rule 14a-8, February 21,

As such, the philosophy of the Principles is not merely to respond to the present environmental situation, but to nurture an environmentally responsible economy. For example, CERES intends that corporate signatories will strive to use the best available technology (BAT) in combating environmental pollution from their facilities.[25] Use of BAT is currently required in only specific circumstances under the Clean Air Act, and not at all under most other environmental statutes.

Although the Valdez Principles have been heralded as "a new and aggressive approach to corporate environmental responsibility," the success of the Valdez Principles turns on whether a voluntary code can motivate corporations to shoulder a burden beyond what is legally required. Similarly, the success of the Valdez Principles also depends on whether the disclosure and audit provisions of the Principles can be drafted in a manner to overcome present corporate objections.[26]

1990, Union Pacific Corp.; 1990 SEC No-Act. LEXIS 317, Securities Exchange Act of 1934—Rule 14a-8, February 16, 1990, Dow Chemical Co.; 1990 SEC No-Act. LEXIS 320, Securities Exchange Act of 1934—Rule 14a-8, February 16, 1990, Dow Chemical Co.; 1990 SEC No-Act. LEXIS 262, Securities Exchange Act of 1934—Rule 14a-8, February 9, 1990, Union Carbide Corp.; 1990 SEC No-Act. LEXIS 290, Securities Exchange Act of 1934—Rule 14a-8, February 9, 1990, Union Carbide Corp.; 1990 SEC No-Act. LEXIS 148, Securities Exchange Act of 1934—Rule 14a-8, January 25, 1990, American Express Company; 1989 SEC No-Act. LEXIS 1221, Securities Exchange Act of 1934—Rule 14a-8, December 27, 1989, American Electric Power Co.

[25] *See* Valerie Ann Zondorak, *A New Face in Corporate Environmental Responsibility: The Valdez Principles,* 18 B.C. B.C. ENVTL. AFF. L. REV. 457, n.83 (1991) (citing telephone interview with Michael Fleming, CERES staff member (Jan. 11, 1991)).

[26] According to one commentator:

> Corporate experience in South Africa under the Sullivan Principles indicates that a voluntary code of conduct, demanding responsibility beyond that legally required, may be successful if certain social and economic factors that motivate corporations to comply with the code of conduct are present. In addition, review of various techniques used in social audits suggests that the controversial disclosure and audit standards may be structured in a manner acceptable to both the drafters and the potential signatories.

Valerie Ann Zondorak, *A New Face in Corporate Environmental Responsibility: The Valdez Principles,* 18 B.C. ENVTL. AFF. L. REV. 457 (1991).

Recently there has been a trend toward increased disclosure obligations in environmental legislation. The goal behind such legislation is to harness market forces to encourage voluntary reductions in toxic releases. Disclosures can help inform consumers about environmental risks; consumers can then choose either to avoid the products or to pressure businesses to reduce emissions that pose environmental risks. Examples of this trend include the federal Emergency Planning and Community Right to Know Act (EPCRA), the EPA's Pollution Prevention program, California's Proposition 65, and state property transfer statutes.

Regulations requiring disclosure concerning environmental risks allow consumers and investors to make more informed consumption and investment decisions that create incentives for more responsible corporate behavior. For example, under EPCRA companies that employ more than nine full-time employees and manufacture or use more than the specified threshold quantities of the 300 listed chemicals are required to report chemical releases. These companies must complete and annually file Toxic Chemical Release Forms or Emissions Inventories with the EPA and the states, including estimates of "[t]he annual quantity of the toxic chemical entering each environmental medium" based on "readily available data." EPCRA also requires the EPA to establish and maintain a computer database that includes the national Toxic Release Inventory (TRI) that is available to the public through personal computers.

Congress expected this program to spur voluntary reductions in emissions because the availability of information about releases of toxic chemicals would enable the public to put substantial pressure on companies to reduce emissions. Empirical data indicate that this goal is largely coming to fruition.

EPA and some corporate officials were shocked at the large volume of toxic substance releases reported when EPCRA first came into effect. EPCRA disclosure spurred some companies like Monsanto Corporation to make voluntary reductions. In complying with EPCRA, Monsanto found that its 35 plants released more than 374 million pounds of toxic substances, more than 20 million pounds of it into the air. As a result, Monsanto's chairman pledged to reduce air emissions of hazardous chemicals by 90 percent by the end of 1992. In addition, AT&T established a goal of eliminating all of its toxic air emissions by the year 2000.

By forcing companies to acknowledge the extent of their emissions and the possible resultant health effects while arming environmental groups with information to demand better pollution control programs, EPCRA gave numerous citizens the power to correct environmental problems. For example, a small citizen group in Northern California used TRI information in a successful effort to persuade one company to stop the storage and use of chlorine for treating cooling water at one of its refineries. Similarly, a Massachusetts citizen group used TRI data to influence corporate environmental policy at the Raytheon Company, convincing it to cease using ozone-depleting chemicals to clean computer circuit boards. William K. Reilly indicated, while acting as EPA administrator, that EPCRA's informational approach "could set the pace for a new cooperative way of addressing the nation's environmental goals."[27]

---

[27] Disclosure required by EPCRA also is playing a key role in national pollution prevention efforts. Specifically, the Pollution Prevention Act of 1990 (PPA) declares pollution prevention to be the national policy of the United States. The PPA establishes a hierarchy of preferred options beginning with source reduction and followed by recycling, treatment, and disposal. The TRI is one of the tools used by the PPA to promote voluntary waste reduction. Companies that already file annual TRI reports under EPCRA are required by the PPA to include in their reports descriptions of their source reduction and recycling activities, including estimates of reductions expected to occur over the following two years. This information will be made available to the public. While the EPA does not expect this program to replace "strong regulatory and enforcement programs," the data will "create an improved information base that can facilitate future decisions about regulatory action to prevent pollution."

Disclosure of environmental problems also is required in connection with real estate transactions. Because CERCLA makes property owners liable for cleaning up hazardous substances found on their land, environmental assessments now are performed routinely when real property is sold, making it "difficult for landowners responsible for past contamination to unload their properties on unwitting purchasers." States such as California, Connecticut, Illinois, Indiana, and New Jersey now also require that environmental assessments be conducted or that disclosures of known or suspected environmental contamination problems be made when property is transferred.

New Jersey has one of the most far-reaching statutes requiring disclosure by sellers of industrial property. The New Jersey Industrial Site Recovery Act (ISRA), formerly the Environmental Cleanup Responsibility Act (ECRA), requires that an environmental assessment be conducted with the sale of commercial real estate, and that "cleanup [must] be implemented if the property is found to be contaminated." ISRA requires that the seller declare by approval of the state environmental authorities that there has never been a hazardous waste release or that any release has been cleaned up. If the property is found to be contaminated, a state-approved cleanup plan with adequate financial assurances to support its implementation must accompany the transfer. Additionally, the transfer is voidable by the buyer (but not the state of New Jersey) if the seller does not adhere to the terms of ISRA, and the seller is held strictly liable for all response costs as well as damages for failure to implement the cleanup.

Numerous commentators have argued that "disclosure obligations imposed by environmental and securities laws serve to promote several desirable goals."[28] They assert that, "In addition to encouraging businesses to reduce pollution and to use less hazardous substances in their production processes, disclosure requirements can prevent environmental liabilities from being unfairly transferred to uninformed purchasers or investors."[29] Moreover, it has been repeatedly argued that "the juxtaposition of disclosure obligations in environmental and securities laws may limit the ability of businesses to 'game' regulators by manipulating cost estimates for strategic purposes."[30]

While having an incentive to reassure investors concerning their financial well-being, companies also have an incentive to exaggerate prospective compliance costs when opposing environmental regulators.[31] Thus, it is not infrequent that a business will tell its investors a different story than the company will tell Congress or the regulatory agency in the context of pending environmental legislation or regulation. For example, a company may argue before EPA or Congress that a proposed regulatory or legislative initiative would put it out of business. On the other hand, the same corporation may want to reassure its investors that its financial viability will not be placed at risk by proposed legislation or regulation. By scrutinizing more carefully disclosures made to different regulators and investors, authorities can promote greater consistency in industry estimates of potential compliance costs. The question of whether this consistency leads to better disclosure remains to be tested.[32]

---

[28] *See, e.g.*, Marisa C. Caputo, *SEC Environmental Disclosure Rules and CERCLA Liability*, 4 MD. J. CONTEMP. LEGAL ISSUES 97 (Fall/Winter 1992/93).

[29] *Id.*

[30] *Id.*

[31] *Id.*

[32] Numerous no-action letters have discussed shareholder proposals concerning the CERES principles. *See, e.g.*, 1994 SEC No-Act. LEXIS 839, Securities Exchange Act of 1934—Rule 14a-8(c)(12)(i), December 22, 1994, A. O. Smith Corporation; 1994 SEC No-Act. LEXIS 609, Securities Exchange Act of 1934—Rule 14a-8(c)(10), July 15, 1994, The Procter & Gamble Company; 1994 SEC No-Act. LEXIS 300, Securities Exchange Act of 1934—Rule 14a-8, March 4, 1994, Sigma-Aldrich Corporation; 1994 SEC No-Act. LEXIS 219, Securities Exchange Act of 1934—Rule 14a-8, February 16, 1994, Paul M. Neuhauser, Reconsideration; 1994 SEC No-Act. LEXIS 124, Securities Exchange Act of 1934—Rule

14a-8, January 31, 1994, Amoco Corporation; 1994 SEC No-Act. LEXIS 129, Securities Exchange Act of 1934—Rule 14a-8, January 31, 1994, Allied Signal, Inc.; 1994 SEC No-Act. LEXIS 141, Securities Exchange Act of 1934—Rule 14a-8, January 31, 1994, International Business Machines Corporation; 1994 SEC No-Act. LEXIS 97, Securities Exchange Act of 1934—Rule 14a-8, January 21, 1994, RJR Nabisco, Inc.; 1993 SEC No-Act. LEXIS 1188, Securities Exchange Act of 1934—Rule 14a-8, December 22, 1993, International Paper Company; 1993 SEC No-Act. LEXIS 432, Securities Exchange Act of 1934—Rule 14a 8, March 12, 1993, Eastman Kodak Company; 1993 SEC No-Act. LEXIS 338, Securities Exchange Act of 1934—Rule 14a-8, February 25, 1993, Occidental Petroleum Corporation; 1993 SEC No-Act. LEXIS 282, Securities Exchange Act of 1934—Rule 14a-8(a)(3), February 19, 1993, Cooper Industries; 1993 SEC No-Act. LEXIS 139, Securities Exchange Act of 1934—Rule 14a-8, February 3, 1993, Bristol Myers Squibb Corporation; 1993 SEC No-Act. LEXIS 149, Securities Exchange Act of 1934—Rule 14a-8, February 3, 1993, ITT Corporation; 1993 SEC No-Act. LEXIS 82, Securities Exchange Act of 1934—Rule 14a-8, January 27, 1993, Eastman Kodak Company; 1993 SEC No-Act. LEXIS 97, Securities Exchange Act of 1934—Rule 14a-8, January 27, 1993, Eastman Kodak Company; 1993 SEC No-Act. LEXIS 100, Securities Exchange Act of 1934—Rule 14a-8, January 27, 1993, Eastman Kodak Company; 1993 SEC No-Act. LEXIS 102, Securities Exchange Act of 1934—Rule 14a-8, January 27, 1993, Eastman Kodak Company; 1993 SEC No-Act. LEXIS 78, Securities Exchange Act of 1934—Rule 14a-8(c)(7), January 26, 1993, R. R. Donnelley & Sons Company; 1993 SEC No-Act. LEXIS 80, Securities Exchange Act 1934—Rule 14a-8(c)(3), January 26, 1993, McDonald's Corporation; 1993 SEC No-Act. LEXIS 85, Securities Exchange Act of 1934—Rule 14a-8, January 26, 1993, Louisiana-Pacific Corporation; 1993 SEC No-Act. LEXIS 91, Securities Exchange Act of 1934—Rule 14a-8, January 26, 1993, Homestake Mining Company; 1993 SEC No-Act. LEXIS 104, Securities Exchange Act of 1934—Rule 14a-8, January 26, 1993, Eastman Kodak Company; 1993 SEC No-Act. LEXIS 112, Securities Exchange Act of 1934—Rule 14a-8, January 26, 1993, Texas Instruments Incorporated; 1993 SEC No-Act. LEXIS 44, Securities Exchange Act of 1934—Rule 14a-8, January 13, 1993, American Express Company; 1993 SEC No-Act. LEXIS 34, Securities Exchange Act of 1934—Rule 14a-8, January 11, 1993, Exxon Corporation; 1993 SEC No-Act. LEXIS 2, Securities Exchange Act of 1934—Rule 14a-8, January 4, 1993, Westinghouse Electric Corporation; 1992 SEC No-Act. LEXIS 1081, Securities Exchange Act of 1934—Rule 14a-8, December 2, 1992, TEXFI Industries, Inc.; 1992 SEC No-Act. LEXIS 489, Securities Exchange Act of 1934—Rule 14a-8, April 3, 1992, McDonald's Corporation; 1992 SEC No-Act. LEXIS 469, Securities Exchange Act of 1934—Rule 14a-8, March 27, 1992, Union Carbide Corporation; 1992 SEC No-Act. LEXIS 479, Securities Exchange Act of 1934—Rule 14a-8, March 25, 1992, PepsiCo, Inc.; 1992 SEC No-Act. LEXIS 409, Securities Exchange Act of 1934—Rule 14a-8, March 23, 1992, Newmont Mining Corporation; 1992 SEC No-Act. LEXIS 356, Securities Exchange Act of 1934—Rule 14a-8, March 13, 1992, Texaco, Inc.; 1992 SEC No-Act. LEXIS 322, Securities Exchange Act of 1934—Rule 14a-8, March 9, 1992, Burlington Resources, Inc.; 1992 SEC No-Act. LEXIS 325, Securities Exchange Act of 1934—Rule 14a-8, March 6, 1992, Weyerhaeuser Company; 1992 SEC No-Act. LEXIS 309, Securities Exchange Act of 1934—Rule 14a-8, March 3, 1992, American Electric Power Company; 1992 SEC No-Act. LEXIS 265, Securities Exchange Act of 1934—Rule 14a-8, February 25, 1992, United Technologies Corporation; 1992 SEC No-Act. LEXIS 213, Securities Exchange Act of 1934—Rule 14a-8, February 19, 1992, Eastman Kodak Company; 1992 SEC No-Act. LEXIS 138, Securities Exchange Act of 1934—Rule 14a,

February 4, 1992, B. M. Squibb Co.; 1992 SEC No-Act. LEXIS 72, Securities Exchange Act of 1934—Rule 14a-8, January 17, 1992, AT&T Company; 1992 SEC No-Act. LEXIS 33, Securities Exchange Act of 1934—Rule 14a-8, January 14, 1992, Eastman Kodak Company; 1992 SEC No-Act. LEXIS 39, Securities Exchange Act of 1934—Rule 14a-8, January 14, 1992, E. I. Du Pont De Nemours and Company; 1991 SEC No-Act. LEXIS 1361, Securities Exchange Act of 1934—Rule 14a-8, December 12, 1991, Emerson Electric Co.; 1991 SEC No-Act. LEXIS 583, Securities Exchange Act of 1934—Rule 14a-8, April 4, 1991, McDonald's Corporation; 1991 SEC No-Act. LEXIS 496, Securities Exchange Act of 1934—Rule 14a-8, March 28, 1991, Houston Industries Incorporated; 1991 SEC No-Act. LEXIS 477, Securities Exchange Act of 1934—Rule 14a-8, March 15, 1991, McDonald's Corporation; 1991 SEC No-Act. LEXIS 490, Securities Exchange Act of 1934—Rule 14a-8, March 13, 1991, Newmont Mining Corporation; 1991 SEC No-Act. LEXIS 377, Securities Exchange Act of 1934—Rule 14a-8, March 6, 1991, The Dow Chemical Company; 1991 SEC No-Act. LEXIS 404, Securities Exchange Act of 1934—Rule 14a-8, March 6, 1991, Tribune Company; 1991 SEC No-Act. LEXIS 415, Securities Exchange Act of 1934—Rule 14a-8, March 6, 1991, Champion International Corporation; 1991 SEC No-Act. LEXIS 430, Securities Exchange Act of 1934—Rule 14a-8, March 6, 1991, Texaco, Inc.; 1991 SEC No-Act. LEXIS 384, Securities Exchange Act of 1934—Rule 14a-8, March 1, 1991, Amoco Corporation; 1991 SEC No-Act. LEXIS 397, Securities Exchange Act of 1934—Rule 14a-8, March 1, 1991, Exxon Corporation; 1991 SEC No-Act. LEXIS 347, Securities Exchange Act of 1934—Rule 14a-8(c)(7), February 27, 1991, E. I. Du Pont De Nemours and Company; 1991 SEC No-Act. LEXIS 382, Securities Exchange Act of 1934—Rule 14a-8, February 27, 1991, Eastman Kodak Company; 1991 SEC No-Act. LEXIS 393, Securities Exchange Act of 1934—Rule 14a-8, February 27, 1991, Eastman Kodak Company; 1991 SEC No-Act. LEXIS 412, Securities Exchange Act of 1934—Rule 14a-8, February 27, 1991, Chevron Corporation; 1991 SEC No-Act. LEXIS 202, Securities Exchange Act of 1934—Rule 14a-8(a)(3)(i), February 13, 1991, Chevron Corporation; 1991 SEC No-Act. LEXIS 212, Securities Exchange Act of 1934—Rule 14a-8(a)(3)(i), February 13, 1991, Chevron Corporation; 1991 SEC No-Act. LEXIS 213, Securities Exchange Act of 1934—Rule 14a-8(a)(3)(i), February 13, 1991, Chevron Corporation; 1991 SEC No-Act. LEXIS 250, Securities Exchange Act of 1934—Rule 14a-8(a)(3), February 13, 1991, ARCO; 1991 SEC No-Act. LEXIS 142, Securities Exchange Act of 1934—Rule 14a-8, January 25, 1991, General Electric Company; 1991 SEC No-Act. LEXIS 146, Securities Exchange Act of 1934—Rule 14a-8, January 25, 1991, Union Carbide Corporation; 1991 SEC No-Act. LEXIS 107, Securities Exchange Act of 1934—Rule 14a-8, January 22, 1991, The New York Times Co.; 1991 SEC No-Act. LEXIS 30, Securities Exchange Act of 1934—Rule 14a-8, January 4, 1991, Washington Post Co.; 1991 SEC No-Act. LEXIS 24, Securities Exchange Act of 1934—Rule 14a-8, January 3, 1991, American Electric Power; 1991 SEC No-Act. LEXIS 26, Securities Exchange Act of 1934—Rule 14a-8, January 3, 1991, IBM The Valdez Principles; 1990 SEC No-Act. LEXIS 1358, Securities Exchange Act of 1934—Rule 14a-8, December 26, 1990, The Mead Corp.; 1990 SEC No-Act. LEXIS 1316, Securities Exchange Act of 1934—Rule 14a-8, December 18, 1990, American Electric Power Co.; 1990 SEC No-Act. LEXIS 1167, Securities Exchange Act of 1934—Rule 14a-8, October 5, 1990, Centerior Energy Corp.; 1990 SEC No-Act. LEXIS 494, Securities Exchange Act of 1934—Rule 14a-8, March 19, 1990, Eastman Kodak Company; 1990 SEC No-Act. LEXIS 476, Securities Exchange Act of 1934—Rule 14a-8, March 13, 1990, Aetna Life & Casualty Co.; 1990 SEC No-Act. LEXIS 477, Securities Exchange Act of 1934—Rule 14a-8, March 12, 1990, Sun Co., Inc.; 1989 SEC No-Act. LEXIS 204, Securities Exchange Act of 1934—Section 17—Rule 17a-5, February 10, 1989, Ceres

§ 6.3 Proxy Activity Concerning CERES

Proponents of environmental questions have been increasingly successful in forcing issuers to include their proposals in proxy materials. A review of SEC staff no-action letters reveals an increased reluctance on the part of the staff to take a no-action enforcement provision when the proposal relates to an environmental matter.[33] For example, on July 15, 1994, the SEC informed The Procter & Gamble Company that it could not exclude a shareholder proposal requesting that the company endorse the CERES principles from proxy materials.[34] The SEC staff explained that it was "unable to concur in [the Company's] view that the proposal may be excluded pursuant to Rule 14a-8(c)(10)."[35] While the rule allows the omission of a proposal that has been rendered moot, based on the information provided, the staff said it was "unable to conclude that the Company's existing policies, practices and procedures with respect to the environment substantially implement the CERES principles."[36]

Whether the staff grants the no-action relief ultimately depends on the nature of the proposal, the company's rationale for excluding it, and whether proper procedure has been followed by all parties involved with the proposal.

Partners, Inc.; 1984 SEC No-Act. LEXIS 1707, Securities Act of 1933—Section 4(2), Feb 16, 1984, Courseware, Inc.

[33] *See* Donald A. Carr & Stuart J. Kaswell, *Publicly Traded Securities, in* BNA'S ENVIRONMENTAL DUE DILIGENCE GUIDE §171.

[34] The Procter & Gamble Company, 1994 SEC No-Act. LEXIS 609 (July 15, 1994).

[35] *Id.* at 1.

[36] *Id.*

§ 6.4 Responsible Care

The Chemical Manufacturer's Association (CMA) "Responsible Care" (RC) program requires that member companies engage in continual efforts to improve people's security and to protect the environment.[37] The Chemical Manufacturers Association Responsible Care program is a "broad industry commitment to improve performance through a process that ensures responsiveness to the public's concerns." Responsible Care is unique because the bylaws require the CMA member companies to participate in the initiative, and the public is directly involved in developing the program elements through a public advisory panel.[38]

Responsible Care was born out of the Bhopal incident.[39] In order to ease uncertainties created by Bhopal and the public concern that such an accident could happen in the United States as well, the Chemical Manufacturers Association launched the Community Awareness and Emergency Response (CAER) program in April 1985, less than half a year after the Indian disaster. CMA knew Congress would react to the disaster, but it also knew the public would not wait for congressional action and neither should CMA. As CMA developed CAER, it stayed in close contact with Congress and EPA. The congressional establishment of local emergency planning committees and state emergency response commissions grew out of that relationship.

---

[37] For a discussion of the application of Responsible Care in Europe by members of the Chemical Industries Association (CIA), *see* Peter Simmons and Brian Wynne, *Responsible Care: Trust, Credibility, and Environmental Management, in* KURT FISCHER AND JOHAN SCHOT, ENVIRONMENTAL STRATEGIES FOR INDUSTRY: INTERNATIONAL PERSPECTIVES ON RESEARCH NEEDS AND POLICY IMPLICATIONS (Island Press 1993). CIA launched Responsible Care in March 1989, adapting the program from the experience of the Canadian Chemical Producers Association, which began in 1985.

[38] *See* Kenneth R. Myers, *What It Means to Be 'Green'; Environmental Movement Impacts Public, Corporate Policy,* THE LEGAL INTELLIGENCER § 2 (November 16, 1994).

[39] Ronald Begley and Marjorie Coeyman, *After Bhopal: A CMA 'War Room,' New Programs, and Changed Habits,* CHEMICAL WEEK 32 (December 7, 1994).

Community emergency responders found they were ill prepared to handle major chemical emergencies, thus making the CAER project more crucial.[40] While almost all U.S. plants had emergency response plans in place, almost none were sharing them with their communities. In Bhopal, it was clear that residents did not know what to do in an emergency. Plant managers initially balked at the idea of telling their communities about potential risks at their plant sites, but they discovered their communities had an open-minded interest in cooperatively developing emergency response plans that respected companies' needs to keep some information confidential and communities' needs to be safe. "The plant managers learned that no one wanted them to leave town. They just wanted to know they and their families were safe."

In developing CAER, CMA set out to identify the practices of the companies that were among the "best of the best" and package them and spread them out in the industry to raise performance. That approach was extended to the subsequent Responsible Care codes. Responsible Care was adopted by the United States industry in 1988.[41]

Responsible Care has caused an "escalation" of industry expenditures on health, safety, and environmental protection. CMA estimates "3%-4% of sales—or about $12 billion—is spent by the industry on health, safety, and environmental protection, compared with maybe 1% or less pre-Bhopal. It's the number one item on the agenda of every CEO I know in the industry."[42]

---

[40] CAER became an international model for companies around the world interested in responding to community concerns. For example, the Canadian Chemical Producers Association (Ottawa) imported the program. After finding it too narrow, the Canadian companies, with the help of companies that operated in the U.S. as well as in Canada, broadened the CAER program into Responsible Care.

[41] Ronald Begley and Marjorie Coeyman, *After Bhopal: A CMA 'War Room,' New Programs, and Changed Habits*, CHEMICAL WEEK 32 (December 7, 1994).

[42] *Id.*

§ 6.5 Business for Social Responsibility

Business for Social Responsibility (BSR) is a Washington, D.C., based trade association for American businesses concerned with responsible businesses practices including an empowered workforce, childcare, community outreach/involvement, product development, sourcing, environment, and commercial success. According to BSR's promotional literature:

> BSR is a broad-based membership organization which promotes responsible commercial success. It will enable businesses to secure a competitive advantage by demonstrating concern about the consequences of businesses decisions. Launched in 1992, BSR has grown to 850 member companies. Each company is working on different issues and making efforts to improve and/or incorporate additional practices. Socially responsible business practices do not have a set definition, but reflect continuous improvement. BSR's definition of corporate responsibility involves business making decisions which support commercial success of the enterprise while taking into consideration and fairly addressing the legitimate claims of all who are impacted by the decision—investors, employees, customers, suppliers, communities, the environment, and future generations. BSR wants to both inspire companies to behave in a responsible manner and demonstrate ways to do so.

With BSR's expanding membership has come some criticism not only of the BSR program but also of BSR's claims that "[s]omething strange and wonderful is taking place in business."[43] Many have begun to question whether "[s]lowly, but ever so surely, companies of all size and sectors are discovering that they function best when they merge business interest with the interests of customers, employees, suppliers, neighbors, investors, and other groups affected, directly or indirectly, by their companies' operations."[44] As one political scientist explained, "whether Americans are

---

[43] See JOEL MAKOWER AND BUSINESS FOR SOCIAL RESPONSIBILITY, BEYOND THE BOTTOM LINE 1 (1994).

[44] *Id.* For a critical review of this book, *see* Brett D. Fromson, *Putting Social Responsibility to Work for Your Business,* WASH. POST, December 18, 1994, at Book World 8 (book review).

environmentalist in any enduring sense remains problematic."[45] According to theories critical of BSR's conclusions, "[a] critical look at American culture raises profound questions about the public strength and endurance of American environmentalism and warns that public opinion polls are not yet prophecy."[46]

§ 6.6 Green Seal

Green Seal is a nonprofit organization that awards its certification mark to products that meet strict environmental standards set on a category-by-category basis.[47] Green Seal has issued standards for tissue paper, re-refined engine oil, water-efficient fixtures, printing and writing paper, and compact fluorescent light bulbs,[48] and intends to develop standards for house paints, laundry detergents, household cleaners, batteries, windows, and major appliances.[49]

In its product certification procedure, rather than focus on a single product attribute, Green Seal identifies and evaluates the significant environmental impacts of a product and its packaging from the manufacturing stage through consumer use to recycling or disposal.[50] Manufacturers pay a fee for the testing of their products, as well as a licensing fee for use of the Green Seal logo and an annual monitoring fee.[51] The Green Seal certification mark must be accompanied by a detailed explanation of the certification, such as the amount of recycled and postconsumer material, absence of bleach, or similar characteristics.[52]

---

[45] Rosenbaum, at 341.

[46] *Id.* at 342.

[47] U.S. Guide, ¶ 612. Green Seal sometimes is assisted in its standard-setting process by advisory panels composed of business, government, academia, and public interest representatives. The proposed standards are released for public comment, and revised periodically. "Some of the impacts that Green Seal may evaluate in establishing a standard are toxic chemical pollution; energy consumption; depletion and pollution of water resources; harm to fish, wildlife and natural areas; waste of natural resources; destruction of the earth's atmosphere; and global warming."

[48] *Id.*

[49] *Id.*

[50] *Id.*

[51] *Id.*

[52] *Id.*

Green Seal has also established what it calls its Environmental Partners Program. The purpose of the Environmental Partners Program is to "promote environmentally responsible business practices, including the purchase, sale, and use of products and services which are less damaging to the environment." Environmental partners must agree to purchase Green Seal certified products and other products less damaging to the environment. This program requires that companies recycle office materials, assist Green Seal in setting environmental product standards, appoint an Environmental Partners Liaison, and contribute to an annual report that is released to the public.

§ 6.7 Strategies for Today's Environmental Partnership (STEP)

Recognizing that "One of the most significant long-term trends affecting the future vitality of the petroleum industry is the public's concerns about the environment," American Petroleum Institute (API)[53] member companies developed a positive, forward looking strategy called STEP: Strategies for Today's Environmental Partnership. The STEP program was launched in November 1992 and "serves as an umbrella for API's environmental programs, industry-sponsored initiatives, and individual company programs."[54] The STEP program aims to address public concerns by improving the industry's environmental, health, and safety performance; documenting performance improvements; and communicating them to the public.[55]

API launched the STEP program in November 1993, announcing it at the API annual meeting. STEP is modeled after the chemical industry's Responsible Care program.[56] The STEP program was designed to demonstrate that the oil industry "is serious about protecting the

[53] The address and telephone number of the American Petroleum Institute (API) American Petroleum Institute are as follows: 1220 L Street, NW, Washington, DC 20005; Tel: 202/682-8128.

[54] *See Update on Corporate Environment Reporting,* BUSINESS & THE ENVIRONMENT (August 1994).

[55] *See* FDCH Congressional Testimony, September 14, 1994.

[56] *US Petroleum Industry Launches Its Version of "Responsible Care" Program,* BUSINESS & THE ENVIRONMENT (February 1993).

environment."[57] The program established seven sets of management practices including a set on pollution prevention. API members support and participate in this program.[58] STEP is built on eleven guiding principles that were added to API's bylaws. These include preventing pollution, promoting safe operating practices, addressing community concerns, maintaining crisis readiness, reducing product hazards, conserving resources, and achieving reasonable government policies. "STEP . . . does for our industry what we would like to see the government do more often—it sets environmental goals and then lets the industry find the best way to achieve those goals. Government policy should be based on reality, not politics."[59]

The first API report issued under STEP (Petroleum Industry Environmental Performance 1992) offered forty pages of data on chemical releases, oil spills, workplace safety, and environmental expenditures in the U.S. for the industry as a whole. The initial report, thus, provided a base of indicators on chemical releases, oil spills, workplace safety, and environmental expenditures.

The second report was issued in July 1994. It covered the same four categories and added two new sections with data on refinery residuals and used oil. API refined certain categories and conducted some of the data-gathering a bit differently in the second report.[60] The new report included data on refinery residuals and used oil collection.

The July 1994 report showed that refinery residuals (wastes and other by-products) were 3.4 million tons less in 1991 than the year before. Used oil collection centers operating at 7,000 sites in 1992 took in more than 7 million gallons of used oil; environmental expenditures by the domestic petroleum industry rose from \$7.8 billion in 1990 to \$10.5 billion in 1992, and job-related injuries and illnesses among petroleum workers fell by 26 percent between 1977 and 1992.

---

[57] Other competitor programs include: the Public Environmental Reporting Initiative (PERI), a voluntary private sector effort "aimed at improving and expanding corporate environmental reporting," and the Coalition for Environmentally Responsible Economies (CERES), a general code of environmental conduct for business. Sun was the first Fortune 500 company and is the only oil company to endorse the CERES Principles.

[58] *See* FDCH Congressional Testimony, September 14, 1994.

[59] *API Declaring new Victory over Pollution,* 9 WASTE TREATMENT TECHNOLOGY NEWS (July 1994).

[60] *Update on Corporate Environmental Reporting,* 5 BUSINESS & THE ENVIRONMENT (August 1994).

All chemical release information in the report was based on the Environmental Protection Agency's annual Toxic Release Inventory (TRI), which covers more than 300 chemicals. For refineries, releases and transfers of TRI chemicals in 1991 were 98,359,970 pounds, down 27 percent from 135,247,928 pounds in 1988. For all industries, releases and transfers of TRI chemicals in 1991 were 4,458,805,714 pounds, down 31 percent from 6,426,484,611 pounds in 1988. Petroleum refineries generate slightly more than 2 percent of the chemical releases that the EPA monitors through the TRI. The API also noted, based on EPA data, that more than 60 percent of the by-products of refining are recycled. The API also analyzed the total number and volume of oil spills from 1983 to 1992, during which time approximately 78,000 spills of crude oil and petroleum products (a total of 64.9 million gallons) occurred in U.S. waters. The API noted that more than half of the spills occurring between 1983 and 1991 (during which time the Coast Guard stopped tracking facility ownership) were from vessels and facilities owned by the U.S. government, the armed forces, other industries, and private citizens, rather than the companies that constitute the U.S. petroleum industry.

For the second half of the reporting period more small spills and fewer large spills occurred. The API may attribute the increase in small spills to better reporting. This shift in spill size caused a drop from a high of 14 million gallons in 1989 to 1 million gallons in 1992. From 1983 to 1992, three-fourths of the oil spill occurred during transportation by tankers, barges, pipelines, railroad, or tank truck.[61]

The API analyzed U.S. government statistics on chemical releases from refineries, oil spills, and job-related illnesses and injuries and conducted the first API survey of environmental expenditures since 1984 in the hopes that energy and environmental policymakers would use this information as a yardstick to assess the performance of the petroleum industry and to measure its progress.[62]

---

[61] *See Environment Assessed in Second API Report*, ENHANCED ENERGY RECOVERY & REFINING NEWS (July 1994). For more information, *see generally Petroleum Industry Environmental Performance, Second Annual Report*, published by the American Petroleum Institute. To order by telephone, dial 202/682-8375.

[62] *US Petroleum Industry Launches Its Version of "Responsible Care" Program*, BUSINESS & THE ENVIRONMENT (Feb. 1993).

The utility of API's report is, however, somewhat in question, since unlike the Canadian chemical industry's sector-wide report, the API report does not provide a list of the companies reporting or a cross-reference that links the companies with the reportable substances their facilities emit.[63]

§ 6.8 The Public Environmental Reporting Initiative (PERI)

Nine corporations developed an alternative framework for reporting on environmental policies, performance, and practices.[64] The original volunteer group that developed the PERI guidelines consisted of Amoco Corp., Dow Chemical Co., E.I. Du Pont De Nemours and Co., IBM Corp., Northern Telecom, Phillips Petroleum Co., Polaroid Corp., Rockwell International, and United Technologies.[65]

PERI, led by environment health and safety officers from Amoco, Phillips Petroleum, Dow Chemical, DuPont, and others, developed its program by circulating a draft report to scores of companies for comment.[66]

---

[63] *Update on Corporate Environmental Reporting*, 5 BUSINESS & THE ENVIRONMENT, (Aug. 1994).

[64] Other groups that offered advice on reporting environmental information prior to PERI include: Business Council for Sustainable Development, Coalition for Environmentally Responsible Economies, Council on Economic Priorities, World Industry Council for the Environment, Investor Responsibility Research Center, and Global Environmental Management Initiative.

Among industry's voluntary codes of conduct are the International Chamber of Commerce's Business Charter for Sustainable Development, which calls for greater openness and the periodic provision of "appropriate information" to stakeholders, and the Coalition for Environmentally Responsible Economies (CERES) principles, which call on signatory companies to complete a standardized annual report on their environmental performance. STEP is Strategies for Today's Environmental Partnership, a program designed to demonstrate that the oil industry "is serious about protecting the environment."

Other guidelines developed by industry groups include the Global Environmental Management Initiative (GEMI) in the United States; and in Europe, the new European Chemical Industry Council's (CEFIC) sectoral environmental reporting guidelines, which, in the future, are likely to be followed by the majority of chemical companies in producing reports. National industry associations such as the Confederation of British Industry (CBI) also have produced guidelines for public disclosure of environmental information.

[65] PR News, October 17, 1994.

[66] The PERI guidelines were prepared by a group of executives from such companies as IBM, DuPont, Polaroid, Dow Chemical, and Rockwell International. L. A. Mitchell, *State Working to Lure Firms in Environmental Technology*, ARIZ. BUS. GAZETTE (December 15, 1994).

PERI developed flexible guidelines that are designed to integrate reporting for Responsible Care, the Environmental Protection Agency's 33/50 program, and other initiatives.[67] The goal of PERI is to promote meaningful public disclosure of businesses' environmental performance. This reporting would focus on progress in reducing environmental releases, implementation of emergency response programs, environmental compliance, and product stewardship. The PERI effort was designed to be flexible enough to enable any company, regardless of its size or industry, to communicate environmental results clearly and concisely.

The industrial group publicly released guidelines for corporate reporting of environmental matters. PERI is committed to publicly disclosing corporate environmental performance information. The PERI guidelines reflect significant input from representatives of CERES based on joint work with them over the last two years. Many companies, including IBM, now report publicly their corporate environmental information pursuant to PERI.[68]

Reports issued under the PERI guidelines will cover nine areas: (1) company profile, (2) environmental policy, (3) environmental management structure, (4) environmental releases, (5) environmental risk management, (6) environmental compliance, (7) product stewardship or product life-cycle management, (8) employee recognition, and (9) stakeholder involvement. They are described in greater detail as follows:

1. A company profile including its major lines of business and the nature of the environmental impact of company operations;
2. Description of the company's environmental policies;
3. Description of the level of management accountability for the company's policies and programs, and a description of the company's environmental organization;
4. Quantification of the amount of emissions, effluents, or wastes released to the environment;

---

[67] *See Industry Standardizes Environment Reports,* Chemical Week (June 30, 1993).

[68] *See* 1994 SEC No-Act. LEXIS 141, *7.

5. Environmental Risk Management, which includes
    - an environmental audit program,
    - its frequency and scope,
    - remediation programs in place or being planned,
    - environmental emergency response programs,
    - workplace hazards and controls;
6. Environmental compliance: specific information on the company's compliance with laws and regulations;
7. Description of product stewardship or product life-cycle management that focuses on reducing the health and environmental impact of products from manufacturing through disposal;
8. Description of employee recognition for programs that encourage environmental excellence;
9. Description of stakeholder involvement including a statement that describes the company's efforts to involve other stakeholders in its environmental efforts and to relate to the communities in which it operates.

Reports issued in compliance with the PERI guidelines address virtually every aspect of the company's environmental programs, practices, procedures, and performance.[69]

The PERI guidelines were developed after discussions with environmental groups, investor groups, other companies, and CERES representatives. The PERI report is a cooperative effort in which the company can contribute to the final result. In contrast, certain other voluntary reports, such as the CERES report, have been criticized because they give no assurance that the basic principles will not be changed in the future without the consent of a subscribing company. In other words, many companies are concerned because to join CERES or Responsible Care, a company undertakes an open-ended commitment, the substance of which is determined by others. Many boards of directors do not believe that entering into such a commitment is a prudent course of action, since the board of directors of any corporation has a duty to represent all stockholders of the company and not just those who have a special interest or concern.[70]

---

[69] *See* 1994 SEC No-Act. LEXIS 124, *19.
[70] 1994 SEC No-Act. LEXIS 141.

The advantage of following the PERI guidelines is that customers, shareholders, and suppliers can compare the company's efforts with its past efforts and with those of other companies. The companies that helped form the PERI guidelines had been publishing their own environmental reports for at least a couple of years. "But not all were balanced . . . Our idea was to come up with a comprehensive, balanced report." The PERI guidelines, which are copyrighted, can be used by any company at no cost. PERI makes free informational packets available.[71] Since the program is voluntary and no group administers the program, there is no information on how many corporations besides these original nine have chosen to use the PERI guidelines.[72]

One of the changes to the draft PERI guidelines was to make them more applicable internationally. The International Standards Organization plans to include environmental reporting standards in its environmental management practices standards. The World Industry Conference on the Environment and the Global Environmental Management Initiative also are developing reporting guidelines.[73]

The impact of voluntary reporting programs like PERI has been questioned. Some suggest that such programs are "likely to be superseded by more binding initiatives," possibly instituted by the European Community or the International Standards Organization. Others assert that the reports would be more valuable if companies set environmental goals and reported on their successes or failures in reaching them. For example, it was suggested that firms report, for example, how much energy they used and how much hazardous waste they produced per unit of production.[74]

Another concern with programs such as the PERI report card has been the cost of producing the environmental reports. For firms the size of PERI charter members, the costs range from $50,000 to $200,000.[75]

---

[71] L. A. Mitchell, *Execs Unveil Guide To Telling 'Green' Story; Reports Pushed as Relations Tool,* ARIZ. BUS. GAZETTE (Dec. 15, 1994).

[72] PR News, October 17, 1994.

[73] PR News , October 17, 1994.

[74] Liz Skinner, Greenwire, December 20, 1993.

[75] *Id.*

While no one can predict when universal standards for environmental reports (such as the ISO 14000) will be instituted, or by whom, many say it is only a matter of time before these voluntary reports become compulsory. If they do, companies that have an environmental reporting framework in place are likely to have a tremendous leg up on their competitors in the corporate race to stay ahead of regulations.[76]

§ 6.9 EPA's Voluntary Pollution Reduction Programs

EPA has been active in a number of voluntary partnerships to achieve environmental improvement. The common theme of these programs has been to promote pollution prevention among traditional manufacturers, service providers, and other government agencies. The following is a sampling of several voluntary programs sponsored by EPA.[77]

§ 6.9.1 The 33/50 Program

The free flow of information between the government and the public is essential to a democratic society. Improvements in the national information infrastructure provide a tremendous opportunity to improve the delivery of government information to the taxpayers who paid for its collection and to provide that information at an equitable price.

Some of the most powerful examples of the power inherent in information collection and dissemination come from the experience of federal agencies. For example, the Emergency Planning and Community Right-to-Know Act of 1986 established a Toxic Release Inventory (TRI), which requires industries to report their estimated total releases of toxic chemicals into the environment. The Environmental Protection Agency has used a variety of means for making the data available to the public, including a collaborative effort involving the agency, the nonprofit community, and philanthropy. This effort involved making the TRI available through an online service called RTK NET (the Right-to-Know Computer Network), operated by OMB Watch and Unison Institute.

---

[76] *Id.*

[77] The following examples have been taken from a variety of internal EPA source documents.

As a result of the TRI program, EPA and industry developed the "33/50" program, in which CEOs of major corporations set a voluntary goal of reducing their pollution by 33 percent by 1992 and 50 percent by 1995.[78] The 33/50 (Industrial Toxics) Project, which was announced in February 1991, remains one of the major components of EPA's pollution prevention strategy. The project established a national goal to reduce releases and offsite transfers of seventeen chemicals by one-third by 1992 and one-half by 1995 with emphasis on the use of pollution prevention techniques. The baseline for these reduction goals was the 1988 TRI. Then EPA Administrator Bill Reilly asked companies that produced or used these chemicals to participate in the program by making voluntary commitments to reduce their releases to all environmental media. Progress in achieving the program goals was monitored through the use of information reported to the TRI.[79]

EPA encouraged participation in the early reduction program through the adoption of pollution prevention measures. EPA defines pollution prevention as "the use of materials, processes, practices, or products that avoid, reduce or eliminate wastes or toxic releases, through activities such as toxic use reduction, source reduction and closed-loop recycling."[80]

EPA continues to emphasize its commitment in the long term to urge companies to participate in pollution prevention programs such as "33/50," and to continue to find new ways to use and emit less polluting and lower toxicity compounds.

The focus of the national strategy is to reduce the ongoing generation of toxic pollution in any form (air emissions, wastewater discharges, hazardous waste, runoff, or fugitive releases) through reduction in the use of toxic substances, process changes, and product changes. EPA and the Great Lakes states agreed to implement this effort to reduce the levels of toxic substances found in the Great Lakes basin by promoting pollution prevention activities to significantly reduce or eliminate the use and/or release of toxic substances at the source, with a special focus on reducing

---

[78] Because of RTK NET's success, EPA is seeking to expand the information available on the service. *See* 58 Fed. Reg. 49,025, 49,033 (1993).

[79] 56 Fed. Reg. 27,338 (1991).

[80] *Id.* at 27,340.

or eliminating persistent bioaccumulative toxic substances. EPA is currently promoting pollution prevention through a number of regulatory and nonregulatory activities.

For example, EPA is implementing the national 33/50 program in the Great Lakes system. Under this national program, EPA has received voluntary commitments from industry to reduce the emission of seventeen priority pollutants by 33 percent by the end of 1992 and by 50 percent by the end of 1995. EPA has also been working with utilities located within the Great Lakes basin to accelerate the phase-out of transformers containing PCBs.[81] Additionally, EPA and the Great Lakes states have undertaken many pollution prevention activities including "clean sweeps," which not only provide for the collection and environmentally safe disposal of contaminants but also provide educational fact sheets to participants on how to prevent future pollution. These pollution prevention activities are designed to complement other ongoing efforts to reduce toxics in the Great Lakes system.

The only requirement to participate in the program is a letter with numerical reduction commitments. Over 1,200 United States manufacturers are currently participating in the program. TRI data indicate that the program exceeded its 1992 reduction by 40 percent or 600 million pounds.

The success of voluntary programs such as EPA's 33/50 program suggests that many companies may welcome the opportunity to take the initiative in setting broader and more far-reaching elective goals that make environmental and economic sense for particular facilities. As in the Statement of Environmental Principles, facilities would be encouraged to develop additional objectives beyond those EPA may identify as national goals.[82] As EPA explained:

> Environmental leadership in today's world is not simply a matter of how companies manage their facilities. To demonstrate leadership, companies must show that they are making environmental

[81] 58 Fed. Reg. 20,802, 20,827 (1993).

[82] EPA recognized that many states have already established pollution prevention planning requirements for various categories of industrial facilities. EPA hoped to avoid duplication of state requirements in developing the federal programs. 58 Fed. Reg. 4802, 4807 (1993).

> considerations a more integral and salient feature of the development and implementation of their business plans. A leading company will consider environmental factors in deciding what products it will manufacture, what materials it will use, what industrial processes it will adopt, and even how it will approach financial practices (*e.g.*, cost accounting) and institutional culture (*e.g.*, employee incentives).
>
> An increasing number of companies are beginning to formalize their commitment to the environment by incorporating a Design for Environment (DfE) ethic into their business operations. The Design for Environment concept is analogous to other design improvement initiatives recently undertaken by companies that seek to improve the quality of the products they make, the processes they use, and other aspects of their operations. A company committed to Design for Environment will seek every available opportunity to produce safer products, use less polluting materials, implement production processes that minimize health and environmental impacts, and devise institutional systems that encourage environmental stewardship. The Environmental Leadership Program should recognize but not dictate these actions, acknowledge the industry pacesetters, and encourage others to follow their lead, thus advancing the state of the art.[83]

EPA sought to incorporate into its voluntary pollution prevention programs "clear encouragement of environmentally conscious product design, marketing, and raw material use, without being either prescriptive or administratively burdensome."[84] Rather, EPA stated that it was "interested in general principles that allow ample flexibility to accommodate the wide variety in the design, manufacture, and use of products."[85]

---

[83] *Id.*
[84] *Id.*
[85] *Id.*

§ 6.9.2 WASTEWI$E

WASTEWI$E promotes cost-effective steps to reduce solid waste from business. Companies fill out brief registration forms and then conduct a facility-wide waste assessment. Companies then commit to achieving reduction goals in three areas: reducing waste generation, recycling, and buying or manufacturing recycled products. Participants are encouraged to prepare annual progress reports on achieving their goals. EPA provides public recognition and technical assistance for member companies.

§ 6.9.3 Energy Star

Energy Star Computers reduces air pollution emissions from electricity generation for computers. Energy Star Computer companies have designed computers and printers that automatically shut down when not in use. This feature cuts electricity use in half. EPA has signed partnership agreements with most major computer manufacturers. The U.S. government as the largest purchaser of computer equipment in the world is a participant in this program. This particular program is akin to the McDonald's waste reduction and recycling programs in that a major player (the U.S. government) has the purchasing power to influence the way manufacturers and suppliers do business, in this case, the development of energy-efficient computers.

§ 6.9.4 Design for the Environment

Design for the Environment (DfE) is a cooperative effort between EPA and industry to promote consideration of environmental impacts at the earliest stages of product design. Initial projects include evaluating alternative dry-cleaning processes and environmentally safe substitutes for toxic chemicals used in printing processes. A new focus is a joint project between the accounting and insurance industries to integrate environmental considerations into capital budgeting and cost accounting systems.

Industry, consumers, and the government have all shifted the focus away from the traditional command-and-control, end-of-the-pipeline solutions to a more holistic approach. This approach has many different elements. Union Carbide, in part, focuses on integrating environmental policies into the daily work environment through a variety of measures and holds employees accountable for facility compliance. API's STEPS program sets up an industry standard that association members must comply with as a condition of membership. Hyde Corporation employs pollution prevention measures to save money. Frigidaire offers energy-efficient appliances to consumers. Consumers make environmentally responsible investments and integrate other lifestyle changes into their daily routines. EPA sponsors a program to promote consideration of environmental impacts early in product design. Both industry and consumers have integrated environmental concern into their own behavior patterns. Whether it be to gain competitive advantage, open new markets, or just to feel good, an environmental ethic or consensus has emerged.

§ 6.9.5 Permits Improvement Team

EPA programs also has been involved in projects that emphasize streamlining and process design. The Permits Improvement Team (PIT) mission is to examine and recommend improvements in all permitting programs and procedures. The PIT is developing recommendations in the areas of streamlining, permit alternatives, public participation, pollution prevention,[86] improved program training, and performance measures.

§ 6.9.6 The Environmental Leadership Program

The Environmental Leadership Program (ELP) grew out of a desire to test innovative compliance approaches such as third-party auditing. ELP allows facilities to identify ways to streamline reporting requirements and reduce compliance inspections, without sacrificing environmental and public health protection. Facilities will use innovative management techniques such as environmental auditing and pollution prevention to

[86] The term "pollution prevention" means source reduction as defined under the Pollution Prevention Act of 1990 (42 U.S.C.A. §13101 *et seq.*). Source reduction is further defined under §6603(5)(A) which (i) reduces the amount of any . . . pollutant . . . released into the environment . . . prior to recycling, treatment, or disposal; and (ii) reduces the hazards to public health and the environment associated with the release of such . . . pollutants . . . .

reduce the burden of paperwork and inspections on the facilities, while enhancing compliance with existing environmental laws. At the completion of these one-year pilot projects, the lessons learned from these projects will be applied to others.

The ELP is intended to encourage a long-term commitment by corporations and facilities to high quality environmental management practices, pollution prevention, and sustainable development.[87] According to EPA, "A program of this type has exciting potential for several reasons. First, American businesses—often applying principles of Total Quality Management—see competitive and other advantages in moving beyond compliance to reduce pollution at the source, and in using energy and other resources more efficiently."[88] EPA stated that "[t]he extensive and dramatic participation of companies in EPA's targeted voluntary toxics reduction ("33/50") and energy efficiency programs (Green Lights, Energy Star Computers, and others) is a sign of how much environmental improvement can result when progressive companies push the state of the art without waiting for regulations."[89] Thus, EPA concluded that, "with positive recognition and appropriate incentives from the Federal Government, these companies may decide to accelerate their development of breakthrough approaches to environmentally sound production and management."[90]

Moreover, EPA asserted that "if ultimate success in achieving sustainable development depends on providing for environmental concerns in the design stage of strategies and projects, then it is critical to encourage marketplace innovations and leadership from our business community."[91] James Gustave Speth, while President of the World Resources Institute, made that point quite clearly in the Foreword to *Beyond Compliance: A New Industry View of the Environment,* WRI's study of the voluntary efforts of progressive companies:

---

[87] *See* 58 Fed. Reg. 16,826 (1993); 58 Fed. Reg. 4802 (1993).

[88] 58 Fed. Reg. at 4802.

[89] *Id.*

[90] *Id.*

[91] *Id.*

> Important as pressure from environmentalists and governmental direction are to stimulating change, in the end only the corporate community can efficiently provide the necessary organization, technology, and financial resources needed to design and implement change on the scale required. Companies that are trying to be leaders on a new path to a sustainable future merit our encouragement and support, just as the inevitable backsliders deserve a vigorous shove onto the trail.[92]

EPA's environmental leadership program is designed to "advance pollution prevention by encouraging—but not dictating—superior environmental management systems that lead companies to design and develop their own innovative ideas and breakthrough improvements."[93] Encouraging a CEO-level focus through a comprehensive national program that emphasizes risk reduction priorities and measurable results can help set the pace for these voluntary efforts.

Toward this end, on January 15, 1993, EPA published a Federal Register notice requesting comment on the possible creation of a national voluntary program to encourage and publicly recognize environmental leadership and to promote pollution prevention in the manufacturing sector.[94] The agency requested responses to fifty-six specific questions about the structure of such a program, possible goals and measures, the need for incentives, the role of compliance screening, and other related issues. Two basic components were proposed for the "Environmental Leadership Program" (ELP)—a "Corporate Statement of Environmental Principles" and a "Model Facility" program.

The agency received a wide variety of comments on the original proposal from industry, states, environmental groups, and other non-governmental organizations. In addition, the agency held a public meeting on May 6, 1993, in Washington, D.C., and received additional comments from thirty groups. While no true consensus emerged on the best structure or goals for the program, the comments clearly indicated an interest in a

---

[92] B. Smart, Beyond Compliance: A New Industry View of the Environment, at X (World Resources Institute, Washington, DC, 1992).

[93] *Id.*

[94] 58 Fed. Reg. 4802 (1993).

voluntary program to recognize environmental excellence. After extensive review and analysis of the comments by the Office of Enforcement, the Office of Pollution Prevention and Toxics, and the Administrator's Pollution Prevention Policy Staff, the agency refined the goals and the immediate focus of the project.

In the January 28, 1994, Federal Register, the EPA Administrator announced the agency's intent to further develop the ELP concept, initially through a small number (3-5) of voluntary, facility-based pilot projects.[95] The new Office of Compliance (OC), within the reorganized Office of Enforcement and Compliance Assurance (OECA), will coordinate the "model facility" pilot project effort with significant regional and state partner involvement. At the same time, EPA opted not to further develop its own "Corporate Statement of Environmental Principles," but rather to work cooperatively with organizations that have developed their own corporate or industry-specific codes.[96]

EPA's pilot project effort is distinguished from a possible future, full-scale Environmental Leadership Program. A major goal of the pilot projects is to further explore possible components of a full-scale program. Options were raised during public comment on various program elements, including the review and selection process, recognition mechanisms, and other possible incentives for facilities. The pilot projects will explore these options, and will have a definite life span of about 12-18 months. At the end of this time, EPA will determine if a full-scale program is feasible, and if implementing such a program can help improve environmental compliance.

The EPA pilot projects, therefore, represent the experimental first step in the evolution of the ELP. The pilot phase is also an excellent opportunity to strengthen partnerships between government, industry groups, and regulated entities as a prelude to more extensive emphasis by the agency on voluntary compliance initiatives. The pilot projects, and any future leadership program, will be a vehicle for facilities to continue building positive, proactive relationships with EPA and state and local agencies.

---

[95] 59 Fed. Reg. 4066 (1994).

[96] The Office of Pollution Prevention and Toxics will continue to lead any future agency involvement in this area. *See* 59 Fed. Reg. 32,062 (1994).

EPA expects a number of potential benefits to facilities that are selected for pilot projects. The agency will publicly recognize those facilities that demonstrate outstanding environmental management practices, and also provide them with an opportunity to help shape the possible future, full-scale leadership program.[97] EPA will use the pilot projects to evaluate recognition mechanisms and other incentives that could be offered in a full-scale program. While mechanisms for recognition will be determined in discussions with each facility, they may include press releases, letters to community groups, local and state agencies, and/or site visits by EPA officials. It is important to note that any future program would offer recognition and other incentives on a continuing basis (similar to the OSHA Voluntary Protection Program), not as a one-time award.

The pilots represent an opportunity for facilities to inform and directly participate in EPA's effort to reassess its environmental auditing policy. In addition, it is anticipated that the projects selected will generate empirical data useful for evaluating EPA's compliance policies and spur the development of methodologies for evaluating compliance behavior. Finally, EPA is interested in discussing possible policy modifications and other incentives that could help facilities overcome barriers to self-monitoring and compliance efforts.[98]

---

[97] *Id.* at 32,063.

[98] Guidance on environmental auditing and state-of-the-art environmental management practices is available from many sources:

> The EPA Environmental Auditing Policy Statement, 51 Fed. Reg. 25,004 (July 9, 1986), which includes a discussion of elements of an effective auditing program;
>
> The *Draft Corporate Sentencing Guidelines for Environmental Violations,* 24 Env't Rep. (BNA) No. 30, at 1378 (Nov. 26, 1993), which includes a discussion of "Minimum Factors for Demonstrating a Commitment to Environmental Compliance" in Part D.

Voluntary standards on environmental management systems and environmental auditing may also provide guidance to facilities interested in preparing pilot project proposals. EPA is participating in work groups organized by the International Organization for Standards (ISO) and the National Sanitation Foundation (NSF International). The American Society for Testing and Materials (ASTM) is coordinating U.S. participation in ISO Technical Committee 207 (TC-207) on environmental management systems (EMS).

Similarly, initiatives such as the Global Environmental Management Initiative (GEMI), the Responsible Care Program, and the CERES principles have been developed in the private sector and by nongovernmental groups. These and other private sector efforts may be useful for facilities interested in submitting pilot proposals, and facilities are encouraged to develop proposals with industry and trade association involvement and support.

§ 6.9.7 The XL Community Pilot Program

EPA's XL community pilot program invites local governments, communities, states, federally recognized tribes, and others to join together to conduct a limited number of pilot projects that demonstrate and assess the merits of community-designed and directed strategies for achieving greater environmental quality and sustainable economic goals.[99] These pilots will build on the experience gained in the administration's Empowerment Zone and Ecosystem Management Initiatives. Both of these efforts encourage participants to achieve their community's economic development and environmental quality goals by using community-based planning to mobilize the resources necessary to sustain economic and environmental health. XL community pilot programs will integrate the mutually supportive goals of economic development and environmental protection at the community level utilizing full public participation.

Under XL, EPA will:

✓ assist local communities to achieve superior environmental performance.

✓ offer assistance in one or more areas, depending on the nature of the proposed project—flexibility in the implementation of environmental regulations, technical support, coordination of federal programs, the identification of government resources, and other types of support identified on a case-by-case basis—in exchange for a commitment on the part of the community entity to go beyond the environmental results of compliance and achieve greater environmental quality than would have been realized under traditional approaches.

✓ respond to requests for regulatory flexibility to support local communities' efforts to create innovative, alternative environmental management strategies that are supportive of community economic goals.

[99] 60 Fed. Reg. at 63,712 (Nov. 1, 1995).

The purpose of XL is to explore alternative approaches for achieving greater environmental results than would have been obtained under existing approaches to environmental protection. Ideally, XL community pilot projects are consistent with and help to establish long-range community environmental goals and bring together groups such as facilities, community organizations, and governments at all levels to achieve the goals of greater environmental quality consistent with economic development. As such, proposals that demonstrate the greatest support from community stakeholders and are consistent with a broader community vision or plan will be given preference in the selection process. Proposals were invited from a range of community entities[100] and should be designed around a defined geographic area.[101]

Factors in determining eligibility for the XL pilot project include

✓ Environmental Results

✓ Stakeholder Involvement, Support, and Capacity for Community Participation

✓ Economic Opportunity

✓ Feasibility

✓ Transferability

✓ Monitoring, Reporting and Evaluation

✓ Equitable Distribution of Environmental Risks

✓ Community Planning

✓ Innovative Approaches/Multimedia Focus/Pollution Prevention

✓ Enforcement and Compliance History

✓ Relationship of XL Community Pilots to Other Community-Based Reinvention Efforts

---

[100] Community entities include, but are not limited to, local governments, tribal governments, regional area consortia/governments, councils of government, private non-profit citizen/neighborhood/community organizations, nonprofit educational institutions, Empowerment Zones and Enterprise Communities designated under the administration's Community Empowerment Initiative, and other local entities either public or private.

[101] Geographic areas could include urban and rural areas, political jurisdictions, tribal lands, and ecologically defined areas such as watersheds and ecosystems, among others. EPA encourages community groups within the same geographic area whose project objectives are similar to consolidate their proposals.

ELP differs from the XL programs being announced today in that the XL programs include flexibility in existing regulations in exchange for the attainment of environmental results beyond what would have been achieved through full compliance with those regulations. ELP projects, on the other hand, work to achieve improvements in environmental quality within existing regulatory requirements. EPA expects that compliance-oriented ELP projects may include regulatory innovations, and that some projects conducted pursuant to today's notice will also address compliance systems.

§ 6.9.8 Common Sense Initiative

On July 22, 1994, EPA Administrator Browner announced the new environmental policy Common Sense Initiative (CSI). This initiative is designed to shift environmental protection from the current "pollutant-by-pollutant, end-of-pipeline, command-and-control" approach to an "industry-by-industry, multimedia, prevention-oriented" approach. Six pilot industries were identified for CSI:

- ✓ auto manufacturing,
- ✓ computers and electronics,
- ✓ iron and steel,
- ✓ metal finishing and plating,
- ✓ petroleum refining, and
- ✓ printing.

The Common Sense Initiative was launched to move EPA beyond the traditional medium-by-medium approach to environmental management to a systematic, sector-based approach. Each of the six industry sectors is directed by a consensus-based, multistakeholder advisory subcommittee, with CSI as a whole directed by the Common Sense Initiative Council operating under the federal Advisory Committee Act. The purpose of CSI is to recommend changes in environmental regulations, statutes, and programs that will result in "cleaner, cheaper, and smarter" outcomes for entire industries. Such changes, when accepted and promulgated, will lead to permanent adjustments to current programs.

Each of the CSI sector-specific subcommittees is developing a plan covering a broad spectrum of activities including (but not limited to) regulations, pollution prevention, reporting requirements and public access to data, permitting innovative compliance assistance and enforcement, and innovative technology. In some cases, these plans will include projects that meet the criteria outlined today for regulatory reinvention pilots.[102]

EPA'S Common Sense Initiative requires that EPA and outside groups undertake a cooperative investigation of cheaper ways to provide cleaner outcomes. CSI is designed to remedy the unintended consequence of EPA's historical approach to regulation, which has created a situation in which any given regulated entity may find itself bound by a number of environmental requirements that are not coordinated to achieve efficiency at the point of compliance. A firm may find itself responding to air, water, land, and toxic-substance controls that appear duplicative or unnecessarily burdensome. As a result of the regulatory burden, however, these firms are discouraged from making investments that might very well protect the environment better at a lower overall cost. EPA is looking for solutions that make sense nationally as well as locally—solutions where real people make real investments to protect the environment.

EPA staff has asked industry, environmentalists, state and local governments, and others to work together under CSI to learn just how multiple environmental requirements are working at their points of intersection. When this process turns up ways to improve the environmental performance of these industries by providing cheaper, more flexible ways to comply with the requirements, EPA will work to bring about this "cheaper" solution. In return, the agency will look to these industries to identify even more effective strategies to control pollution in their plants and communities, so that the public benefits from a "cleaner" solution as well. By pursuing this "cheaper, cleaner" strategy with these initial industries, EPA hopes to learn how to better effect national environmental controls that also make sense locally. Some of the areas the Common Sense Initiative will explore are

- ✓ pollution prevention,
- ✓ alternative compliance mechanisms,
- ✓ reform of data management and recordkeeping (especially introduction of electronic reporting),
- ✓ review and reform of existing regulations,
- ✓ coordinating multiple new rulemakings across program lines,

[102] Firms or other project sponsors in CSI industries are encouraged by EPA to develop XL projects.

✓ multimedia permitting, and
✓ the introduction of improved environmental technology.

§ 6.10 United Nations Environment Programme (UNEP)

In August, the United Nations Environment Programme (UNEP) issued a code of ethics to guide the private sector in enhancing safety in the international trade in chemicals. The code is a voluntary guide to standards of conduct in chemical production and management, with the purpose of reducing health and environmental risks. The three-part code describes the chemical industry's commitment to improved health, safety, and environmental protection. The second section details general commitments and specific guidance in eight areas, while the third covers periodic monitoring and follow up of all parties' voluntary compliance.[103]

To develop "The Code of Ethics on the International Trade in Chemicals," UNEP convened a series of informal meetings from May 1992 to April 1994. Representatives from industry, consumer groups, unions, and government reviewed existing industry initiatives and programs. According to UNEP, "industry and governments should be encouraged to cooperate with NGOs to ensure that the code's standards of conduct are adhered to; while those groups should serve as watchdogs with a view to improving industry's performance."[104] UNEP will provide progress reports on the code's implementation.[105]

---

[103] *Are Voluntary Environmental Characters Effective,* 5 BUSINESS & THE ENVIRONMENT, (September 1994).

[104] *Id.*

[105] *Id.* A copy of the code (in Arabic, Chinese, English, French, Russian, or Spanish) is available from UNEP Environmental Law and Institutions Program Activity Center (ELI/PAC), P.O. Box 30552, Nairobi, Kenya. Tel: +254 2 62 3493; Fax: +254 2 230198.

Environmental Codes of Conduct

Form 6.1
Code of the Valdez/CERES Principles

1. Protection of the Biosphere.

We will minimize and strive to eliminate the release of any pollutant that may cause environmental damage to the air, water, or earth or its inhabitants. We will safeguard habitats in rivers, lakes, wetlands, coastal zones and oceans and will minimize contributing to the greenhouse effect, depletion of the ozone layer, acid rain, or smog.

2. Sustainable Use of Natural Resources.

We will make sustainable use of renewable natural resources, such as water, soils and forests. We will conserve nonrenewable natural resources through efficient use and careful planning. We will protect wildlife habitat, open spaces and wilderness, while preserving biodiversity.

3. Reduction and Disposal of Waste.

We will minimize the creation of waste, especially hazardous waste, and wherever possible recycle materials. We will dispose of all wastes through safe and responsible methods.

4. Wise Use of Energy.

We will make every effort to use environmentally safe and sustainable energy sources to meet our needs. We will invest in improved energy efficiency and conservation in our operations. We will maximize the energy efficiency of products we produce or sell.

5. Risk Reduction.

We will minimize the environmental, health and safety risks to our employees and the communities in which we operate by employing safe technologies and operating procedures and by being constantly prepared for emergencies.

6. Marketing of Safe Products and Services.

We will sell products or services that minimize adverse environmental impacts and that are safe as consumers commonly use them. We will inform consumers of the environmental impacts of our products or services.

7. Damage Compensation.

We will take responsibility for any harm we cause to the environment by making every effort to fully restore the environment and to compensate those persons who are adversely affected.

8. Disclosure.

We will disclose to our employees and to the public incidents relating to our operations that cause environmental harm or pose health or safety hazards. We will disclose potential environmental, health or safety hazards posed by our operations, and we will not take any action against employees who report any condition that creates a danger to the environment or poses health and safety hazards.

9. Environmental Directors and Managers.

At least one member of the Board of Directors will be a person qualified to represent environmental interests. We will commit management resources to implement these Principles, including the funding of an office of vice president for environmental affairs or an equivalent executive position, reporting directly to the CEO, to monitor and report upon our implementation efforts.

10. Assessment and Annual Audit.

We will conduct and make public an annual self-evaluation of our progress in implementing these Principles and in complying with all applicable laws and regulations throughout our worldwide operations. We will work toward the timely creation of independent environmental audit procedures which we will complete annually and make available to the public.

Form 6.2
Mission Statement of the Business for Social Responsibility

Business for Social Responsibility is an alliance of businesses that defines, support, and promotes responsible business policies and practices that benefit not only our companies, but also our employees and communities, our economy and environment. BSR seeks to reshape the way business does business so that members and other companies can best address the many problems and opportunities that face business and society.

Form 6.3
Statement of Values of the Business for Social Responsibility

Prosperity and Well-Being. Responsible policies and practices enhance both the commercial success of businesses and the well-being of people in communities where we conduct business and throughout the world.
Responsibility and Leadership. The impact and resources of business create responsibilities to our employees, customers, business partners, communities, and the environment, as well as to our shareholders.
Stewardship and Sustainability. Our business and our society can best prosper when all citizens have access to quality education, health care, housing, and a healthy and safe environment, as well as the opportunities created by our market economy.
Diversity. We recognize, value, and benefit from the diversity of our employees, neighbors and customers.
Partnership and Advocacy. Members of the business community should partner with one another and with representatives of those impacted by business decisions to define, implement, and promote policies of responsible corporate citizenship.
Evolving Goals. Socially responsible corporate goals, policies, and practices are not static; companies need continually to review the standards, needs and concerns of their various stakeholders.
Universal Commitment. Responsible policies and practices should be implemented wherever businesses do business, in the United States and throughout the world.

Form 6.4
American Petroleum Institute Environmental Mission and Guiding Environmental Principles

The members of the American Petroleum Institute are dedicated to continuous efforts to improve the compatibility of our operations with the environment while economically developing energy resources and supplying high quality products and services to consumers. The members recognize the importance of efficiently meeting society's needs and our responsibility to work with the public, the government, and others to develop and to use natural resources in an environmentally sound manner while protecting the health and safety of our employees and the public. To meet these responsibilities, API members pledge to manage our businesses according to these principles:

✓ To recognize and to respond to community concerns about our raw materials, products and operations.

✓ To operate our plants and facilities, and to handle our raw materials and products in a manner that protects the environment, and the safety and health of our employees and the public.

✓ To make safety, health and environmental considerations a priority in our planning, and our development of new products and processes.

✓ To advise promptly, appropriate officials, employees, customers and the public of information on significant industry-related safety, health and environmental hazards, and to recommend protective measures.

✓ To counsel customers, transporters and others in the safe use, transportation and disposal of our raw materials, products and waste materials.

✓ To economically develop and produce natural resources and to conserve those resources by using energy efficiently.

✓ To extend knowledge by conducting or supporting research on the safety, health and environmental effects of our raw materials, products, processes and waste materials.

✓ To commit to reduce overall emission and waste generation.

✓ To work with others to resolve problems created by handling and disposal of hazardous substances from our operations.

✓ To participate with government and others in creating responsible laws, regulations and standards to safeguard the community, workplace and environment.

✓ To promote these principles and practices by sharing experiences and offering assistance to others who produce, handle, use, transport or dispose of similar raw materials, petroleum products and wastes.

Chapter 7

Business Transfer Statutes and the Duty to Audit

§ 7.1 Overview

While few statutes require that industry conduct comprehensive environmental audits as a condition of transferring property, many states have enacted business transfer statutes that indirectly encourage auditing prior to or during transfer of properties that may contain environmental hazards. These environmental statutes restrict, in varying degrees, the transfer of hazardous waste sites—and sites containing other environmental perils—by requiring that certain environmental disclosures be made concerning the condition of the property prior to transfer. The requirements of the laws vary from obtaining state approval of a transaction to close, terminate, or transfer an industrial establishment to simply notifying and disclosing the existence of a hazardous waste site to state or local authorities or to the party that will acquire the new possessory interest in the property. The sanctions for noncompliance include avoidance of the transaction and civil penalties, sometimes especially severe.[1]

---

[1] But note that these statutes conflict with the common law notion of caveat emptor; as such, the purchaser may be expected to know all that is visible at time of purchase. *See* Smith Land & Improvement Corp. v. Celotex Corp., 851 F.2d 86 (3d Cir. 1988) (presence of asbestos and hazardous waste disposal sites readily visible), *cert. denied,* 488 U.S. 1029 (1989); Philadelphia Elec. Co. v. Hercules, Inc. 762 F.2d 303 (3rd Cir. 1985), *cert. denied,* 474 U.S. 980 (1985) (site contamination polluting nearby stream); Clauser v. Taylor, 112 P.2d 661 (Calif. 1941) (recision allowed for failure to disclose matter outside buyers' ability to observe). *Cf.,* CALIF. CIV. CODE § 0795.

The following table illustrates states with specialized environmental lien, disclosure, or business transfer statutes:

State	Mini-CERCLA	Lien Law	Deed	Misc. Envtl. Disc.	Hazardous Waste Disclosure
Alabama	X	—	—	—	—
Alaska	X	X	—	—	—
Arkansas	X	SUPER	—	—	—
California	X	X	X	X	X
Colorado	X	X	X	—	DEED
Connecticut	X	SUPER	—	—	X
Delaware	X	—	X	—	X
Dist. of Col.	—	—	—	—	—
Florida	X	—	—	—	X
Georgia	X	—	X	—	X
Hawaii	X	—	—	—	—
Idaho	X	—	—	—	—
Illinois	X	SUPER	X	X	X
Indiana	X	X	X	—	X
Iowa	X	X	—	—	X
Kansas	X	—	—	—	—
Kentucky	X	X	X	—	X
Louisiana	X	SUPER	—	—	—
Maine	X	SUPER	—	—	—
Maryland	X	X	X	X	X
Massachusetts	X	SUPER	X	X	X
Michigan	X	SUPER	X	X	X
Minnesota	X	X	X	—	X

State	Mini-CERCLA	Lien Law	Deed	Misc. Envtl. Disc.	Hazardous Waste Disclosure
Mississippi	X	—	—	—	—
Missouri	X	—	X	—	X
Montana	X	X	—	X	—
Nebraska	X	—	X	—	X
Nevada	X	—	—	—	—
New Hampshire	X	SUPER	X	—	X
New Jersey	X	SUPER	—	—	ECRA/ISRA
New York	X	X	X	—	X
North Carolina	X	—	X	—	X
North Dakota	X	—	X	—	X
Ohio	X	X	—	X	X
Oklahoma	X	—	—	—	—
Oregon	X	X	X	—	X
Pennsylvania	X	X	X	—	X
Rhode Island	X	—	X	X	X
South Carolina	X	—	—	—	—
South Dakota	X	X	—	—	—
Tennessee	X	X	X	—	— DEED
Texas	X	X	X	X	X
Utah	X	—	—	—	X
Vermont	X	X	X	—	X

State	Mini-CERCLA	Lien Law	Deed	Misc. Envtl. Disc.	Hazardous Waste Disclosure
Virginia	X	X	—	X	X
Washington	X	—	—	—	—
West					
Virginia	X	—	X	—	X
Wisconsin	X	SUPER	—	X	X
Wyoming	X	—	—		—

§ 7.2 The New Jersey Environmental Cleanup Responsibility Act (ECRA) and ECRA-Type Laws

The New Jersey Environmental Cleanup Responsibility Act (ECRA) was one of the first of the state statutes. The ECRA imposed preconditions upon the consummation of a transaction to close, terminate, or transfer the operations of an industrial establishment in New Jersey. The preconditions included submitting to the New Jersey Department of Environmental Protection a notification form, a site evaluation form, and a declaration essentially stating that the property was not contaminated at the time of the transfer or, alternatively, a cleanup plan backed by financial security, which the department must approve. The sanctions for noncompliance with the act were severe: avoidance of the transaction and imposition of damages and stringent, escalating penalties.[2]

[2] Under ISRA, only private parties have the power to void a transaction. Although it did under ECRA, ISRA took away the state's authority to void a transaction. The legislative history indicated that since private parties had attempted to use this remedy and the state had not, the legislature decided to continue the remedy for private parties but not for the state.

From the beginning, New Jersey's Department of Environmental Protection and Energy (NJDEPE) was criticized for its handling of the program. These attacks gained in severity as New Jersey's economy lagged behind the nation in the most recent and continuing recession. After many months of effort in New Jersey's legislature, Senate S.1070 passed the Senate and Assembly on June 10, 1993, and was promptly signed by Governor Florio on June 16, 1993. Thus, New Jersey enacted ISRA to replace the ECRA program.[3]

The changes to ISRA were intended to improve the efficiency of the program and better balance legitimate business concerns with protection of the environment. Many of the studies required by ISRA are certain to continue the debate on environmental policy both in and out of New Jersey. As these studies are delivered, the legislature will revisit ISRA itself, conceivably as early as 1994. NJDEPE's actual implementation of ISRA will then be clear, and the legislature should be able to assess its actual impact on, and reception by, the regulated community.

---

[3] N.J. STAT. ANN. § 13:1K-7, amended by ISRA Section 2. The legislature clearly announced its probusiness purposes in enacting the changes:

> . . . [I]t is in interest of the environment and the State's economic health to promote certainty in the regulatory process . . . to create a more efficient regulatory structure and to allow greater privatization. . . .
>
> . . . it is the policy of this State to protect the public health, safety, and the environment, to promote efficient and timely cleanups and to reduce any unnecessary financial burden of remediating contaminated sites; . . . by streamlining the regulatory process, by establishing summary administrative procedures for industrial establishments that have previously undergone environmental review, and reducing oversight . . . where less...review will assure the same degree . . . of protection to public health, safety and the environment; and that new procedures . . . shall be designed to guard against redundancy . . . and minimize governmental involvement in certain business transactions.

The legislative history contained several other statements of note (unless otherwise noted all citations will be to the Assembly Policy and Rules Committee Statement to S.1070 of June 3, 1993, and will be cited as "APRC Statement"):

> 1. [U]ntil regulations are adopted . . . [NJDEPE shall] act reasonably in the interim period when reviewing applications and petitions and in all other interactions with the public." APRC Statement p. 17, 7th bullet.
>
> 2. NJDEPE should: "[a]llow cleanups that do not remediate property to pristine levels, provided that appropriate and DEPE approved engineering or institutional controls are implemented." APRC Statement p. 17, 8th bullet.
>
> 3. NJDEPE should: "[p]rovide, to the greatest extent possible, finality to compliance with ISRA." APRC Statement p. 17 last bullet.

The impact of ISRA's changes may not be limited to New Jersey. Because of ISRA's continued extraterritorial effect, out-of-state counselors involved in New Jersey transactions must be educated as to the implications of ISRA. In addition, other state legislatures may consider New Jersey's experience under ISRA when evaluating their own environmental laws and policies. The Connecticut Transfer of Hazardous Waste Establishment Act[4] also ties environmental cleanups to business transactions involving the transfers of real property. The intent of the act is to protect purchasers by compelling full disclosure of the environmental conditions of properties before transfer. Unlike the New Jersey ECRA, the Connecticut act does not require the removal of any hazardous waste from the property and does not permit the avoidance of the transaction if the act is violated. The Connecticut act permits the transferee of the property to recover damages from the transferor. Additionally, the transferor is strictly liable for all cleanup and removal costs and for all direct damages resulting from a violation.

Another ECRA-inspired law is the Illinois Responsible Property Transfer Act of 1988.[5] Like the Connecticut Transfer of Hazardous Waste Establishment Act,[6] the Illinois act is intended to inform parties involved in certain real property transactions of existing environmental liabilities associated with the property, as well as previous use and the environmental condition of the property. The Illinois act imposes stiff civil penalties of up to $1,000 for each day a violation continues. Moreover, if a person knowingly gives false information in the disclosure statement, the person is fined $10,000 for each day the violation continues.

The following states have statutes affecting the transfer of hazardous waste sites and requiring notice and disclosure: Illinois,[7] Massachusetts,[8] Minnesota,[9] Missouri,[10] New York,[11] Pennsylvania[12] and West Virginia.[13]

---

[4] CONN. GEN. STAT. §§ 22a-134-22a-134e.
[5] ILL. ANN. STAT. ch. 30, para. 901.
[6] CONN. GEN. STAT. §§ 22a-134-22a-134e.
[7] ILL. ANN. STAT. ch. 111, para. 1021 § 1021(m), (n).
[8] MASS. ANN. LAWS ch. 21c, § 7.
[9] MINN. STAT ANN. § 115B.16, subd. 2.
[10] MO. ANN. STAT. § 260.465(2).
[11] N.Y. ENVTL. CONSV. LAW § 27-1317.
[12] 35 PA. CONS. STAT. ANN. § 6018.405.
[13] W.VA. CODE § 20-5E-20(a).

Additionally, the following states have statutes requiring approval of state environmental officials of transfers: Colorado,[14] Iowa,[15] Missouri,[16] and New York.[17]

§ 7.2.1 New Jersey Environmental Cleanup Responsibility Act

Enacted in 1983,[18] ECRA represented the New Jersey Legislature's response to the problem of abandoned hazardous waste sites. *New Jersey Department of Environmental Protection v. Ventron,*[19] a suit brought by NJDEPE against the current owners of a decaying chemical plant and their predecessors in interest, illustrated the difficulties that governmental agencies experienced in cleaning up abandoned facilities. NJDEPE was compelled to engage in expensive and time-consuming litigation to remedy the environmental damage at the plant. In similar situations, state and federal agencies were forced to initiate comprehensive cleanup efforts and legal actions against previous owners. ECRA's sponsor, State Senator Raymond Lesniak, explained that contaminated property "leav[es] a blight on our State and health hazards and economic burdens to our residents."[20] ECRA was passed in response to *Ventron*-type problems at hazardous industrial facilities.

ECRA's primary purpose was to impose liability for cleanup costs on property owners without protracted litigation. As courts have recognized, "[t]he essential goal of ECRA is to secure the cleanup of contaminated industrial sites at the earliest possible date."[21] This motive was expressed in

---

[14] COLO. REV. STAT. ANN. § 25-15-206(1).

[15] IOWA CODE ANN. §445B.430(1), (2).

[16] MO. ANN. STAT. § 260.465(1).

[17] N.Y. ENVTL. CONSERV. LAW § 27-1317,

[18] The statute became effective on December 31, 1983.

[19] 468 A.2d 150 (N.J. 1983).

[20] *See* Raymond Lesniak, *ECRA Is Coming,* 104 N.J. LAW. 41 (1983).

[21] Dixon Venture v. J. Dixon Crucible, 561 A.2d 663 (N.J. Super. Ct. App. Div. 1989), *aff'd as modified,* 584 A.2d 797 (N.J. 1991).

the legislature's finding that the generation, handling, storage, and disposal of hazardous substances and wastes pose an inherent danger of exposing citizens, property, and natural resources of this state to substantial risk of harm or degradation; that the closing of operations and the transfer of real property utilized for the . . . hazardous substances and wastes should be conducted in a rational and orderly way, so as to minimize potential risks; and that it is necessary to impose a precondition on any closure or transfer of these operations by requiring the adequate preparation and implementation of acceptable cleanup procedures.[22] As the court observed in *Superior Air Products v. NL Indus.*,[23] ECRA imposes "a self-executing duty to remediate without the necessity and delay" of a determination of liability through litigation.[24]

ECRA required all "industrial establishments"[25] to obtain prior NJDEPE approval of either a negative declaration that the site is free from possible contamination or a cleanup plan as a precondition to the sale, closure, or transfer of business operations.[26] Any "owner or operator" of an industrial establishment planning to sell or close operations must notify NJDEPE within five days after public release of its decision to close or within five days of the execution of any agreement of sale or option to buy, as applicable.[27] The NJDEPE regulations describe the type of information to be submitted, including inventories of hazardous substances; description

---

[22] N.J. STAT. ANN. § 13:1K-7.

[23] 522 A.2d 1025 (N.J. Super. App. Div. 1987).

[24] *Id.*

[25] An "industrial establishment" is defined as any place of business engaged in operations that involve the generation, manufacture, refining, transportation, treatment, storage, handling, or disposal of hazardous substances or wastes on-site, above or below ground, having a [specified] Standard Industrial Classification number. N.J. Stat. Ann. § 13:1K-9. A Standard Industrial Classification (SIC) number is assigned by the federal Office of Management and Budget and describes the type of operation. For example, number 22 refers to textile mill products and number 23 refers to apparel and similar materials. ECRA initially included SIC designation codes covering a wide range of businesses. In 1986, pursuant to its authority under the statute, NJDEPE adopted exemptions for several SIC codes originally included in ECRA.

[26] N.J. STAT. ANN. § 13:1K-9.

[27] *Id.*

of current operations; copies of all soil, groundwater, and surface water sampling results; and details of the impending transaction.[28]

After conducting a preliminary site inspection of all industrial establishments that have notified the department, NJDEPE issues a preliminary report describing site conditions and providing guidance for ECRA compliance. The department supplements all initial notice submissions with its own review of available records of relevant federal, state, county, and municipal agencies. Following approval and implementation of an appropriate sampling plan, the owner or operator of the establishment must submit either a cleanup plan or a negative declaration. All cleanup plans must be accompanied by financial security, guaranteeing the performance of the cleanup plan, in an amount equal to the cost estimate approved by NJDEPE for the plan's implementation.[29]

ECRA provided stringent penalties for failure to comply with the act or regulations. Failure to comply allows NJDEPE or the transferee to void the sale or transfer of the industrial establishment or real property.[30] The transferee is also entitled to recover damages due to the voiding of the transaction and the transferor is held to be strictly liable for all cleanup and removal costs.[31] In addition, NJDEPE may seek penalties of a maximum of $25,000 per day from any person giving false information or otherwise failing to comply with ECRA.[32] Any corporate officer who knowingly directs or authorizes violations of the act will be personally liable for ECRA penalties.[33]

ECRA served as a major incentive for businesses to plan and conduct environmentally safe industrial operations. In its six years of existence, ECRA was responsible for more than $200 million dollars in privately funded cleanups.[34]

---

[28] N.J. ADMIN. CODE § 7:1-3.7(d).

[29] N.J. STAT. ANN. § 13:1K-9(a)(2) and (b)(3).

[30] *Id.* § 13:1K-13(a) and (b).

[31] *Id.* § 13:1K-13(a).

[32] *Id.* § 13:1K-13(c).

[33] *Id.*

[34] *See* Motiuk and Sheridan, *New Jersey's ECRA: Problems, Policies, Future Trends,* 21 ENV'T REP. (BNA) No. 13, at 549 (July 27, 1990). "Already some 370 sites have been cleaned up or investigated under the program, and companies have agreed to provide $475 million to clean up another 480 sites in coming years." Barrett, *New Jersey, After Federal Leadership Waned, Has Become Environmental Protection Pioneer,* WALL ST. J., July 5, 1988, p. 44, col. 4.

ECRA was the first state law that imposed wide-ranging liability on owners and operators of commercial and industrial real estate prior to transfer or closing of business operations. As the New Jersey Supreme Court has noted, ECRA is quite unlike other environmental regimes in that it uses market forces to bring about the reversal of environmental pollution. It recognized that some environmental conditions not posing an imminent hazard to air or water resources of the state may safely attend economic activities.[35] Inspired by ECRA, Connecticut,[36] Iowa,[37] Missouri,[38] and Illinois[39] passed similar statutes.[40]

Dissatisfied with the pace of cleanup actions, New Jersey elected to regulate directly all business transactions involving the "closing, terminating, or transferring" of operations of industrial establishments that may pose an environmental risk by imposing *preconditions* upon the consummation of such transactions. New Jersey enacted the ECRA, effective in 1983, and rules implementing ECRA.[41] The preconditions upon the consummation of a closure, transfer, or termination of operations include the execution of a cleanup plan approved by the NJDEPE and the approval by the NJDEPE of a declaration that there has not been a discharge of hazardous substances or waste or, if there has been, that the property has been cleaned up pursuant to NJDEPE procedures.[42]

ISRA continues to apply to industrial establishments, essentially as previously defined under ECRA (*i.e.*, a place of business with a subject Standard Industrial Classification [SIC] number denoting hazardous substances or wastes). ISRA is triggered when the owner or operator of an

---

[35] Dixon Venture v. Joseph Dixon Crucible Co., 584 A.2d 797 (N.J. 1991) (per curiam).

[36] CONN. GEN. STAT. ANN. §§ 22a-134.

[37] Iowa Code Ann. § 445B(1), (2).

[38] Mo. Ann. Stat. § 260.465B(1), (2).

[39] Ill. Ann. Stat. ch. 30, para. 901.

[40] *See* Motiuk and Sheridan, *New Jersey's ECRA: Problems, Policies, Future Trends*, 21 ENV'T REP. (BNA) No. 13, at 549 (July 27, 1990). State legislatures in New York, Pennsylvania, and Massachusetts considered ECRA-type legislation.

[41] N.J. STAT. ANN. § 13:1K-6; N.J. ADMIN. CODE § 7:26B-1.1 (Supp. 8-21-89).

[42] N.J. STAT. ANN. § 13:K-9.

industrial establishment is involved in either "closing operations" or "transferring ownership or operations." These triggers are not dramatically different from ECRA's recent regulations.

Under ECRA and the rules governing the implementation of ECRA, sanctions for noncompliance could be severe—avoidance of the transaction and imposition of damages and stringent, escalating penalties.[43] This remedy existed in no other state statute requiring environmental compliance on the happening of a business deal. It provided for the broad power of a transferee to void a transfer for any violation of ECRA.

This provision has been changed in ISRA to permit a right to void only (1) if the transferor fails to perform a remediation and obtain NJDEPE approval and (2) after the transferee gives notice to the transferor of that failure, affording the transferor a reasonable period of time to comply (*i.e.*, a cure period). Although arguments have been made that this remedy should be eliminated, the legislature recognized that some actions have been filed by transferees seeking to void transactions and that transferees might need such protection in some circumstances.

NJDEPE has lost its separate right to void a transfer for a failure to file a negative declaration or cleanup plan. The legislature was convinced that the threat of voiding was not a weapon needed by NJDEPE because it had never sought to use it. It is not clear whether this loss applies retroactively to violations occurring prior to ISRA's effective date.

---

[43] ECRA provided that

> [f]ailure of the transferor to comply with . . . this act is grounds for voiding the sale or transfer of an industrial establishment or any real property utilized in connection therewith by the transferee, entitles the transferee to recover damages from the transferor, and renders the owner or operator of the industrial establishment strictly liable, without regard to fault, for all cleanup and removal costs and for all direct and indirect damages resulting from the failure to implement the cleanup plan.

Thus, if the owner or operator of an industrial establishment failed to comply with ECRA, the NJDEPE or the purchaser could have voided the transaction. The purchaser is entitled to recover damages resulting from avoidance of the transaction. Additionally, the owner or operator is strictly liable for cleanup and removal costs and damages caused by not implementing, or by implementing improperly, the cleanup plan.

IRSA ended this state right and determined that only private parties may sue to void the transaction.

NJDEPE retains its power to fine up to $25,000 per day. With the loss of NJDEPE's right to void, the regulated community may be more willing to engage in transactions that strain past interpretations of what events necessitate compliance with ISRA. NJDEPE has a Remediation Guarantee Fund that it can use to remediate a site when someone required to post a remediation funding source fails to do required work.[44]

NJDEPE can assert a lien against all property owned by the person required to do the work and the discharger, to the same extent as the Spill Act allows under N.J.S.A.[45] It can also seek treble damages.[46] Under ISRA Section 48, there will be a study by NJDEPE and the attorney general of key concepts of environmental liability. A report is due to the legislature within 9 months.[47]

ISRA contains two separate amnesty provisions designed to allow past violations to be corrected without fines and penalties.[48] Those guilty of ECRA violations occurring before the effective date of ISRA are given one year to seek amnesty.[49] To be eligible, violators must enter into an ACO or

[44] ISRA § 45.

[45] N.J. Admin. Code § 58:10-23.11f.

[46] ISRA §§ 45.e, 45.f.

[47] N.J. Admin. Code § 7:26B-9.2(b).

[48] ISRA § 21.

[49] ECRA provides stringent civil penalties for submitting false information or failing to comply with ECRA:

> [a]ny person who knowingly gives or causes to be given any false information or who fails to comply with the provisions of this act is liable for a penalty of not more than $25,000.00 for each offense. If the violation is of a continuing nature, each day during which it continues shall constitute an additional and separate offense. Penalties shall be collected in a civil action by a summary proceeding. . . . Any officer or management official of an industrial establishment who knowingly directs or authorizes the violation of any provisions of this act shall be personally liable for the penalties. . . .

N.J. STAT. ANN. § 13:1K-13(c); *see also* N.J. ADMIN. CODE § 7:26B-9.3(a)-(c). Thus, failure to comply with ECRA may result in stringent, escalating civil penalties. Moreover, any officer or manager who knowingly participates in violating ECRA is liable for the penalties. N.J. ADMIN. CODE § 7:26B-9.3(c).

Memorandum of Agreement (MOA), which will govern their belated compliance with ISRA. If the party thereafter fails to honor the ACO or MOA, the NJDEPE can enforce penalties from the date of the original violation.

Amnesty does not appear to resolve other enforcement issues resulting from past violations (*e.g.*, rights of transferees). Also, in order to get amnesty a person has to remediate the site. This remediation may be more costly and difficult than it would have been at the time of the violation and may require dealing with adverse interests of current owner and operator in order to perform as required. Conversely, it may be easier if the site has been through ECRA since the violation by reason of some other trigger.

Information tendered in the amnesty process cannot be used in a criminal matter. NJDEPE has yet to adopt regulations on amnesty.[50] ECRA may void the transaction if the owner or operator does not submit a negative declaration or cleanup plan.[51]

When ISRA is triggered, the owner or operator must follow a specific process. Five days after (1) the actual closing of operations or release of its decision to close, or (2) execution of an agreement to transfer, the owner or operator must provide written notice to NJDEPE. This notice is equivalent to the present ECRA general information submission.[52] Subsequent to this initial notice, the owner or operator must proceed to remediate the site (which includes investigation of the site).[53] No specific period is set, but regulations will likely provide NJDEPE's target schedule. Remediation includes preliminary assessment (equivalent to ECRA's site evaluation submission), site investigation and remedial investigation (collectively

---

[50] Spill Act, ISRA Section 39. There is a separate amnesty procedure for certain violations of the Spill Act (unreported discharges) if a directive has not yet issued, and if that person enters into an MOA or ACO to remedy the discharge. Information tendered in this process also cannot be used in a criminal matter. However, this immunity may be of no value if the violation of the Spill Act was also a violation of federal law. Given NJDEPE's recent interpretations concerning discharge reporting obligations, this provision may be very useful to eliminate this liability. Once the amnesty period passes, NJDEPE enforcement will be more strenuous.

[51] N.J. STAT. ANN. § 13:1K-13(b); N.J. ADMIN. CODE § 7:26B-9.1(b).

[52] ISRA § 4.a.

[53] ISRA § 4.b.(1).

equivalent to ECRA's use of successive sampling plans to characterize a site's areas of concerns), and, when needed, a remedial action workplan (like ECRA's cleanup plan) and remedial action (the cleanup itself).

When a site is clean, an owner or operator can submit a negative declaration.[54] Upon approval, a deal that has not yet closed (*e.g.*, under a remediation agreement) can proceed.[55] When a site is not clean, an owner or operator can submit a remedial action workplan. Once it is approved, and the remediation funding source is posted, the deal can occur. When all of the remediation at a contaminated site has been completed, NJDEPE may issue a "no further action letter."

Concomitantly with the heavy burden put upon businesses and individuals, the NJDEPE has had to cope with an enormous workload in administering ECRA. Recognizing that many operations fall within the broad industry groupings specified in ECRA, the NJDEPE has exempted various subgroups within the interdicted primary SIC groupings that do not pose a significant threat to the environment. Some of the operations and transactions not subject to ECRA are the following: any business engaged principally in agricultural production or distribution; "[a] sale or transfer of assets of an industrial establishment . . . in the ordinary course of business"; and 43 subgroups within SIC numbers 22-39, 46-49, 51, or 76, such as a water supply and an irrigation system.[56]

The NJDEPE has adopted the practice of issuing applicability determinations, or letters of nonapplicability, which have become a safe harbor for many business transactions. Out of 5,000 of these requests processed in 1987, 93 percent resulted in exemptions or findings of nonapplicability. The NJDEPE now routinely processes applications within a few days, and generally in no more than three weeks in the most complicated situations. A person seeking an applicability determination must submit a required form and fee to the NJDEPE, must permit entrance to and inspection of the property, and must show the NJDEPE that ECRA does not apply.[57]

Although ISRA is silent on the process, presumably NJDEPE will continue its helpful process of issuing letters of nonapplicability. Letters of

---

[54] ISRA § 4.d.
[55] ISRA § 4.c.
[56] N.J. ADMIN. CODE. § 7:26B-1.8(a)(3), (6), (b).
[57] *Id.* § 7:26B-1.9 (Supp. 8-21-89).

nonapplicability explain NJDEPE's position as to whether a particular transaction or event triggers the need to comply with ECRA.

Even after granting exemptions, the NJDEPE faces numerous case filings a year, which require NJDEPE notification and individual compliance. An owner or operation should consider the following: (1) seeking an exemption for a low-volume generator; (2) filing a negative declaration, if possible; and (3) seeking an administrative consent order, which will permit a transaction to proceed, pending ECRA compliance, but places the applicant at the mercy of the NJDEPE, which can require a substantial bond or other form of financial security.

§ 7.2.2 Connecticut Transfer of Hazardous Waste Establishment Act

Connecticut also ties environmental cleanups to business transactions involving property transfers. Basically the Connecticut Transfer of Hazardous Waste Establishment Act[58] is intended to protect purchasers by compelling full disclosure of the environmental conditions of properties before transfer. Since the underlying environmental risks exist independently under CERCLA and other environmental laws, the practical effect of the act is to bring to focus environmental considerations that should be routinely addressed in all transactions.

The Connecticut Transfer of Hazardous Waste Establishment Act applies to the "transfer of [an] establishment," defined as

> the transfer of any operations which involve the generation, recycling, reclamation, reuse, transportation, treatment, storage, handling or disposal of hazardous waste, or any other transaction or proceeding through which an establishment undergoes a change in ownership, including, but not limited to, sale of stock in the form of a statutory merger or consolidation, sale of the controlling share of the assets, the conveyance of real property, change of corporate identity or financial reorganization, but excluding corporation reorganization not substantially affecting the ownership of the establishment.[59]

Thus, this broad term includes the sale of stock through a merger or consolidation, the disposition of the controlling share of assets, a sale of

---

[58] CONN. GEN. STAT. ANN. §§ 22a-134 to 22a-134e.

[59] *Id.* § 22a-134(1).

real property, a change in corporate identity, and a financial reorganization significantly affecting the ownership of the establishment.

The definition of "establishment" is not tied to the SIC system and is as follows: "any establishment [that since] May 1, 1967, generated more than one hundred kilograms of hazardous waste per month or [that] recycled, reclaimed, reused, stored, handled, treated, transported or disposed of hazardous waste generated by another" entity.[60]

Unlike the New Jersey program, this state act defines "hazardous waste" as "any waste material, except by-product material, source material or special nuclear material, . . . which may pose a present or potential hazard to human health or environment when improperly disposed of, treated, stored, transported, or otherwise managed."[61] The definition excludes sewage.[62] Thus, the act applies to, for example, a sale of stock substantially affecting the ownership of a business engaged in the transportation or handling of hazardous waste since May 1967.

Again unlike the New Jersey laws, the Connecticut act does not specifically authorize the voiding of the underlying transaction in the event of noncompliance with the act. The Connecticut act also differs in several other significant respects from those of New Jersey:

1. the act does not apply to closures or termination of leases;
2. the act deals only with hazardous wastes, not hazardous substances or petroleum products;
3. excludes low-volume generators directly from its coverage; and
4. has relatively simple reporting requirements.

When applicable, the act also requires the transferor to submit a "negative declaration" to both the transferee and the Connecticut Commissioner of Environmental Protection. A "negative declaration" is

> a written declaration on a form prescribed by the commissioner stating (1) that there has been no discharge, spillage, uncontrolled loss, seepage or filtration of hazardous waste on-site, or that any such discharge, spillage, uncontrolled loss, seepage or filtration has been cleaned up in accordance to procedures approved by the

---

[60] *Id.* § 22a-134(3).
[61] *Id.* § 22a-115(1).
[62] *Id.* § 22a-134(4).

commissioner or determined by him to pose no threat to human health or safety or the environment which would warrant containment and removal or other mitigation measures and (2) that any hazardous waste which remains on-site is being managed in accordance with [the Act] and regulations. . . .[63]

Again in contrast to the early New Jersey ECRA law, the Connecticut act does not require that all hazardous substances or wastes be removed from the property. All that is necessary in order to make a negative declaration is that the seller provide assurances that any hazardous waste is being properly managed.[64]

If the transferor is unable to make a negative declaration, the transferor, before a transfer, must certify to the Connecticut Commissioner of Environmental Protection that, "to the extent necessary to minimize or mitigate a threat to human health or the environment," the transferor "will contain, remove, or otherwise mitigate" the effects of any hazardous wastes according to a schedule and procedures approved by the commissioner and embodied in an administrative order, stipulated judgment, or consent agreement.[65]

If the transferor violates the act, the transferee may recover damages from the transferor.[66] Furthermore, the transferor is strictly liable for all cleanup and removal costs and for all direct damages resulting from the violation.[67]

If a person knowingly gives false information or does not comply with the act, the person is civilly liable for up to $100,000.[68] In contrast to the New Jersey laws, the Connecticut act does not empower the commissioner to void the transaction in the event of noncompliance with the act.

---

[63] *Id.* § 22a-134(2), (5).

[64] *Id.* § 22a-134(5).

[65] *Id.* § 22a-134a(c).

[66] *Id.* § 22a-134b.

[67] *Id.*

[68] *Id.* § 22a-134d.

§ 7.2.3 Illinois Responsible Property Transfer Act of 1988

The last of the ECRA-inspired laws was the Illinois Responsible Property Transfer Act of 1988.[69] The act is intended to see that parties involved in certain real property transactions know about existing environmental liabilities related to the properties, as well as the prior use and environmental condition of the properties.[70]

The Illinois Responsible Property Transfer Act applies to certain "transfers of real property . . . after January 1, 1990."[71] The act defines a "transfer" as "any conveyance of an interest in real property," such as by deed; a lease of at least forty years; an assignment of at least 25 percent of the beneficial interest in a state land trust; or "a mortgage, trust deed, or collateral assignment of a beneficial interest in a" state land trust.[72]

The act applies to transfers of "real property," including improvements thereon, located in Illinois and containing at least one "facility" subject to reporting requirements under the Emergency Planning and Community Right-to-Know Act of 1986,[73] and regulations thereunder or a storage tank requiring notification under the Solid Waste Disposal Act.[74] Thus, the real property must be located in Illinois and be subject to reporting under the federal laws, the Emergency Planning and Community Right-to-Know Act, or the Solid Waste Disposal Act.

The Illinois act requires the transferor, within thirty days after execution of any written contract and not later than thirty days before the transfer of real property, to deliver to the transferee and to any lender a disclosure statement in a form prescribed by Illinois.[75] A transferee or a lender may void a duty "to accept a transfer or finance a transfer . . . yet to be" consummated if the disclosure statement reveals environmental defects previously unknown to the transferee or lender or if the transferor fails to make the required disclosure, unless the transferee or lender waives the

---

69 Ill. Ann. Stat. ch. 30, para. 901.
70 *Id.* at ch. 30, para. 902.
71 *Id.* at ch. 30, para. 904, § 4(a).
72 *Id.* ch. 30, para. 903, § 3(g).
73 42 U.S.C. § 11022.
74 42 U.S.C. § 6991.
75 Ill. Ann. Stat. ch. 30, para. 904, § 4(a).

right.[76] However, the transferee or lender must act to void the transaction within ten days after requesting or receiving the disclosure statement.[77]

Both the transferor and the transferee are responsible for seeing that the disclosure statement is recorded in the county where the property is located within thirty days after transfer of the real property.[78] Both the transferor and the transferee may be required to file the disclosure statement with the Environmental Protection Agency.[79]

To enforce the act, the act imposes stiff civil penalties for violations. A person violating the duty to disclose is civilly liable for up to $1,000 for the violation and $1,000 for each day the violation continues.[80] A person who knowingly makes a false statement in the disclosure statement is civilly liable for up to $10,000 for the violation and $10,000 for each day the violation continues.[81] Failure to record a disclosure statement results in joint and several civil penalty of up to $10,000 against the transferor and the transferee.[82]

§ 7.3 Other State Notification Laws

The following table illustrates the states having environmental disclosure laws pertaining to hazardous wastes and other environmental hazards. The table also delineates which states place deed restrictions on certain properties for environmental hazards.

State	Citation	Description of Miscellaneous Environmental Disclosure Requirements
Alabama	—	—
Alaska	—	—

[76] *Id.* ch. 30, para. 904, § 4(c).
[77] *Id.*
[78] *Id.* ch. 30, para. 906, § 6.
[79] *Id.*
[80] *Id.* at ch. 30, para. 907, § 7.
[81] *Id.* at ch. 30, para. 907, § 7(b).
[82] *Id.* at ch. 30, para. 907(c).

State	Citation	Description of Miscellaneous Environmental Disclosure Requirements
Arkansas	—	—
California	CAL. CIV. CODE §1102 *et seq.*	"Real Estate Transfer Disclosure Statement" to be exchanged when real property with one to four dwelling units is transferred, inquires into the seller's knowledge of asbestos, radon gas, lead-based paint, and fuel or chemical tanks.
Colorado	—	—
Connecticut	—	—
Delaware	—	—
Dist. of Col.	—	—
Florida	—	—
Georgia	—	—
Hawaii	—	—
Idaho	—	—
Illinois	ILL. REV. STAT. ch. 765, para. §77 *et seq.* ILL. REV. STAT. ch. 410, para. §45 *et seq.*	Sellers of real property required to disclose the condition of the property, including whether the seller is aware of "unsafe concentrations" of lead paint. Authorizes the Illinois Department of Public Health to order an owner of residential property to remove, replace, or permanently cover the lead-bearing substances if it determines that the lead-bearing substances present on the property may be hazardous to children or if it learns that an individual has a level of lead in the blood in excess of permissible health limits.
Indiana	—	—
Iowa	—	—
Kansas	—	—
Kentucky	—	—

State	Citation	Description of Miscellaneous Environmental Disclosure Requirements
Louisiana	—	—
Maine	—	—
Maryland	M.D. CODE ANN, REAL PROP. § 10-702	Disclosure form requires seller to disclose knowledge of asbestos, radon, USTs, and lead paint.
Massachusetts	MASS. GEN. LAWS ANN. ch. 111, 190 *et seq.*	Massachusetts has an extensive lead poisoning prevention and control program. Owners of residential buildings must sensor lead-based materials and disclose the presence of lead to a prospective buyer.
Michigan	MICH. COMP. LAWS ANN. § 565.957	Requires sellers of real property to disclose any knowledge of "environmental problems."
Minnesota	—	—
Mississippi	—	—
Missouri	—	—
Montana	MONT. CODE ANN. § 75-3-600	Requires seller to disclose knowledge of radon.
Nebraska	—	—
Nevada	—	—
New Hampshire	—	—
New Jersey	—	—
New York	—	—
North Carolina	—	—
North Dakota	—	—
Ohio	OHIO REV. CODE ANN. § 5302.30	Requires sellers of real property to disclose the condition of the property, including the presence of lead-based paint.

State	Citation	Description of Miscellaneous Environmental Disclosure Requirements
Oklahoma	—	—
Oregon	—	—
Pennsylvania	—	—
Rhode Island	R.I. GEN. LAWS §5-20.8-2	Sellers required to disclose knowledge of asbestos, lead paint, and radon.
South Carolina	—	—
South Dakota	—	—
Tennessee	—	—
Texas	TEX. PROP. CODE ANN. § 5.003	Disclosure form requires sellers to disclose knowledge of radon, lead, and asbestos.
Utah	—	—
Vermont	—	—
Virginia	VA. CODE ANN. § 55-519 VA. CODE ANN. §36-107.1	Disclosure form may require disclosure of seller's knowledge of asbestos, radon, and USTs. Owners of property who have been notified by local building officials or local health officials of levels of unacceptable levels of lead-based paint required to notify prospective purchasers in writing. Failure to comply with the notice requirements may subject the seller to a civil penalty of $1000.
Washington	—	—
West Virginia	—	—

State	Citation	Description of Miscellaneous Environmental Disclosure Requirements
Wisconsin	WIS. STAT. ANN. §7099.03	"Real Estate Condition Report" covers the owner's knowledge of the presence of various hazards including asbestos and USTs.
	WIS. STAT. ANN. §254.166	Authorizes the Department of Health and Social Services to order the owner of property determined to be a lead hazard to reduce or eliminate imminent lead hazards within 5 days and other lead hazards within 30 days. There are also other notification statutes that are not triggered by a transfer of the property.
Wyoming	—	—

§ 7.3.1 California

The California Health and Safety Code[83] requires that until July 1, 1996, "any owner of nonresidential real property who knows, or has reasonable cause to believe, that any release of hazardous substance has come to be located on or beneath that real property must, prior to the sale, lease, or rental of the real property by that owner, give written notice of that condition to the buyer, lessee, or renter of the real property."[84] Failure of the owner to provide written notice to the buyer, lessee, or renter shall subject the owner to actual damages and any other remedies provided by law. In addition, where the owner has actual knowledge of the presence of any release of a material amount of a hazardous substance and knowingly and willfully fails to provide written notice to the buyer, lessee, or renter, the owner will be held liable for a civil penalty not to exceed $ 5,000 for each separate violation.

---

[83] CAL HEALTH & SAF. CODE § 25359.7.

[84] *Id.* § 25359.7(a).

In addition, the California code requires that any lessee or renter of real property who knows or has reasonable cause to believe that any release of a hazardous substance has come or will come to be located on or beneath that real property will, within a reasonable period of time, either prior to the release or following the discovery by the lessee or renter of the presence or believed presence of the hazardous substance release, give written notice of that condition to the owner of the real property or to the lessor under the lessee's or renter's lease or rental agreement. A lessee or renter who fails to provide written notice, when required, to the owner or lessor may be subject to actual damages as well as any other remedy provided by law. If the lessee or renter has knowledge of the presence of a release of a material amount of a hazardous substance, or of a hazardous substance release that is required to be reported to a state or local agency pursuant to law, on or under the real property leased or rented by the lessee or renter and knowingly and willfully fails to provide written notice, when required, to the owner or lessor, both of the following apply:

✓ Owner's or lessor's written notice to the lessee or renter, under the lessee's or renter's lease or rental agreement, except this does not apply to lessees and renters of property used exclusively for residential purposes; and

✓ The lessee or renter is liable for a civil penalty not to exceed $5,000 for each separate violation.

A lessee or renter may cure a default under the lessee's or renter's lease or rental agreement, by promptly commencing and completing the removal of, or taking other appropriate remedial action with respect to, the hazardous substance release. The removal or remedial action must automatically be conducted in accordance with all applicable laws and regulations and in a manner which is reasonably acceptable to, and which is approved in writing by, the owner or lessor. Curing the default will not, however, relieve the lessee or renter of any liability for actual damages or for any civil penalty for a violation.[85]

---

[85] For further discussion of the California rules, *see* Handling real property sales transactions, CEB Action Guide; Minimizing toxics liability when buying real property and businesses, CEB Action Guide, Spring 1992; Taking a closer look: Significant new California legislation relating to real property enacted in 1987, 11 CEB Real Prop L. Rep. 37; Taking a closer look: significant new California legislation enacted in 1988, 12 CEB Real Prop L. Rep. 55.

§ 7.3.2 Colorado[86]

The Colorado Hazardous Waste Act[87] covers hazardous waste disposal sites.[88] The act is triggered by a "substantial change" in ownership, design, or operation of a hazardous waste disposal site,[89] hazardous waste incinerator, or incinerator site.[90] Change in ownership includes assignment or transfer of the certificate of designation.[91]

When the owner intends to make such a "substantial change" he must submit an application[92] to the Board of County Commissioners or governing body of the municipality and request approval of the change before it becomes effective.[93] Approval of a substantial change may be made only if all of the factors in section 25-15-203 are satisfied.[94]

Colorado state regulations require that, within sixty days after closure of a hazardous waste disposal facility, the owner or operator must record a statement in the land records notifying potential purchasers that the property was once used to manage hazardous waste.[95]

§ 7.3.3 Connecticut

The Connecticut Transfer Act[96] requires transferors of facilities that produce more than 100 kilograms each month of hazardous waste or that deal with hazardous wastes generated by other entities to certify to the

---

[86] COLO. REV. STAT. §§ 25-15-101, *et seq.* (1993).

[87] *Id.*

[88] *Id.* at § 206(1).

[89] *Id.*

[90] COLO. REV. STAT. § 25-15-507.

[91] *Id.*

[92] The application must be accompanied by a fee not to exceed $10,000, which may be partially or totally refunded. *Id.* at § 206(3). The purpose of the fee is to assist in funding the review of the substantial change. *Id.*

[93] *Id.* at § 206(1). However, if a hazardous waste disposal site has been designated by the council pursuant to section 25-15-217, any substantial change is to be approved by the council until January 1, 1986, and changes thereafter are to be approved by the department.

[94] *Id.* at § 206(2).

[95] COLO. REV. STAT. § 25-25-303; 6 Colo. Code Regs. 1007-3 §264.119.

[96] CONN. GEN. STAT. ANN. § 22a-454(b).

transferee that either: (1) there has been no discharge of hazardous waste at the property or (2) if a discharge existed, it has been cleaned up. The certification is to be filed with the state, and if the transferor cannot certify to one of the above two items, either the transferor or transferee must certify that it will clean up the site.[97]

§ 7.3.4 Illinois

In Illinois, a person may not transfer an interest in or use real property previously "used as a hazardous waste disposal site" unless the person notifies, in writing, the Illinois Environmental Protection Agency of the transfer and the transferee of any restrictions the agency imposes upon the use of the property.[98] The state is required to impose conditions "reasonably necessary to protect public health and the environment." These may include permanently prohibiting use of the property to prevent an "unreasonable risk of injury to human health or to the environment."[99]

§ 7.3.5 Indiana

The Indiana Environmental Hazardous Disclosure and Responsible Party Transfer Law[100] requires that the transferor complete and record a disclosure[101] statement concerning the environmental[102] history of the

---

[97] *See* I. LEO MOTIUK ET AL., *Environmental Transfer Law Update: New Jersey and the Nation,* TOXICS L. REP. (BNA) 487, 488 (1992).

Note also that the Governor of Connecticut signed the Landowner Liability Act, Public Act. No. 93-375, on June 30, 1993, effective as of that date, which establishes an innocent landowner defense to avoid liability for spills or discharges on property. New Jersey's ISRA provides for a similar innocent purchaser defense.

[98] ILL. ANN. STAT. at ch. 111 1/2, para. 1021, § 1021(m), (n).

[99] *Id.* at ch. 111 1/1, para. 1039(g) (Smith-Hurd 1988 & Supp. 1992).

In addition, as discussed *supra,* the Illinois' Responsible Property Transfer Act, ILL. ANN. STAT. ch. 30, 1/2 901, requires that the transferor provide notice, in the form of a disclosure statement, to the transferee and lender. The transferee or lender is released from any obligation to the transfer prior to closing, if the transferor does not provide the disclosure statement. Motiuk et al., *supra,* note 99 at 488.

[100] IND. CODE § 13-7-22.5-1 *et seq.*

[101] *Id.* § 13-7-22.5 -1 defines a "Disclosure document" as "a document that sets forth certain information about a property that is to be transferred."

[102] *Id.* § 13-7-22.5 -1.5 of the act defines an "Environmental defect" as an

property[103] upon transfer.[104] This disclosure statement must be provided to

---

environmentally related commission, omission, activity, or condition that:

1. Constitutes a material violation of an environmental statute, regulation, or ordinance;
2. Would require remedial activity under an environmental statute, regulation, or ordinance;
3. Presents a substantial endangerment to:
   (A) The public health;
   (B) The public welfare; or
   (C) The environment;
4. Would have a material, adverse effect on the market value of the property or of an abutting property; or
5. Would prevent or materially interfere with another party's ability to obtain a permit or license that is required under an environmental statute, regulation, or ordinance to operate the property or a facility or process on the property.

[103] *Id.* § 13-7-22.5-6 of the statute defines "Property" as:

a specific and identifiable parcel of real property, including any improvement that:

1. Contains one (1) or more facilities that are subject to reporting under Section 312 of the federal Emergency Planning and Community Right-to-Know Act of 1986 (42 U.S.C. 11022);
2. Is the site of one (1) or more underground storage tanks for which notification is required under 42 U.S.C. 6991a and IC 13-7-20-13(8)(A); or
3. Is listed on the Comprehensive Environmental Response, Compensation and Liability Information System (CERCLIS) in accordance with Section 116 of CERCLA (42 U.S.C. 9616).

The term does not include property that has been subject to bonding or other financial assurances released by the appropriate governmental agency after compliance with applicable state laws.

[104] *Id.* § 13-7-22.5-7 defines "Transfer" as:

(a) As used in this chapter, "transfer" means a conveyance of an interest in property by any of the following:

1. A deed or other instrument of conveyance of fee title to property.
2. A lease whose term, if all options were exercised, would be more than forty (40) years.
3. An assignment of more than twenty-five percent (25%) of the beneficial interest in a land trust.
4. A collateral assignment of a beneficial interest in a land trust.
5. An installment contract for the sale of property.
6. A mortgage or trust deed.
7. A lease of any duration that includes an option to purchase.

(b) The term does not include a conveyance of an interest in property by any of the following:

1. A deed or trust document, which without additional consideration, confirms, corrects, modifies, or supplements a deed or trust document that was previously recorded.
2. A deed or trust document that, without additional consideration, changes title to property without changing beneficial interest.

the transferee[105] and lender thirty days prior to the transfer.[106] The transferee or lender[107] is released from any obligation pertaining to the transfer if the

---

3. A tax deed or a deed from a county transferring property the county received under IC 6-1.1-25-5.5.
4. An instrument of release of an interest in property that is security for a debt or other obligation.
5. A deed of partition.
6. A conveyance occurring as a result of the foreclosure of a mortgage or other lien on real property.
7. An easement.
8. A conveyance of an interest in minerals, gas, or oil (including a lease).
9. A conveyance by operation of law upon the death of a joint tenant with right of survivorship.
10. An inheritance or devise.
11. A deed in lieu of foreclosure.
12. A Uniform Commercial Code sale or other foreclosure of a collateral assignment of a beneficial interest in a land trust.
13. A deed that conveys fee title under an installment contract for the sale of property.
14. A deed that conveys fee title under an exercise of an option to purchase contained in a lease of property.

[105] *Id.* § 13-7-22.5-8(a) defines a "Transferee "as:
1. A buyer, mortgagee, grantee, or lessee of real property;
2. An assignee of an interest of more than twenty-five percent (25%) in a land trust; or
3. In the case of a transfer to the trustee of a land trust, the owners of the beneficial interest of the land trust.

(b) The term includes a prospective transferee.

[106] "Transferor" is defined as:
1. A seller, grantor, mortgagor, or lessor of real property;
2. An assignor of an interest of more than twenty-five percent (25%) in a land trust; or
3. In the case of a transfer by the trustee of a land trust, the owner of the beneficial interest of the land trust.

(b) The term includes a prospective transferor.

Id. § 13-7-22.5-9.

[107] *Id.* § 13-7-22.5-3 defines a "Lender" as "a person that provides loans secured by an interest in property or by an assignment of beneficial interest in a land trust." § 13-7-22.5 -2 defines a "Land trust" as a trust that is established under terms providing that:
1. The trustee holds legal or equitable title to property;
2. The beneficiary has the power to manage the trust property, including the power to direct the trustee to sell the property; and
3. The trustee may sell the trust property:

(A) Only at the direction of the beneficiary or other person; or
(B) After a period of time stipulated in the terms of the trust.

statement discloses an environmental defect in the property.[108]

A buyer of property who finances the purchase of the property through a mortgage loan is not required to deliver a disclosure document to the mortgagee that provides that mortgage loan.[109] A person who lends money and takes a mortgage on property to secure that loan is not required to record a disclosure document concerning the property in the office of the recorder of the county in which the property is located, or to file a copy of the disclosure document with the department of environmental management.

A party to a transfer of property may bring a civil action against another party to the transfer of property to recover consequential damages based upon a violation of this act. The party may also recover reasonable costs and attorney's fees.[110]

It is unlawful for a transferor to knowingly make a false statement in a disclosure document.[111] Each day that the transferor knows of the falsity of the statement made in the disclosure document but fails to correct that statement through the filing, recording, and delivery of a corrected disclosure statement constitutes a separate infraction.[112]

A person who is responsible for filing a disclosure document in the office of the county recorder and who fails to record the disclosure document is also guilty of an unlawful act. However, the failure of a transferee to record a disclosure document within the period allowed is not an infraction under this section if the disclosure document:

1. Was not delivered to the transferee within the time allowed under section 10 [IC 13-7-22.5-10] of this chapter; or
2. Contains one or more false statements about substantive matters.[113]

---

[108] *Id. See generally Seller Beware: The Indiana Responsible Party Transfer Law*, 24 IND. L. REV. 761 (1991); *Indiana Environmental Law: An Examination of 1989 Legislation*, 23 IND. L. REV. 329 (1990); *Indiana Responsible Property Transfer Law*, 34 RES GESTAE 78 (1990); *Look before you leap into real estate deals: Indiana's Responsible Property Transfer Law*, 33 RES GESTAE 264 (1989).

[109] IND. CODE ANN. § 13-7-22.5-19.5 (Burns 1993).

[110] *Id.* § 13-7-22.5-21 (Burns 1993).

[111] *Id.* § 13-7-22.5-18 (Burns 1993).

[112] *Id.*

[113] *Id.* § 13-7-22.5-19 (Burns 1993).

If a recorded disclosure document reports the existence of an environmental defect on a property, a person who has a financial interest in the property may record, in the same recorder's office in which the disclosure document is recorded, a document that reports that the environmental defect has been eliminated from the property. A document so filed must be certified by a registered professional engineer who does not have a financial interest in the property.[114]

The Indiana Environmental Hazardous Disclosure and Responsible Party Transfer Law (RPLA)[115] covers the same property as the Illinois Responsible Party Transfer Act except that it adds real property listed on the Comprehensive Environmental Response, Compensation and Liability Information System (CERCLIS).[116]

RPLA is triggered upon the transfer of any real property after December 31, 1989,[117] that meets the requirements above. Transfer is defined as the conveyance of any interest in real property and includes deeds, leases over forty years, assignments of more than 25 percent of the beneficial interest in a land trust, installation contracts, and leases with the option to purchase.[118] Foreclosures, deeds in lieu of foreclosure, easements, inheritance or devise are some transactions specifically excluded from the definition of transfer.[119]

Once a transaction triggers RPLA, a disclosure document must be completed and delivered to the transferee.[120] The transferor must disclose potential environmental liability, identify the property and its characteristics, the type of transfer, uses of the property by the transferor as well as previous owners and lessees of the transferor, and any release of hazardous substances or petroleum required to be reported by state or federal law.[121] The disclosure document must be delivered at least 30 days

---

[114] *Id.* § 13-7-22.5-22 (Burns 1993).

[115] *Id.*

[116] IND. CODE ANN. § 13-7-22.5-6.

[117] *Id.* § 13-7-22.5-6.

[118] *Id.* § 13-7-22.5-7.

[119] *Id.*

[120] *Id.* A transferee is defined as a buyer, mortgagee, grantee, lessee, assignee of more than 25 percent interest in a land trust, or trustee who is the owner of the beneficial interest of a land trust. *Id.* at § -22.5-8(a).

[121] IND. CODE ANN. § 13-7-22.5-15.

prior to the transfer[122] but can be waived if all parties execute written statements that they are aware of the purpose and intent of the disclosure document.[123] Even if the thirty day period is waived, the disclosure document must still be delivered to all parties to the transfer on or before the transfer becomes final.[124]

If environmental defects are revealed in the disclosure document that were not known to the parties to the transfer, then any party, including the lender, may void the transfer before it becomes final.[125] In addition, if the transferor does not provide a disclosure document, then any party who did not receive the document may demand one.[126] If the party does not receive the disclosure within ten days of the demand, or receives it within ten days but it reveals environmental defects, the party may void the transfer or obligation to finance the transfer.[127] A party may not void the transaction due to a transferee's failure to respond to a demand or for environmental defects once the transfer is final.[128]

After delivery to the transferee and lender of the disclosure document, the disclosure must be recorded in the county where the property is located.[129] In addition, a copy of the document must be filed within thirty days after the transfer with the Indiana Department of Environmental Management.[130] The transferor and transferee are jointly responsible for recording the document.[131]

---

[122] *Id.*

[123] *Id.* at § 13-7-22.5-10(a).

[124] *Id.*

[125] *Id.* at § 13-7-22.5-11. An "environmental defect" is a commission, omission, activity, or condition that (1) is a material violation of an environmental statute, regulation, or ordinance; (2) would require remedial activity; (3) substantially endangers public health, welfare, or the environment; (4) would materially affect the market value of the property or adjacent property; or (5) would prevent or interfere with the ability of another party to receive a required permit or license. *See id.* at § 13-7-22.5-1.5.

[126] *Id.* at § 13-7-22.5-12.

[127] *Id.* The same remedy is available in the case of a transferor who waives the thirty day requirement but fails to provide the disclosure document before the date the transfer is to become final.*Id.* at § 13-7-22.5-13.

[128] *Id.* at § 13-7-22.5-14.

[129] *Id.* at § 13-7-22.5-16(a).

[130] *Id.*

[131] *Id.* at § 13-7-22.5-16(c).

Numerous penalties are available under RPLA. Failure to prepare and provide the disclosure document is a Class B offense punishable by up to a $1,000 civil penalty and court costs.[132] Failure to record the disclosure document is a Class A offense.[133] A transferee does not violate the recordation requirement, however, if the disclosure document was not delivered during the required time period, or contains false statements with respect to substantive matters.[134] Knowingly making a false statement in the disclosure document is also a Class A offense.[135] The state is authorized to bring suit in the judicial circuit where the violation took place[136] and may impose a civil penalty of up to $10,000 as well as court costs.[137]

§ 7.3.6 Iowa

Iowa law provides that "a person shall not substantially change the manner in which a hazardous waste or hazardous substance disposal site on the registry pursuant to section 455B.426 is used without the written approval of the director."[138] In addition, a person can "not sell, convey, or transfer title to a hazardous waste or hazardous substance disposal site which is on the registry pursuant to section 455B.426 without the written approval of the director," who must respond to a request for a change of ownership within thirty days of its receipt. Decisions of the state concerning the use or transfer of a hazardous waste or hazardous substance disposal site may be appealed.

If the state has reason to believe that a person has not complied with this requirement, or that there is an imminent danger that the statute will be violated, "the director may institute a civil action in district court for injunctive relief to prevent the violation and for the assessment of a civil

---

[132] IND. CODE ANN. § 34-4-32-4.
[133] *Id.* at § 34-4-32-19.
[134] *Id.*
[135] *Id.* at § 34-4-32-18.
[136] *Id.* at § 34-4-32-1.
[137] *Id.* at § 34-4-32-4.
[138] IOWA CODE § 455B.430.

penalty, not to exceed one thousand dollars per day for each day of violation."[139]

Immediately upon the listing of real property in the registry of hazardous waste or hazardous substance disposal sites, a person liable for cleanup costs must submit to the state a report consisting of documentation of the responsible person's liabilities and assets, including, if filed, a copy of the annual report submitted to the secretary of state pursuant to chapter 490. A subsequent report must be submitted annually on April 15 for the period the site remains on the registry.[140]

---

[139] *Id.* Moneys collected under this subsection must, however, be deposited in the remedial fund.

[140] The Iowa code dictates:

> With each declaration of value submitted to the county recorder under chapter 428A, there shall also be submitted a statement that no known wells are situated on the property, or if known wells are situated on the property, the statement must state the approximate location of each known well and its status with respect to section 159.29 or 455B.190. The statement shall also state that no known disposal site for solid waste, as defined in section 455B.301, which has been deemed to be potentially hazardous by the department of natural resources, exists on the property, or if such a known disposal site does exist, the location of the site on the property. The statement shall additionally state that no known underground storage tank, as defined in section 455B.471, subsection 11, exists on the property, or if a known underground storage tank does exist, the type and size of the tank, and any known substance in the tank. The statement shall also state that no known hazardous waste as defined in section 455B.411, subsection 3, or listed by the department pursuant to section 455B.412, subsection 2, or section 455B.464, exists on the property, or if known hazardous waste does exist, that the waste is being managed in accordance with rules adopted by the department of natural resources. The statement shall be signed by at least one of the sellers or their agents. The county recorder shall refuse to record any deed, instrument, or writing for which a declaration of value is required under chapter 428A unless the statement required by this section has been submitted to the county recorder. A buyer of property shall be provided with a copy of the statement submitted, and, following the fulfillment of this provision, if the statement submitted reveals no well, disposal site, underground storage tank, or hazardous waste on the property, the county recorder may destroy the statement. The land application of sludges or soils resulting from the remediation of underground storage tank releases accomplished in compliance with department of natural resources rules without a permit is not required to be reported as the disposal of solid waste or hazardous waste.

§ 7.3.7 Massachusetts

In Massachusetts, no transfer or lease of land upon which or in which hazardous waste has been disposed may occur "until notice of such disposal is recorded in the registry of deeds" or in the land court for the district where the property is located.[141] The act also limits the use of such land to a hazardous waste disposal facility until notice of such disposal is rccordcd.[142]

§ 7.3.8 Michigan

Michigan's Environmental Response Act (MERA)[143] requires written notification to the transferee of releases at the site. The transferor must provide the location, amount, and date of each release. MERA[144] is very broad because it covers "all real property."[145] If a person transfers an interest in real property and has actual or constructive knowledge there has been a release of a reportable quantity of a hazardous substance, the disclosure requirements of MERA are triggered.[146]

If a declaration of value is not required, the above information shall be submitted on a separate form. The director of the department of natural resources shall prescribe the form of the statement and the separate form to be supplied by each county recorder in the state. The county recorder shall transmit the statements to the department of natural resources at times directed by the director of the department.

The owner of the property is responsible for the accuracy of the information submitted on the form. The owner's agent shall not be liable for the accuracy of information provided by the owner of the property. The provisions of this paragraph do not limit liability which may be imposed under a contract or under any other law.

[141] MASS. ANN. LAWS ch. 21C, § 7.

[142] Id. *See also* Hemingway Transport v. Kahn, 73 B.R. 494 (Bankr. D. Mass. 1987) (purchaser not entitled to recision due to seller's failure to record a notice of disposal of hazardous waste as required by state law).

[143] MICH. COMP. LAWS ANN. § 299.610c.

[144] *Id.*

[145] *Id.*

[146] *Id.*

In order to comply with MERA, prior to the transfer, the transferor must provide written notice to the transferee of the nature and extent of any such release.[147] If the instrument of conveyance must be recorded, then the notice must also be recorded.[148] If the site is subject to a cleanup plan that allows cleanup to less than a residential cleanup standard (*i.e.*, commercial or industrial cleanup standards), notice of remediation must be recorded with the deed.[149] The act makes no provisions for remedies available for violations of these requirements.

§ 7.3.9 Minnesota[150]

Minnesota requires pretransfer recording of disclosure statements at sites of hazardous disposal facilities.[151] If the owner of real property knows, or should have known, that property was used as a hazardous waste disposal facility site or was "subject to extensive contamination by release of a hazardous substance," the owner, before transferring ownership, is required to record an affidavit about the use and condition of the property.[152] The owner must record, in the county where the property is located, an affidavit informing the potential transferee of the prior use, contamination, and any restrictions upon the use of the property.[153] Failure to record the required "affidavit . . . does not affect or prevent any transfer of ownership of the property."[154] Any person who knowingly does not record the affidavit is, however, "liable . . . for any release or threatened release of any hazardous substance from a facility located on" the property.[155]

---

[147] *Id.*
[148] *Id.*
[149] MICH. COMP. LAWS § 324.20120b.
[150] MINN. STAT. § 115B.01.
[151] *See* § 115B.16.
[152] MINN. STAT. ANN. § 115B.16, subd. 2.
[153] *Id.*
[154] *Id.*
[155] *Id.* § 115B.16, subd. 4(b).

§ 7.3.10 Missouri[156]

Missouri law dictates that "[n]o person may substantially change the manner in which an abandoned or uncontrolled hazardous waste disposal site on the registry prepared and maintained by the department pursuant to section 260.440 is used without the written approval of the director."[157] The law also states that "[n]o person may sell, convey or transfer title to an abandoned or uncontrolled hazardous waste disposal site which is on the registry prepared and maintained by the department pursuant to section 260.440 without disclosing to the buyer early in the negotiation process that the site is on the registry, specifying applicable use restrictions and providing all registry information for the site."[158]

The seller is obligated under the statute to notify the buyer that he may be assuming liability for any remedial action at the site; provided, however, the sale, conveyance, or transfer of property shall not absolve any person responsible for site contamination, including the seller, of liability for any remedial action at the site. The seller must also notify the department of the transfer of ownership within thirty days after the transfer. The transfer does not, however, relieve any person responsible for the environmental damage of liability for remedial measures.[159] If the director believes that a violation has occurred, or may occur, the director may seek a civil injunction to prevent such violation and a civil penalty of up to one thousand dollars per day for each day the violation continues.[160]

Decisions of the director concerning the use of an abandoned or uncontrolled hazardous waste site may be appealed to the commission. In addition, if the department has reason to believe that the notice provisions of the section have been violated, or are in imminent danger of being violated, it may institute a civil action in any court of competent jurisdiction for injunctive relief to prevent such violation and for the assessment of a civil penalty not to exceed one thousand dollars per day for each day of violation.[161]

---

156 MO. ANN. STAT. § 260.465.1.

157 *Id.*

158 *Id.*

159 *Id.*

160 *Id.* § 260.475(4).

161 *Id.*

§ 7.3.11 New York

In New York, a person may not "substantially change" the way "in which an inactive hazardous waste disposal site . . . is used" without notifying the New York Department of Environmental Conservation (NYDEC) at least sixty days before the change.[162] A "substantial change" includes constructing a facility or paving a parking lot upon the property.[163] A "substantial change" also includes transferring title to or leasing the property.

In addition, the New York Inactive Hazardous Waste Disposal Sites Rules (N.Y. Rules)[164] apply to inactive hazardous waste disposal sites listed on the state's registry of those sites.[165] The N.Y. Rules are triggered by the transfer of title or lease of the property, or before any "substantial change" in the use of the property.[166]

Prior to transferring, leasing, or substantially changing the use of a listed site, the owner must notify the NYDEC and the Commissioner of Health (the Commissioner) in writing by certified mail.[167] If the Commissioner issues a declaration that there is a condition on the property dangerous to life or health, the owner may not transfer, lease, or substantially change its use without prior written approval of both the Commissioner and the NYDEC.[168] The Commissioner and NYDEC are required to approve or disapprove within 60 days following notification by the owner.[169]

---

[162] N.Y. ENVTL. CONSERV. LAW § 27-1317.
[163] *Id.*
[164] *Id.*
[165] *Id.* at § 375.9(a).
[166] *Id.*
[167] *Id.*
[168] *Id.* at § 375.9(b).
[169] *Id.*

§ 7.3.12 Pennsylvania

The Pennsylvania Solid Waste Management Act[170] and the Pennsylvania Hazardous Sites Act[171] require notice when property that has been exposed to hazardous substances is transferred. A grantor in a deed or other transfer instrument must disclose in the document the fact that real property has been used as a hazardous waste disposal site.[172] The law only applies if hazardous waste presently is being disposed of on the property or if the grantor has ever disposed of, or has actual knowledge of, any previous disposal.[173]

§ 7.3.13 Rhode Island[174]

The Rhode Island Hazardous Waste Management Act (RIHWMA)[175] disclosure requirements are triggered by the transfer of real property.[176] The act requires that "as soon as practicable," but not later than before a transfer agreement is signed, the seller must deliver a written disclosure to the buyer.[177] The disclosure must include information on hazardous waste on the property.[178] The act applies to any real property.[179]

§ 7.3.14 South Dakota

South Dakota[180] requires that transferors of residential properties complete a disclosure statement notifying the transferee of environmental conditions at the property.

---

[170] 35 PA. CONS. STAT. ANN. § 6018.405.

[171] *Id.*§ 6020.512.

[172] *Id.* § 6018.405.

[173] Pennsylvania is also currently considering several bills that impose additional ISRA-like obligations.

[174] R.I. GEN. LAWS § 23-19.1-6 (1993).

[175] *Id.*

[176] *Id.*

[177] *Id.*

[178] *Id.*

[179] *Id.*

[180] S.D. CODIFIED LAWS ANN. § 43-4-37 through 43-4-44, effective July 1, 1993.

§ 7.3.15 West Virginia

In West Virginia, a grantor in a deed or other transfer instrument or a lessor in a lease agreement must disclose in the document the fact that the real property or the subsurface thereof has been "used for the storage, treatment, or disposal of hazardous waste."[181] The act applies only to a grantor or lessor who has an interest in the property at the time it is used for such purpose or who has actual knowledge of the previous use for such purpose.[182] Additionally, the grantee or the lessee must disclose, in writing and no later than the transfer or lease, to the grantor or the lessee any plan to use the property for the purpose of storage, treatment, or disposal of hazardous waste.[183]

---

[181] W. VA. CODE § 20-5E-20(a). The code states:

(a) The grantor in any deed or other instrument of conveyance or any lessor in any lease or other instrument whereby any real property is let for a period of time shall disclose in such deed, lease or other instrument the fact that such property or the subsurface of such property (whether or not the grantor or lessor is at the time of such conveyance or lease the owner of such subsurface) was used for the storage, treatment or disposal of hazardous waste. The provisions of this subsection shall only apply to those grantors or lessors who owned or had an interest in the real property when the same or the subsurface thereof was used for the purpose of storage, treatment or disposal of hazardous waste or who have actual knowledge that such real property or the subsurface thereof was used for such purpose or purposes at any time prior thereto.

(b) Any grantee of real estate or of any substrata underlying said real estate or any lessee for a term who intends to use the real estate conveyed or let or any substrata underlying the same for the purpose of storing, treating or disposing of hazardous waste shall disclose in writing at the time of such conveyance or lease or within thirty days prior thereto such fact to the grantor or lessor of such real estate or substrata. Such disclosure shall describe the proposed location upon said property of the site to be used for the storage, treatment or disposal of hazardous waste, the identity of such waste, the proposed method of storage, treatment or disposal to be used with respect to such waste and any and all other information required by rules and regulations of the director.

[182] *Id.*

[183] W. VA. CODE § 20-5E-20(b).

§ 7.3.16 Washington[184]

The Washington Superfund[185] covers nonresidential real property on which a significant release of a hazardous substance has been found to have occurred by the Washington Department of Ecology (the Department).[186] The statute's disclosure requirements are triggered by the sale of the property found to have a significant release by the Department requiring presale noticc.[187] Thc sellcr must inform the purchaser of any releases of hazardous substances on the property over the past twenty years prior to the sale of the property.[188] In addition, the owner of the property must file a notice in the property records that indicates that a release occurred on the property, the date of the release, and that further information can be obtained from the Department.[189] Unless the parties provide otherwise, the seller is liable for damages to the buyer for failure to inform the purchaser of a release.[190]

§ 7.4 State Statutes Requiring Approval by State Environmental Officials

The following table contains a brief description of certain state statutes requiring approval by state environmental officials.

State Statutes	Approval Requirement
Colorado State Hazardous Waste Siting Act, COLO. REV. STAT. ANN. § 25-15-206(1).	Requires that "the board of county commissioners or the governing body of the municipality" approve a "substantial change in the ownership of a hazardous waste disposal site" before it becomes effective.

[184] Senate Bill No. 6085, enacted Oct. 16, 1987, to be codified.
[185] *Id.*
[186] *Id.*
[187] *Id.*
[188] *Id.*
[189] *Id.*
[190] *Id.*

State Statutes	Approval Requirement
Iowa § 455B.430.	Requires that the Iowa Director of the Department of Natural Resources approve any substantial change in the way "in which a hazardous waste or hazardous substance disposal site . . . is used" or any sale of such property. The director may seek an injunction to prevent the violation and a civil penalty of up to $1,000 per day for each day the violation continues.
MO. ANN. STAT. § 260.465(1).	Requires that the Missouri Director of the Department of Natural Resources approve any substantial change in the way "in which an abandoned or uncontrolled hazardous waste disposal site . . . is used." Allows them to seek a civil injunction to prevent violation and a civil penalty of up to $1,000 per day for each day the violation continues.
N.Y. ENVTL. CONSERV. LAW § 27-1317.	Requires prior state approval if there is a "substantial change" in "an inactive hazardous waste disposal site, for which the commissioner of health [has issued] a declaration." The commissioner must disapprove a change of use if the new use will hamper substantially a proposed, pending, or completed remedial action at the site or will harm the environment or health.

§ 7.5 General Notice Laws Under the RCRA Program and State Equivalents

RCRA § 3004[191] requires that the owner of a hazardous waste facility provide notification to the general public so as to alert a potential purchaser that the property was previously used to manage hazardous waste.[192] The

---

[191] 42 U.S.C. § 6924; 40 C.F.R. § 264.1199(b)(1).

[192] *See* ALA. ADMIN. CODE r. 14-5.07; ALASKA ADMIN, CODE tit. 18, §62.410; ARIZ. COMP. ADMIN, R. & REG. 18-8-264; Reg. Ark. Dep. of Pol'n Ctrl. & Ecol. § 15; FLA. ADMIN. CODE ANN. 391-3-11-.10; Idaho Dept. Health & Welfare, Div. of Envtl. Rules & Reg., § 01.5232; Me. Dep. Env't Prot., Rules for Haz. Waste Mgmt., ch. 855, § 80(H); MD. REGS. CODE tit. 10, § 51.05.07(j)(1); Miss. Dep't Nat. Res. Haz. Waste Mgmt. Regs. pt. 264; Mont. Admin. Rules § 16.44.702; NEB. ADMIN RULES & REG. tit. 128, ch. 21001; NEV. ADMIN. CODE ch. 444, § 444.9025; N.M. Haz. Waste Mgmt. Reg., r.501; N.D. ADMIN CODE § 33-24-05-68; OHIO ADMIN. CODE § 3745-55-19; Okla. Rules & Reg. for Indus. Waste Mgmt. § 210; S.D. ADMIN. R. § 74:28:25:01; TENN. COMP. R. & REG. ch. 1200, § 1-11-.05(7)(j); TEX. ADMIN. CODE tit. 31, § 335.125; UTAH CODE ANN. tit. 10, § 6606(5); Va. Haz. Waste Mgmt.

notice must be placed in a deed or other document that would allow discovery during a typical title search. Penalties for failure to provide this type of notice may be assessed up to $25,000.

Since the RCRA program is delegable to the states, many states have enacted restrictions similar to RCRA § 3004 in their respective statutes governing hazardous waste management. These statutes must be at least as stringent as the federal rule, but some go further.[193]

The following chart lists the applicable state statutes requiring that some form of general notice be placed on the public record or in the chain of title to alert prospective purchasers and the public of a site's status as formerly managing hazardous waste:

State	Statute	Code
Alabama	ALA. CODE § 22-30-9	ALA. ADMIN. CODE r. 14-5.07
Alaska	ALASKA STAT. § 46.03.299	ALASKA ADMIN. CODE tit. 18, § 62.410
Arizona	ARIZ. REV. STAT. ANN. § 49-922	ARIZ. COMP. ADMIN. R. & REGS. § 18-8-264
Arkansas	ARK. CODE ANN. § 82-4204	Reg. Ark. Dept. of Pollution Control 7 ecology, § 15
California	CALIF. HEALTH & SAFETY CODE § 25245	CALIF. CODE REG. § 264.119
Colorado	COLO. REV. STAT. § 25-15-208	6 COLO. CODE REGS. §264,119
Connecticut	CONN. GEN. STAT. § 22a-116(d)	CONN. AGENCIES REGS. § 22a-449(c)-29(k)
Delaware	DEL. CODE ANN. tit. 7, § 6305	Regs. Governing Haz. Waste § 264.120
Florida	FLA. STAT. ANN. § 403.704	FLA. ADMIN. CODE. ANN. 391-3-11-.10

Reg. § 9.6(j); WIS. ADMIN. CODE, Sec. Nat. Res. § 181.42(9)(i). Hawaii and Vermont have no implementing regulations.

[193] Note that the federal notice requirement is only applicable in states that have not had the RCRA program delegated to them.

State	Statute	Code
Georgia	—	—
Hawaii	HAW. REV. STAT. § 342	No implementing regulations
Idaho	IDAHO CODE § 39-4405	Idaho Dept. Health & Welfare, Div. of Envt. Rules & Reg., § 01.5232
Illinois	ILL. REV. STAT., ch. 111 1/2, para. 1021 (n) and para 1039(g)	ILL. ADMIN. CODE tit. 35, §724.219
Indiana	IND. CODE § 13-7-8.5-5.14	IND. ADMIN. CODE tit. 320, r. 4.1-21-9
Iowa	IOWA CODE ANN. § 455 B.481	IOWA ADMIN. CODE r. 141 (455B)
Kansas	KAN. STAT. ANN. § 65-3431	KAN. ADMIN. REGS. § 28-31-8(c)
Kentucky	KY. REV. STAT. ANN. § 244.866 (4)	401 KY. ADMIN. REGS. § 35.070(10(2)
Louisiana	LA. REV. STAT. ANN. § 1136	LA. ADMIN. CODE tit. 33, § 3525
Maine	ME. REV. STAT. ANN. tit 38, § 1303-A	Maine Dept. Envt. Prot., Rules for Haz. Waste Mgmt., ch. 855 (80(H).
Maryland	MD. CODE ANN., HEALTH-ENVTL. § 7-208	MD. REGS. CODE tit. 10, § 51.05.07(J)(1)
Massachusetts	MASS. GEN. LAWS ANN., ch. 21 C, § 4(g)	MASS. REGS. CODE tit. 310, § 30.595
Michigan	MICH. COMP. LAWS ANN. § 115B.16	MICH. ADMIN. CODE r 299.9613
Minnesota	MINN. STAT. ANN. § 115B.16	MINN. 7045.0496
Mississippi	MISS. CODE ANN. § 17.17.27(1)	Miss. Dept. Nat. Res. Haz. Waste Mgmt. Regs. part 264
Missouri	MO. ANN. STAT. § 260.370	MO. CODE REGS. tit. 10, § 25-7.264(G)
Montana	MONT. CODE ANN. § 75-10-405	MONT. ADMIN. R. 16.44.702
Nebraska	—	NEB. ADMIN R. & REGS. tit. 128, ch.21001
Nevada	NEV. REV. STAT. § 459.490	NEV. ADMIN. CODE ch. 444, § 444.9025
New Hampshire	N.H. REV. STAT. ANN. § 147-A:3	Dept. of Health 7 Welfare, Div. Pub. Health Haz. Waste Rules, § 1905.08
New Jersey	N.J. STAT. ANN. § 13:1E-56	N.J. ADMIN. CODE tit. 7, § 26-9.9(n)

State	Statute	Code
New Mexico	N.M. STAT. ANN. § 74-4-4	N.M. Haz. Waste. Mgmt. Regd, r.501
New York	N.Y. ENVTL. CONSERV. LAW. § 27-0918	N.Y. COMP. CODES R. & REGS. tit. 6, § 373-2.7(i)(2)
North Carolina	N.C. GEN. STAT. § 130A-294	N.C. ADMIN. CODE tit. 10, r. 10F.0032(h)
North Dakota	N.D. CENT. CODE § 23-20.30-03	N.D. ADMIN. CODE § 33-24-05-68
Ohio	OHIO REV. CODE ANN. § 3734.12(D)(8)	OHIO ADMIN. CODE § 3745-55-19
Oklahoma	OKLA. STAT. ANN. tit. 63, § 1-2004	Okla. Rules & Reg. for Industrial Waste Mgmt. § 210
Oregon	ORE. REV. STAT. § 466.015	ORE. ADMIN. R. 340-104-001
Pennsylvania	35 PA. CONS. STAT. § 6020.512	25 PA. CODE § 75.265(0)(20)
Rhode Island	R.I. GEN. LAWS § 23-19.1-6	Rules & Reg. for Haz. Waste Gen., Treatment, Storage & Disposal, § 9.16
South Carolina	S.C. CODE ANN. § 44-56-30	S.C. CODE REGS. 61-79.264.120
South Dakota	S.D. CODIFIED LAWS. ANN. § 44-56-30	S.D. ADMIN. R. 74:28:25:01
Tennessee	TENN. CODE ANN. § 68-46-107	TENN. COMP. R. & REGS. ch. 1200-1-11-.05(7)(j)
Texas	TEX. REV. CIV. STAT. ANN. art. 4477-7, § 4(c)	TEX. ADMIN. CODE tit. 31, § 335.125
Utah	UTAH CODE ANN. § 26-14-6	—
Vermont	VT. STAT. ANN. tit.10, § 6606(5)	—
Virginia	VA. CODE ANN. § 10-266	Va. Haz. Waste Mgmt. Reg. § 9.6(J)
Washington	WASH. REV. CODE § 70.105.130	WASH. ADMIN. CODE § 173-303-610(10)
West Virginia	W. VA. CODE § 20-5E-6	W. VA. REG. tit.47, § 35-15.4
Wisconsin	WIS. STAT. ANN. § 144.62	WIS. ADMIN. CODE, § NAT. RES. 181.42(9)(i)

Below is a brief description of some of the more stringent state laws requiring recordation of notice that a site was used for hazardous waste disposal.

§ 7.6 Statutes Requiring Notice to Purchasers in the Deed

The following table indicates the states that presently have statutory deed restrictions:

State	Deed Restriction Requirements
Alabama	—
Alaska	—
Arkansas	—
California	Cal. Admin. Code Tit. 22 § 67220.
Colorado	Code of Colo. Regs. Tit. 6 §254.119.
Connecticut	—
Delaware	Del. Code Tit. 7 §9115.
Dist. of Col.	—
Florida	—
Georgia	Ga. Code Ann. §44-5-48. Ga. Code Ann. §12-8-97 et seq
Hawaii	—
Idaho	—
Illinois	Ill. Rev. Stat. Ch. 111 ¶ 11021(n) and 1039(g).
Indiana	Ind. Code §13-7-22.5-16.
Iowa	Iowa Code Ann. §455B.431.
Kansas	—
Kentucky	Ky. Rev. Stat. Ann. §224.580(16).
Louisiana	La. Rev. Stat. Ann. §30:2157

State	Deed Restriction Requirements
Maine	—
Maryland	Md. Real Prop. Code Ann. 10-702.
Massachusetts	Mass. Gen. Laws Ann. Ch. 21C §7.
Michigan	Mich. Stat. Ann. §13.32(10c).
Minnesota	Minn. Stat. Ann. §115B.16.
Mississippi	—
Missouri	Mo. Ann. Stat. §260.440, §260.465.
Montana	—
Nebraska	Neb. Rev. Stat. §81-15102.
Nevada	—
New Hampshire	N.H. Rev. Stat. Ann. §147A:14a.
New Jersey	—
New York	N.Y. Admin. Code Tit. 6 §3732.7(1).
North Carolina	N.C. Gen. Stat. §130A-310.8.
North Dakota	N.D. Cent. Code Ann. §23-20.310 .
Ohio	—
Oklahoma	—
Oregon	Ore. Rev. Stat. §466.360.
Pennsylvania	35 Pa. Cons. Stat. §6018.405, §6020.S12. Pa. Stat. Ann. tit. 35, §6018.405
Rhode Island	R.I. Gen Laws §23-19.1-33.
South Carolina	S.C. Code Ann. §30-5-36
South Dakota	—
Tennessee	Tenn. Code Ann. §68-212-209.
Texas	Tex. Health & Safety Code Ann. §361.181.
Utah	—

State	Deed Restriction Requirements
Vermont	Vt. Stat. Ann. Tit. 10 §1925.
Virginia	—
Washington	—
West Virginia	W.Va. Code §20-5E-20.
Wisconsin	—
Wyoming	—

§ 7.6.1 Delaware[194]

The Delaware Hazardous Substance Cleanup Act{195} (DHSCA) covers facilities and property on which a release of a hazardous substance has occurred.[196] If the secretary determines the release is a threat to public health or the environment, DHSCA's requirements are triggered.[197]

DHSCA requires that the owner of the property place a notice in the real property records in the county where the property is located.[198] The notice must identify the facility[199] and the owner and recorder of the notice,[200] state that a release has occurred[201] and the date of the release,[202] and indicate that further inquiries should be directed to the secretary.[203]

---

194 DEL. CODE ANN. tit. 7, §§ 9115, *et seq.* (1993).
195 *Id.*
196 DEL. CODE. ANN. tit.7, § 9115(a).
197 *Id.*
198 *Id.*
199 *Id.* at § (a)(1).
200 *Id.* at § (2).
201 *Id.* § (3).
202 *Id.* at § (4).
203 *Id.* at § (5).

§ 7.6.2 Georgia[204]

The Georgia Hazardous Waste Management Act (GHWMA)[205] covers real property on which hazardous wastes have been disposed or released in reportable quantities, or is listed on the site inventory and designated as having a known release, but not designated as needing no further action.[206] The act is triggered when the owner knows of disposal, release, or listing.[207]

Once the owner has such knowledge, he or she is required to execute and file an affidavit in the county where the real property is located indicating listing, release, or disposal.[208] The affidavit must state that the property is known to contain hazardous waste, and must be filed within thirty days of when the owner is in receipt of such knowledge.[209]

§ 7.6.3 Louisiana[210]

The Louisiana Solid Waste Management Laws (LSWML)[211] provisions apply to abandoned and inactive hazardous and solid waste disposal sites.[212] The LSWML is triggered when (1) the owner has actual or constructive knowledge the property has been used for disposal and that wastes remain on the property and (2) the property has been identified as an abandoned or inactive disposal site.[213] Once triggered, the owner must record a notice in the mortgage and conveyance records of the parish in which the property is located.[214] The notice must indicate the location of the waste site on the property.[215]

---

[204] GA. CODE ANN. § 12-8-97 *et seq*. (1992).
[205] *Id.*
[206] *Id.* at § 12-8-97(c)(1), (2).
[207] *Id.*
[208] *Id.* at § 12-8-97(c).
[209] *Id.*
[210] LA. REV. STAT. ANN. § 30:2157 (1992).
[211] *Id.*
[212] *Id.*
[213] *Id.*
[214] *Id.*
[215] *Id.*

§ 7.6.4 Massachusetts[216]

The Massachusetts Hazardous Waste Management Act's (MHWMA)[217] provisions apply to land on which hazardous waste has been disposed.[218] If the owner conveys, leases, or devotes the property to any use other than as a facility for disposal the act's requirements are triggered.[219] Once triggered, the owner must record a notice that the land has been used for disposal of hazardous waste in the registry of deeds.[220] If the land is already registered, the notice must be recorded in the registry section of the Land Court for the district prior to conveyance.[221] Violation of MHWMA might result in civil and criminal penalties.[222]

§ 7.6.5 New Hampshire

The New Hampshire Hazardous Waste Management Act allows the state to record any order issued concerning hazardous waste sites with the registry of deeds for the county in which the property subject to the order is located. The state order is considered binding on successors as it "runs with the land."[223]

§ 7.6.6 North Carolina[224]

The North Carolina Solid Waste Management Act (NCSWMA)[225] applies to inactive hazardous waste disposal sites.[226] The sale, lease, conveyance, or transfer of a disposal site is required by official notification from the state.[227] Within 180 days after notification from the department, the

---

[216] Mass Gen. L. ch. 21, §§ 7-9.
[217] *Id.*
[218] *Id.* at § 7.
[219] *Id.*
[220] *Id.*
[221] *Id.*
[222] *Id.* at § 9.
[223] N.H. Rev. Stat. Ann. §147-A:14-a.
[224] N.C. Gen. Stat. § 130A-310.8, *et seq.* (1993).
[225] *Id.*
[226] *Id.* at § 130A-310.8(a).
[227] *Id.* Sites that are undergoing voluntary remedial action are excluded from the disclosure requirements. *Id.* at § 130A-310.8(g).

owner must submit a survey plat of areas the department has designated in the notice.[228] The survey must be prepared and certified by a professional land surveyor and titled "NOTICE OF INACTIVE HAZARDOUS SUBSTANCE OR WASTE DISPOSAL SITE" (the Notice).[229] The Notice must include the size and location of the disposal area[230] and the type, location, and amount of hazardous waste disposed "to the best of the owner's knowledge."[231]

The owner must file a certified copy of thc Noticc in the register of deeds' office in the county where the land is located after the department approves and certifies the Notice.[232] The Notice is then filed in the grantor index under the owner's name.[233] The deed or other instrument of transfer must contain a statement that the property was used as a hazardous waste disposal site and reference the recordation of the Notice whenever an inactive site is sold, leased, conveyed, or transferred.[234]

If the owner does not submit or file the Notice, the secretary may do so and recover costs from the party responsible for contamination.[235] In addition, if the owner who submits and files the Notice is not the responsible party, he or she may recover the reasonable costs for the submission and filing.[236]

§ 7.6.7 Pennsylvania[237]

The Pennsylvania Hazardous Site Act (PHSA)[238] covers hazardous waste disposal sites.[239] PHSA is triggered if the grantor has actual knowledge that hazardous waste is being disposed of on the property or has

---

228 *Id.*
229 *Id.*
230 *Id.* at § 130A-310.8(a)(1).
231 *Id.* at § 130A-310.8(a)(2).
232 *Id.* at § 130A-310.8(b).
233 *Id.* at § 130A-310.8(c).
234 *Id.* at § 130A-310.8(e).
235 *Id.* at § 130A-310.8(d).
236 *Id.*
237 35 PA. STAT. ANN. § 6018.405 (1993).
238 *Id.*
239 *Id.*

been disposed of on the property in the past.[240] Once triggered, the grantor is required to place an acknowledgment of the disposal in the property description of the deed.[241] The acknowledgment should include, but is not limited to, information on the size and location of the disposal site and a description of the hazardous wastes disposed.[242] The acknowledgment must be included in the deed for all conveyances and transfers once the notation is in place.[243]

§ 7.6.8 Rhode Island[244]

In Rhode Island, the state must record in the land records any notices it gives concerning improper disposal of hazardous substances at a site. The notice must be recorded in the land records where the property subject to the state notice is located. The owner can demand recording that the notice from the state was rescinded once the state notice is vacated.[245]

§ 7.6.9 South Carolina[246]

The South Carolina Hazardous Waste Management Act (SCHWMA)[247] applies to permitted hazardous waste storage and disposal sites.[248] SCHWMA is triggered when the property is "sold, leased, conveyed or transferred in any manner."[249] At the time of the sale, lease,

---

[240] *Id.*

[241] *Id.*

[242] *Id.*

[243] *Id.*

[244] R.I. GEN LAWS § 23-19.1-33.

[245] *Id.* Sellers of property are also required to disclose knowledge of asbestos, lead paint, and radon. *See* R.I. GEN. LAWS § 5-20.8-2.

[246] S.C. CODE. ANN. § 30-5-36 (1991).

[247] *Id.*

[248] *Id.*

[249] *Id.*

conveyance, or transfer, the deed must contain the following statement "The real property conveyed or transferred by this instrument has previously been used as a storage or treatment facility for hazardous wastes."[250]

§ 7.6.10 West Virginia[251]

The West Virginia Hazardous Waste Management Act (WVHWMA)[252] applies to real property or subsurface[253] used for the storage, treatment, or disposal of hazardous waste.[254] The notice requirement of WVHWMA is triggered by conveyance or lease of such property.[255] If a grantor or lessor owned the property at the time treatment, storage, or disposal occurred, or has actual knowledge it was used for such purposes in the past,[256] he or she must disclose this knowledge in the deed, other instrument of conveyance, or lease.[257] Disclosure must occur thirty days prior to conveyance.

---

[250] *Id.* The notice can be no smaller than the print in the main body of the document.

[251] W.VA. CODE. § 20-5E-20, *et seq.* (1993).

[252] *Id.*

[253] *Id.* at § 20-5E-20(a). The requirements of WVHWMA apply to grantors or lessors of such property even if they do not own the property at the time of conveyance or lease. *Id.*

[254] *Id.* at §§ 20-5E-20(a), (b).

[255] *Id.* at § 20-5E-20(a).

[256] *Id.*

[257] *Id.* WVHWMA also imposes disclosure requirements on grantees and lessees. If a grantee or lessee intends to use the real estate or substrata for treatment, storage, or disposal he or she must disclose this intent to the grantor or lessor in writing either at the time of the conveyance or lease or within thirty days prior to commencement of such acts. *Id.* at -20(b). This provision is favorable to the grantor or lessor since CERCLA imposes liability on previous owners and present lessors. Thus, a statute requiring notice to former owners or current lessors protects them from CERCLA liability if the grantee or tenant complies with such a provision.

§ 7.7 Statutes That Require the State to Record Notice of Hazardous Waste Contamination

§ 7.7.1 Kentucky

The Kentucky Waste Management Acts (KWMA)[258] was the first state statute to require the state to record disclosure of hazardous waste disposal activities.[259] The KWMA applies to hazardous waste disposal sites.[260] If the state determines that disposal activities have occurred, the notice requirement is triggered.[261] Once triggered, the Natural Resources and Environmental Protection Cabinet (the Cabinet) must record a notice with the local county clerk.[262] The notice must contain a legal description of the property that discloses to any potential transferee that the land was used to dispose hazardous waste and that further information on the site is available from the Cabinet.[263]

§ 7.7.2 Rhode Island[264]

Like Kentucky, under Rhode Island law, the state must record in the land records any notices it gives concerning improper disposal of hazardous substances at a site. The notice must be recorded in the land records where the property subject to the state notice is located. The owner can demand recording that the notice from the state was rescinded once the state notice is vacated.

§ 7.8 Transfers of Federal Property

In 1986, Congress enacted the Superfund Amendments and Reauthorization Act (SARA),[265] which amended CERCLA in a number of

---

[258] KY. REV. STAT. ANN. § 224.876(16).
[259] *Id.*
[260] *Id.*
[261] *Id.*
[262] *Id.*
[263] *Id.*
[264] R.I. GEN LAWS § 23-19.1-33.
[265] PUB. L. NO. 99-499, 100 Stat. 1613.

respects. Among other provisions, SARA added section 120 to CERCLA[266] to address the problem of responsibility for hazardous substances located on federally owned sites. Section 120(a) provides that federal departments, agencies, and instrumentalities are subject to the provisions of CERCLA to the same extent as are private parties.[267] Sections 120(b)-(f) establish a comprehensive program to ensure the identification and cleanup of federal sites. In section 120(h), the provision at issue here, Congress imposed notice and covenant requirements on federal agencies that transfer real property contaminated by hazardous substances. Section 120(h)(1) provides in relevant part:

> Whenever any department, agency, or instrumentality of the United States enters into any contract for the sale or other transfer of real property which is owned by the United States and on which any hazardous substance was stored for one year or more, known to have been released, or disposed of, the head of such department, agency, or instrumentality shall include in such contract notice of the type and quantity of such hazardous substance and notice of the time at which such storage, release, or disposal took place, to the extent such information is available on the basis of a complete search of agency files.[268]

Section 120(h)(2) requires that the EPA issue regulations establishing the "form and manner" of the required notice. Finally, section 120(h)(3) requires that deeds for the transfer of federal real property contain the same notice as that required for contracts to transfer real property, as well as covenants warranting that remedial action has been taken to clean up the site and that the government will take any further remedial action found to be necessary after the date of the transfer.

EPA proposed rules implementing these provisions in January 1988.[269] In this proposal, the EPA, exercising its authority under section 120(h)(2), published a proposed rule implementing the notice requirements

[266] 42 U.S.C. § 9620.

[267] A 1986 amendment to the Comprehensive Environmental Response, Compensation and Liability Act ("CERCLA"), 42 U.S.C. § 9601 *et seq.* (1988), imposes notice and covenant requirements on federal agencies that transfer real property contaminated by hazardous substances.

[268] 42 U.S.C. § 9620(h)(1).

[269] *See* 54 Fed. Reg. 850 (1988).

of section 120(h)(1), stating that in most cases it intended to apply the requirements without regard to whether the hazardous substance activities occurred during the period of federal ownership or prior to the government's acquisition of the property.[270] The EPA expressed concern about the burden of applying the requirements to properties obtained by the government through foreclosure and expressed doubt that Congress intended section 120(h) to apply to such properties, but nonetheless decided that it would be "both prudent and appropriate" to apply the requirements to foreclosed properties, with an exclusion for certain small residential properties.[271] With regard to the latter, the EPA believed that "most small, residential properties would not be the site of any significant hazardous substances activity" and that any health or environmental gains from including them in the regulation "would be minimal."[272] Among other features of the proposed rule, the EPA defined certain terms used in the statute (such as "storage," "release," and "disposal"),[273] but did not define the phrase "transfer of real property," stating only that it presumed Congress meant to incorporate the definition contained in the Federal Property Management Regulations (FPMR).[274] The proposed rule made no specific mention of leases.

In April 1990, fully two years after the statutory deadline,[275] the EPA promulgated its final rule implementing section 120(h)(1).[276] The final rule revised the proposed rule in two respects important here. First, the final rule applied the notice requirements of section 120(h)(1) only to real property on which hazardous substances were stored, released, or disposed of "during the time the property was owned by the United States."[277] The EPA explained that it now believed that Congress was primarily concerned with federal facilities (chiefly military bases and nuclear weapons facilities) whose own operations involved the storage, release, or disposal of hazardous substances and that section 120(h) therefore was not intended to

[270] *See* 53 Fed. Reg. 850, 851 (1988).
[271] *See id.*
[272] *See id.*
[273] *See id.* at 852, 854.
[274] 41 C.F.R. § 101-47. *See* 53 Fed. Reg. at 851.
[275] *See* 42 U.S.C. § 9620(h)(2).
[276] *See* 55 Fed. Reg. 14,208 (1990) (codified at 40 C.F.R. § 373 (1990)).
[277] *Id.* at 14,212; 40 C.F.R. § 373.1.

apply where these activities occurred prior to the government's acquisition of the property.[278] This interpretation, it concluded, was "more appropriate . . . than [EPA's] earlier approach . . . which focused on the manner in which the property was acquired" and would avoid imposing unfair and unmanageable obligations on federal agencies that had no role in the storage, release, or disposal of hazardous substances.[279]

Second, the EPA declined to adopt a general definition of "transfer" in the final rule or to determine the extent to which leases are subject to the notice requirements.[280] In response to commenters who complained that they were unable to find a definition of "transfer" in the FPMR, the EPA stated that it had referred to the FPMR in its proposed rulemaking only in order to ensure that "federal agencies realized that the proposed regulations applied to transfers of property between agencies," as well as to transfers between an agency and a private party and between an agency and a state or local government.[281] Although it recognized that "determining what constitutes a 'transfer' of real property is important for implementing the requirements of section 120(h)," the EPA failed to offer any general definition of the term. With respect to the more specific questions of "whether and to what extent leases and easements should be included among the types of property" subject to the rule, the EPA stated that "these questions involve a complicated area of real property law, and may be affected by specific deed or lease terms and by state common law. Accordingly, EPA has not addressed these issues in the final rule."[282]

Following promulgation of the final rule, certain petitioners, who periodically enter into contracts for the purchase or lease of federal property, filed a petition for review.[283] The petitioners challenged the EPA's final rule on three grounds. First, they argued that the EPA's decision to apply section 120(h)(1) only to real property where hazardous substance activities occurred during the period of government ownership is contrary

[278] *See* 55 Fed. Reg. at 14,210.

[279] *See id.*

[280] *See id.* at 14,208-09.

[281] *Id.* at 14,208.

[282] *Id.* at 14,209.

[283] Hercules Inc. v. United States, 938 F.2d 276 (D. C. Cir. 1991).

to the express terms of the statute, congressional intent, and the statutory scheme taken as a whole. Second, they challenged as arbitrary and capricious the EPA's failure to define the term "transfer" or to include leases among the types of transfers subject to the requirements of section 120(h)(1). Finally, they claimed that the EPA violated the notice and comment requirements of the Administrative Procedure Act (APA)[284] by failing to give adequate notice in the proposed rule that it might adopt the policies on prior ownership and leases embodied in the final rule.[285]

The court rejected petitioners' challenge on the issue of transfers and leases, as well as their contention that the EPA violated the notice and comment provisions of the APA. The court found, however, that the EPA's interpretation of section 120(h) as applying only to properties where contamination occurred during the period of government ownership flew in the face of the plain meaning of the statutory language. Accordingly, the court granted the petition for review in that respect, vacated that portion of the rule, and remanded to the EPA for further rulemaking. In light of the fact that the deadline for issuance of regulations implementing section 120(h)(1) had long since passed, the court said it expected that "the EPA will act expeditiously in issuing a revised rule."[286]

§ 7.9 EPA and HUD Requirements for Disclosure of Known Lead-Based Paint and/or Lead-Based Paint Hazards in Housing

As a result of past federal, state, and local efforts to reduce lead in the environment, the percentage of children with elevated levels of lead in their blood has declined considerably over the last twenty years. Approximately 1.7 million children, however, still have blood-lead levels high enough to raise health concerns. Studies suggest that lead exposure from deteriorated residential lead-based paint, contaminated soil, and lead in dust are among

[284] 5 U.S.C. § 553 (1988).

[285] The EPA stated with respect to leases that the question was better left for case-by-case determination by agencies. "Petitioners challenge[d] each of these features of the EPA's rulemaking and claim[ed] further that the proposed regulation failed to provide adequate notice that the EPA might adopt these positions in its final rule." Hercules Inc. v. United States, 938 F.2d 276 (D.C. Cir. 1991).

[286] Hercules Inc. v. United States, 938 F.2d 276 (D. C. Cir. 1991).

the major existing sources of lead exposure among children in the United States.

Section 1018 of the Residential Lead-Based Paint Hazard Reduction Act of 1992[287] directs EPA and HUD to jointly issue regulations requiring disclosure of known lead-based paint and/or lead-based paint hazards by persons selling or leasing housing constructed before the phaseout of residential lead-based paint use in 1978.[288] Under that authority, EPA and HUD established the following requirements:

1. Sellers and lessors of most residential housing built before 1978 must disclose the presence of known lead-based paint and/or lead-based paint hazards in the housing;
2. sellers and lessors must provide purchasers and lessees with any available records or reports pertaining to the presence of lead-based paint and/or lead-based paint hazards;
3. sellers and lessors must provide purchasers and lessees with a federally approved lead hazard information pamphlet;
4. sellers must provide purchasers with a ten-day opportunity to conduct a risk assessment or inspection for the presence of lead-based paint and/or lead-based paint hazards before the purchaser is obligated under any purchase contract;

[287] 42 U.S.C. 4852d. The Residential Lead-Based Paint Hazard Reduction Act is Title X of the Housing and Community Development Act of 1992, Pub. L. 102-550.

[288] Section 1018 of Title X requires that EPA and HUD promulgate joint regulations for disclosure of any known lead-based paint or any known lead-based paint hazards in target housing offered for sale or lease. (Target housing is defined in section 1004(27) of Title X, section 401(17) of TSCA, and is discussed in Unit IV.C. of this preamble.) Specifically, section 1018 requires the following activities before a purchaser or lessee is obligated under a contract to purchase or lease target housing: (1) Sellers and lessors must provide purchasers and lessees with a lead hazard information pamphlet, as developed under section 406(a) of TSCA; (2) sellers and lessors must disclose the presence of known lead-based paint and/or lead-based paint hazards in such housing and provide purchasers and lessees with any lead hazard evaluation report available to the seller or lessor; (3) sellers must permit purchasers a ten-day opportunity to conduct a risk assessment or inspection for the presence of lead-based paint hazards; and (4) sales contracts must include an attached Lead Warning Statement and acknowledgment, signed by the purchaser. 61 Fed. Reg. at 9065.

5. sales and leasing contracts must include certain disclosure and acknowledgment language; and
6. agents must ensure compliance with these requirements.

Failure to provide the required warnings may constitute a violation of section 1018 of the act; and, thus, may result in civil and criminal penalties and potential triple damages in a private civil suit. The new lead disclosure requirements will be phased in during 1996[289] in the following manner:

1. For owners of more than four residential dwellings, the requirements are applicable on September 6, 1996, and
2. For owners of one to four residential dwellings, the requirements are applicable on December 6, 1996.

According to the government, "These provisions ensure that families receive both specific information on the housing's lead history and general information on lead exposure prevention. With this information, consumers can make more informed decisions concerning home purchase, lease, and maintenance to protect their families from lead hazard exposure."[290]

---

[289] Section 1018 mandated that EPA and HUD promulgate these requirements no later than 2 years after the date of enactment of Title X (October 28, 1994), to take effect 3 years after enactment of Title X (October 28, 1995). Due to promulgation delays, EPA and HUD have revised the effective date provisions for the rule.

[290] 61 Fed. Reg. 9064 (March 6, 1996).

Property Transfer Disclosure Forms

[1] ILL.REV.STAT., Ch. 30, para. 905.
[2] IDAHO CODE § 55-2508 (1994).
[3] MISS. CODE ANN. § 89-1-509 (1995).
[4] S.D. CODIFIED LAWS § 43-4-44 (1995).
[5] TENN. CODE ANN. § 66-5-210 (1995).
[6] REV. CODE WASH. (ARCW) § 64.06.020 (1994).
[7] 61 Fed. Reg. 9064, 9075 (March 6, 1996).
[8] 61 Fed. Reg. 9064, 9076 (March 6, 1996).
[9] 61 Fed. Reg. 9064, 9078 (March 6, 1996).

Form 7.1
Illinois Disclosure Form

ENVIRONMENTAL DISCLOSURE DOCUMENT FOR TRANSFER OF REAL PROPERTY
For Use by County
Recorder's Office

The following information is	County
provided pursuant to the	Date
Responsible Property	Doc. No.
Transfer Act of 1988	Vol.

Seller: .... Page
Buyer: .... Rec'd by:
Document No.: ....

I. PROPERTY IDENTIFICATION:

A. Address of property:... .

Street   City or Village   Township

Permanent Real Estate Index No.:...

B. Legal Description:

Section .... Township .... Range.............

Enter or attach current legal description in this area:

Prepared by:	Return to: ....
name	name
....	....
address	address

LIABILITY DISCLOSURE

Transferors and transferees of real property are advised that their ownership or other control of such property may render them liable for any environmental clean-up costs whether or not they caused or contributed to the presence of environmental problems associated with the property.

C. Property Characteristics:

Lot Size .... Acreage ....

Check all types of improvement and uses that pertain to the property:

.... Apartment building (6 units or less)

.... Commercial apartment (over 6 units)

.... Store, office, commercial building

.... Industrial building

.... Farm, with buildings

.... Other (specify)

II. NATURE OF TRANSFER:

		Yes	No
A.	(1) Is this a transfer by deed or other instrument of conveyance?	.....	.....
	(2) Is this a transfer by assignment of over 25% of beneficial interest of an Illinois land trust?	.....	.....
	(3) A lease exceeding a term of 40 years?	.....	.....
	(4) A mortgage or collateral assignment of beneficial interest?	.....	.....

B. (1) Identify Transferor:

....

Name and Current Address of Transferor

....

Name and Address of Trustee if this is a transfer of beneficial interest of a land trust. Trust No.

(2) Identify person who has completed this form on behalf of the Transferor and who has knowledge of the information contained in this form:

....

....

Name, Position (if any), address, and telephone number.

C. Identify Transferee:

....

Name and Current Address of Transferee

III. NOTIFICATION:

Under the Illinois Environmental Protection Act, owners of real property may be held liable for costs related to the release of hazardous substances.

1. Section 22.2(f) of the Act states:

"Notwithstanding any other provision or rule of law, and subject only to the defenses set forth in subsection (j) of this Section, the following persons shall be liable for all costs of removal or remedial action incurred by the State of Illinois as a result of a release or substantial threat of a release of a hazardous substance:

(1) The owner and operator of a facility or vessel from which there is a release or substantial threat of release of a hazardous substance;

(2) Any person who at the time of disposal, transport, storage or treatment of a hazardous substance owned or operated the facility or vessel used for such disposal, transport, treatment or storage from which there was a release or substantial threat of a release of any such hazardous substance;

(3) Any person who by contract, agreement, or otherwise has arranged with another party or entity for transport, storage, disposal or treatment of hazardous substances owned, controlled or possessed by such person at a facility where there is a release or substantial threat of a release of such hazardous substances; and

(4) Any person who accepts or accepted any hazardous substances for transport to disposal, storage or treatment facilities or sites from which there is a release or a substantial threat of a release of a hazardous substance.

2. Section 4(q) of the Act states:

"The Agency shall have the authority to provide notice to any person who may be liable pursuant to Section 22.2(f) of this Act for a release or a substantial threat of a release of a hazardous substance. Such notice shall include the identified response action and an opportunity for such person to perform the response action."

3. Section 22.2(k) of the Act states:

"If any person who is liable for a release or substantial threat of release of a hazardous substance fails without sufficient cause to provide removal or remedial action upon or in accordance with a notice and request by the agency or upon or in accordance with any order of the Board or any court, such person may be liable to the State for punitive damages in an amount at least equal to, and not more than 3 times, the amount of any costs incurred by the State of Illinois as result of such failure to take such removal or remedial action. The punitive damage imposed by the Board shall be in addition to any costs recovered from such person pursuant to this Section and in addition to any other penalty or relief provided by this Act or any other law."

4. Section 22.18(a) of the Act states:

"Notwithstanding any other provision or rule or law, except as provided otherwise in subsection (b), the owner or operator, or both, of an underground storage tank shall be liable for all costs of preventive action, corrective action and enforcement action incurred by the State of Illinois as a result of a release or a substantial threat of release of petroleum from an underground storage tank."

5. The text of the statutes set out above is subject to change by amendment. Persons using this form may update it to reflect changes in the text of the statutes cited, but no disclosure statement shall be invalid merely because it sets forth an obsolete or superseded version of such text.

IV. ENVIRONMENTAL INFORMATION:

A. Regulatory Information During Current Ownership

1. Has the transferor ever conducted operations on the property which involved the generation, manufacture, processing, transportation, treatment, storage or handling of "hazardous substances," as defined by the Illinois Environmental Protection Act? This question shall not be applicable for consumer goods stored or handled by a retailer in the same form, approximate amount, concentration and manner as they are sold to consumers, provided that such retailer does not engage in any commercial mixing (other than paint mixing or tinting of consumer sized containers), finishing, refinishing, servicing, or cleaning operations on the property.

Yes ....
No ....

2. Has the transferor ever conducted operations on the property which involved the processing, storage or handling of petroleum, other than that which was associated directly with the transferor's vehicle usage?

Yes ....
No ....

3. Has the transferor ever conducted operations on the property which involved the generation, transportation, storage, treatment or disposal of "hazardous or special wastes," as defined by the federal Resource Conservation and Recovery Act and the Illinois Environmental Protection Act?

Yes ....
No ....

4. Are there any of the following specific units (operating or closed) at the property which are or were used by the transferor to manage waste, hazardous wastes, hazardous substances or petroleum?

	Yes	No
Landfill	.....	.....
Surface Impoundment	.....	.....

Land Treatment	.....	.....
Waste Pile	.....	.....
Incinerator	.....	.....
Storage Tank (Above Ground)	.....	.....
Storage Tank (Underground)	.....	.....
Container Storage Area	.....	.....
Injection Wells	.....	.....
Wastewater Treatment Units	.....	.....
Septic Tanks	.....	.....
Transfer Stations	.....	.....
Waste Recycling Operations	.....	.....
Waste Treatment Detoxification	.....	.....
Other Land Disposal Area	.....	.....

If there are "YES" answers to any of the above items and the transfer is other than a mortgage or collateral assignment of beneficial interest, attach a site plan which identifies the location of each unit, such site plan to be filed with the Environmental Protection Agency along with this disclosure document.

5. Has the transferor ever held any of the following in regard to this real property?

a. Permits for discharges of wastewater to waters of the State.
Yes ..... No .....

b. Permits for emissions to the atmosphere.
Yes ..... No .....

c. Permits for any waste storage, waste treatment or waste disposal operation.
Yes ..... No .....

6. Has the transferor had any wastewater discharges (other than sewage) to a publicly owned treatment works?
Yes ..... No .....

7. Has the transferor taken any of the following actions relative to this property?

a. Prepared a Chemical Safety Contingency Plan pursuant to the Illinois Chemical Safety Act.

Yes .....

No .....

b. Filed an Emergency and Hazardous Chemical Inventory Form pursuant to the federal Emergency Planning and Community Right-to-Know Act of 1986.

Yes .....

No .....

c. Filed a Toxic Chemical Release Form pursuant to the federal Emergency Planning and Community Right-to-Know Act of 1986.

Yes .....

No .....

8. Has the transferor or any facility on the property or the property been the subject of any of the following State or federal governmental actions?

a. Written notification regarding known, suspected or alleged contamination on or emanating from the property.

Yes .....

No .....

b. Filing an environmental enforcement case with a court or the Pollution Control Board for which a final order or consent decree was entered.

Yes .....

No .....

c. If item (b) was answered by checking Yes, then indicate whether or not the final order or decree is still in effect for this property.

Yes .....

No .....

9. Environmental Releases During Transferor's Ownership

a. Has any situation occurred at this site which resulted in a reportable "release" of any hazardous substances or petroleum as required under State or federal laws?

Yes ....

No ....

b. Have any hazardous substances or petroleum, which were released, come into direct contact with the ground at this site?

Yes ....

No ....

c. If the answers to questions (a) and (b) are Yes, have any of the following actions or events been associated with a release on the property?

..... Use of a cleanup contractor to remove or treat materials including soils, pavement or other surficial materials

..... Assignment of in-house maintenance staff to remove or treat materials including soils, pavement or other surficial materials

..... Designation, by the IEPA or the IEMA, of the release as "significant" under the Illinois Chemical Safety Act

..... Sampling and analysis of soils

..... Temporary or more long-term monitoring of groundwater at or near the site

..... Impaired usage of an on-site or nearby water well because of offensive characteristics of the water

..... Coping with fumes from subsurface storm drains or inside basements, etc.

..... Signs of substances leaching out of the ground along the base of slopes or at other low points on or immediately adjacent to the site.

10. Is the facility currently operating under a variance granted by the Illinois Pollution Control Board?

Yes ....

No ....

11. Is there any explanation needed for clarification of any of the above answers or responses?

B. Site Information under Other Ownership or Operation

1. Provide the following information about the previous owner or any entity or person the transferor leased the site to or otherwise contracted with for the management of the site or real property:

Name:..

Type of business..

or property usage...

2. If the transferor has knowledge, indicate whether the following existed under prior ownerships, leaseholds granted by the transferor, other contracts for management or use of the facilities or real property:

	YES	NO
Landfill	.....	.....
Surface Impoundment	.....	.....
Land Treatment	.....	.....
Waste Pile	.....	.....
Incinerator	.....	.....
Storage Tank (Above Ground)	.....	.....
Storage Tank (Underground)	.....	....
Container Storage Area	.....	.....
Injection Wells	.....	.....
Wastewater Treatment Units	.....	.....
Septic Tanks	.....	.....
Transfer Stations	.....	.....
Waste Recycling Operations	.....	.....
Waste Treatment Detoxification	.....	.....
Other Land Disposal Area	.....	.....

V. CERTIFICATION:

A. Based on my inquiry of those persons directly responsible for gathering the information, I certify that the information submitted is, to the best of my knowledge and belief, true and accurate.

....
signature

....
type or print name

TRANSFEROR OR TRANSFERORS
(or on behalf of Transferor)

B. This form was delivered to me with all elements completed on

.... 19....

....
signature

....
type or print name

TRANSFEREE OR TRANSFEREES
(or on behalf of Transferee)

C. This form was delivered to me with all elements completed on

.... 19....

....
signature

....
type or print name

Form 7.2
Indiana Disclosure Form

Form of Disclosure Document Pursuant to
Burns Ind. Code Ann. § 13-7-22.5-15 (1993)

The disclosure document delivered by a transferor of property under the Indiana Responsible Property Transfer Law must follow this form:

A WARNING TO THE PARTIES TO A TRANSFER OF PROPERTY: It is highly unlikely that the single act of reading this document would be found to constitute "all appropriate inquiry into the previous ownership and uses of the property" so as to protect you against liability under the "innocent purchaser" provision of the federal Comprehensive Environmental Response, Compensation and Liability Act, 42 U.S.C. 9601(35)(B). You are strongly encouraged not only to read this document carefully but also to take all other actions necessary to the exercise of due diligence in your inquiry into the previous ownership and uses of the property.

ENVIRONMENTAL DISCLOSURE DOCUMENT FOR TRANSFER OF REAL PROPERTY

	For Use by County Recorder's Office
The following information is provided under IC 13-7-22.5, the Responsible Property Transfer Law.	County Date Doc. No. Vol. Page Rec'd by:

I. PROPERTY IDENTIFICATION
A. Address of property:

Street

City or Town   Township

Tax Parcel Identification No. (Key Number):

B. Legal Description:

Section Township Range
Enter or attach complete legal description in this area:

LIABILITY DISCLOSURE

Transferors and transferees of real property are advised that their ownership or other control of such property may render them liable for environmental cleanup costs whether or not they caused or contributed to the presence of environmental problems in association with the property.

C. Property Characteristics:

Lot Size Acreage

Check all types of improvement and uses that pertain to the property:
Apartment building (6 units or less)
Commercial apartment (over 6 units)
Store, office, commercial building
Industrial building
Farm, with buildings
Other (specify)

II. NATURE OF TRANSFER

Yes No

(1) Is this a transfer by deed or other instrument of conveyance of fee title to property?
(2) Is this a transfer by assignment of over 25% of beneficial interest of a land trust?
(3) A lease exceeding a term of 40 years?
(4) A collateral assignment of beneficial interest?
(5) An installment contract for the sale of property?
(6) A mortgage or trust deed?
(7) A lease of any duration that includes an option to purchase?

B. (1) Identify Transferor:
Name and Current Address of
Transferor

Trust No.

Name and Address of Trustee if this is a transfer of beneficial interest of a land trust.

(2) Identify person who has completed this form on behalf of the Transferor and who has knowledge of the information contained in this form:

Name, Position (if any), Telephone No.

and address

C. Identify Transferee:

Name and Current Address of Transferee

III. ENVIRONMENTAL INFORMATION

A. Regulatory Information During Current Ownership

1. Has the transferor ever conducted operations on the property which involved the generation, manufacture, processing, transportation, treatment, storage, or handling of a "hazardous substance," as defined by IC 13-7-8.7-1? This question does not apply to consumer goods stored or handled by a retailer in the same form and approximate amount, concentration, and manner as they are sold to consumers, unless the retailer has engaged in any commercial mixing (other than paint mixing or tinting of consumer sized containers), finishing, refinishing, servicing, or cleaning operations on the property.

Yes No

2. Has the transferor ever conducted operations on the property which involved the processing, storage, or handling of petroleum, other than that which was associated directly with the transferor's vehicle usage?

Yes No

3. Has the transferor ever conducted operations on the property which involved the generation, transportation, storage, treatment, or disposal of "hazardous waste," as defined in IC 13-7-1?

Yes No

4. Are there any of the following specific units (operating or closed) at the property that are used or were used by the transferor to manage hazardous wastes, hazardous substances, or petroleum?

	Yes	No
Landfill		
Surface Impoundment		
Land Application		
Waste Pile		
Incinerator		
Storage Tank (Above Ground)		
Storage Tank (Underground)		
Container Storage Area		
Injection Wells		
Wastewater Treatment Units		
Septic Tanks		
Transfer Stations		
Waste Recycling Operations		
Waste Treatment Detoxification		
Other Land Disposal Area		

If there are "YES" answers to any of the above items and the transfer of property that requires the filing of this document is other than a mortgage or trust deed or a collateral assignment of beneficial interest in a land trust, you must attach to the copies of this document that you file with the county recorder and the department of environmental management a site plan that identifies the location of each unit.

5. Has the transferor ever held any of the following in regard to this real property?
Yes No

(A) Permits for discharges of wastewater to waters of Indiana.
Yes No

(B) Permits for emissions to the atmosphere.
Yes No

(C) Permits for any waste storage, waste treatment, or waste disposal operation.
Yes No

6. Has the transferor ever discharged any wastewater (other than sewage) to a publicly owned treatment works?
Yes No

7. Has the transferor been required to take any of the following actions relative to this property?
Yes No

(A) Filed an emergency and hazardous chemical inventory form pursuant to the federal Emergency Planning and Community Right-to-Know Act of 1986 (42 U.S.C. 11022).
Yes No

(B) Filed a toxic chemical release form pursuant to the federal Emergency Planning and Community Right-to-Know Act of 1986 (42 U.S.C. 11023).
Yes No

8. Has the transferor or any facility on the property or the property been the subject of any of the following state or federal governmental actions?

Yes No

(A) Written notification regarding known, suspected, or alleged contamination on or emanating from the property.

Yes No

(B) Filing an environmental enforcement case with a court or the solid waste management board for which a final order or consent decree was entered.

Yes No

(C) If the answer to question (B) was Yes, then indicate whether or not the final order or decree is still in effect for this property.

Yes No

9. Environmental releases during transferor's ownership.

(A) Has any situation occurred at this site which resulted in a reportable "release" of any hazardous substances or petroleum as required under state or federal laws?

Yes No

(B) Have any hazardous substances or petroleum which were released come into direct contact with the ground at this site?

Yes No

If the answer to question (A) or (B) is yes, have any of the following actions or events been associated with a release on the property?

Yes No

Use of a cleanup contractor to remove or treat materials including soils, pavement, or other surficial materials?

Yes No

Assignment of in-house maintenance staff to remove or treat materials including soils, pavement, or other surficial materials?

Yes No

Sampling and analysis of soils?
Yes No

Temporary or more long term monitoring of groundwater at or near the site?
Yes No

Impaired usage of an onsite or nearby water well because of offensive characteristics of the water?
Yes No

Coping with fumes from subsurface storm drains or inside basements?
Yes No

Signs of substances leaching out of the ground along the base of slopes or at other low points on or immediately adjacent to the site?
Yes No

(C) Is there an environmental defect (as defined in IC 13-7-22.5 - 1.5) on the property that is not reported under question (A) or (B)?
Yes No

If the answer is Yes, describe the environmental defect:

10. Is the facility currently operating under a variance granted by the commissioner of the Indiana department of environmental management?
Yes No

11. Has the transferor ever conducted an activity on the site without obtaining a permit from the U.S. Environmental Protection Agency, the commissioner of the department of environmental management, or another administrative agency or authority with responsibility for the protection of the environment, when such a permit was required by law?

Yes No

If the answer is Yes, describe the activity:

12. Is there any explanation needed for clarification of any of the above answers or responses?

B. Site Information Under Other Ownership Or Operation

1. Provide the following information about the previous owner or about any entity or person to whom the transferor leased the property or with whom the transferor contracted for the management of the property:

Name:

Type of business or property usage

2. If the transferor has knowledge, indicate whether the following existed under prior ownerships, leaseholds granted by the transferor, other contracts for management or use of the property:

	Yes	No
Landfill		
Surface Impoundment		
Land Application		
Waste Pile		
Incinerator		
Storage Tank (Above Ground)		
Storage Tank (Underground)		
Container Storage Area		
Injection Wells		
Wastewater Treatment Units		
Septic Tanks		
Transfer Stations		
Waste Recycling Operations		
Waste Treatment Detoxification		
Other Land Disposal Area		

IV. CERTIFICATION

A. Based on my inquiry of those persons directly responsible for gathering the information, I certify that the information submitted is, to the best of my knowledge and belief, true and accurate.

TRANSFEROR (or on behalf of Transferor)

B. This form was delivered to me with all elements completed on 19

TRANSFEREE (or on behalf of Transferee)

Form 7.3
Idaho Seller Property Disclosure Form

The disclosures required by the provisions of this article pertaining to the property proposed to be transferred are set forth in and shall be made on a copy of the following disclosure form or an alternative form as provided in section 55-2506, Idaho Codc:

SELLER PROPERTY DISCLOSURE FORM

SELLER'S NAME AND ADDRESS: .... Section 55-2501, et seq., Idaho Code, requires Sellers of residential real property to complete a property condition disclosure form.

PURPOSE OF STATEMENT: This is a statement of the conditions and information concerning the property known by the Seller. Unless otherwise advised, the Seller does not possess any expertise in construction, architectural, engineering or any other specific areas related to the construction or condition of the improvements on the property. Other than having lived at or owning the property, the Seller possesses no greater knowledge than that which could be obtained upon a careful inspection of the property by the potential buyer. Unless otherwise advised, the Seller has not conducted any inspection of generally inaccessible areas such as the foundation or roof. It is not a warranty of any kind by the Seller or by any agent representing any Seller in this transaction. It is not a substitute for any inspections. Purchaser is encouraged to obtain his/her own professional inspections.

1. All appliances and service systems included in the sale (such as refrigerator/freezer, range/oven, dishwasher, disposal, hood/fan, central vacuum, microwave oven, trash compactor, smoke detectors, tv antenna/dish, fireplace/wood stove, water heater, garage door opener, pool/hot tub, etc.) are functioning properly except: (please list and explain)..

2. Specify problems with the following:

Basement water.....................................

Foundation..

Roof condition and age........................

Well (type) ....
problem.......................................

Septic system (type) ....
problem.......................................

Plumbing...

Drainage...

Electrical...

Heating...

3. Describe any conditions that may affect your ability to clear title (such as encroachments, easements, zoning violations, lot line disputes, etc.): ....

4. Are you aware of any hazardous materials or pest infestations on the property?...

5. Have any substantial additions or alterations been made without a building permit?...

6. Any other problems, including legal, physical or other not listed above that you know concerning the property:...

The Seller certifies that the information herein is true and correct to the best of Seller's knowledge as of the date signed by the Seller. The Seller is familiar with the residential real property and each act performed in making a disclosure of an item of information is made and performed in good faith.

I/we acknowledge receipt of a copy of this statement.

Seller:	Buyer:
....	....
Date:	Date: ....
....	....
Date:	Date: ....

Form 7.4
Mississippi Form of Seller's Disclosure Statement

The disclosures required by this act pertaining to the property proposed to be transferred are set forth in, and shall be made on a copy of, the following disclosure form:

SELLER'S DISCLOSURE STATEMENT

The following is a Disclosure Statement, made by the Seller, concerning the condition of the property located at: ________________. This Disclosure is not a warranty of any kind by the Seller or any Agent of the Seller in this transaction and is not a substitute for any inspections or warranties the Purchaser may wish to obtain. This statement may be made available to other parties and is to be attached to the Listing Agreements (signed by Owner).

TO THE SELLER: Please complete the following form, including any past history of problems if known. If the condition or question does not apply to your property, mark "N/A." DO NOT LEAVE ANY BLANK SPACES. Attach additional pages if necessary.

APPLIANCES / SYSTEMS

Item	Yes/No	Gas/Elec	Approx Age	Repairs last 2 yrs
Single oven				
Double oven				
Cook top				
Microwave				
Dishwasher				
Disposal				
Trash Compactor				

Item	Yes/No	Gas/Elec	Approx Age	Repairs last 2 yrs
Ice Maker				
Hot Water Heaters				
Refrigerator				
Washer / Dryer				
Central Air				
Heat				
Fireplace				
Chimney				
Other				
Other				

POOL AND POOL EQUIPMENT (Gunite or Vinyl Liner) ....

BURGLAR ALARM YES ..... NO .....

FEE $.........................

APPROXIMATE AGE OF HOUSE: ....
HOMEOWNER'S ASSOCIATION FEE $..........

IF CONDO, MAINTENANCE FEE: $.......... PER ....

TOTAL TAXES: $.................. HOMESTEAD FILED FOR 19.....

AVERAGE ELECTRIC BILL (12 Months) $.....................

AVERAGE GAS BILL (12 Months) $...................

Is there anything normally considered a part of the real estate that you do not plan to leave with the property? (*i.e.,* chandeliers, ceiling fans, bathroom mirrors, etc.)...

Are there any needed repairs of which you are aware? ....
....

Are you aware of any problems which may exist on the property by virtue of prior uses such as, but not limited to, hazardous or toxic waste, asbestos components, lead-based paint, urea-formaldehyde insulation, radon gas, naturally occurring radiation or any past industrial uses of the premises?

YES ..... NO .....

What is the approximate square footage of the heated and cooled living area in house?
....

Where did you get this figure? ....

Are there any wood floors under any carpets?

YES ..... NO .....

Has property previously flooded?

YES ..... NO .....

Is flood insurance required?

YES ..... NO .....

Are you aware of any foundation repairs made in the past?

YES ..... NO .....

If yes, when were they made and who was the contractor? ....

ROOF:

AGE: .....

Are there any leaks?

YES ..... NO .....

Repairs?

....

History of infestation, if any: (termites, carpenter ants, etc.):.........................

....

Any treatments for infestation?

YES ..... NO .....

Any repaired damage?

YES ..... NO .....

Are there any rights of way, easements, or similar matters that may affect our interest in the property?

YES ..... NO ..... UNKNOWN .....

Has there been major damage to the property or any of the structures from fire, wind, floods or any other disaster?

YES ..... NO .....

Please describe: ....

....

Please state any other facts, information or problems (*i.e.,* heating, cooling, electrical, plumbing, mechanical) relating to this property that would be of concern to a buyer:..

....

To the extent of the Seller's knowledge as a property owner, the Seller acknowledges that the information contained above is true and accurate for those areas of the property listed. Owner agrees to save and hold Broker harmless from all claims, disputes, litigations, and/or judgments arising from any incorrect information supplied by Owner, or from any material fact known by Owner which Owner fails to disclose, except that Broker is not held harmless to Owner in claims, disputes, litigations or judgments, arising from conditions of which Broker had actual knowledge.

It is acknowledged and understood by the Owner that this information may be made available to other parties.

....		....	
SELLER	DATE	SELLER	DATE

Form 7.5
South Dakota Property Condition Disclosure Statement

The following form shall be used for the property condition disclosure statement:

SELLER'S PROPERTY CONDITION DISCLOSURE STATEMENT
Seller
Property Address

This Disclosure Statement concerns the real property identified above situated in the City of County of
State of South Dakota.
THIS STATEMENT IS A DISCLOSURE OF THE CONDITION OF THE ABOVE DESCRIBED PROPERTY IN COMPLIANCE WITH § 43-4-38. IT IS NOT A WARRANTY OF ANY KIND BY THE SELLER OR ANY AGENT REPRESENTING ANY PARTY IN THIS TRANSACTION AND IS NOT A SUBSTITUTE FOR ANY INSPECTIONS OR WARRANTIES THE PARTIES MAY WISH TO OBTAIN. Seller hereby authorizes any agent representing any party in this transaction to provide a copy of this statement to any person or entity in connection with any actual or anticipated sale of the property.
IF ANY MATERIAL FACT COMES TO THE ATTENTION OF THE SELLER WHICH WOULD AFFECT ANY STATEMENT MADE BY THE SELLER FOLLOWING THE SIGNING OF THIS STATEMENT AND BEFORE THE TIME OF SETTLEMENT, THE SELLER SHALL NOTIFY THE BUYER AND ANY AGENT REPRESENTING ANY

PARTY TO THIS TRANSACTION IN WRITING OF SUCH MATERIAL FACT OR OTHER INFORMATION.

I. LOT OR TITLE INFORMATION

1. When did you purchase or build the home?

2. Were there any title problems when you purchased the property?

Yes No Unknown

If yes, describe:

3. Are there any unrecorded or recorded liens or financial instruments against the property, other than a first mortgage?

Yes No Unknown

If yes, explain:

4. Are there any easements which have been granted in connection with the property (other than normal utility easements for public water and sewer, gas and electric service, telephone service, cable television service, drainage and sidewalks)?

Yes No Unknown

If yes, explain on separate page and attach hereto.

5. Are there any problems related to establishing the lot lines/boundaries?
Yes No Unknown

Do you have a location survey in your possession or a copy of the recorded plat?

Yes No Unknown

If yes, attach a copy of same.

6. Are you aware of any encroachments from or on adjoining property (*i.e.*, fences, driveway, sheds, outbuildings or other improvements)?

Yes No

If yes, explain:

7. Are you aware of any covenants or restrictions affecting the use of the property in accordance with local law?

Yes No

If yes, attach a copy of the covenants and restrictions, if available.

8. Are you aware of any current or pending litigation, foreclosure, zoning, building code or restrictive covenant violation notices, mechanics' liens, judgments, special assessments, zoning changes or changes that could affect your property?

Yes No Unknown

If yes, explain:

9. Is the property currently occupied by the owner?

Yes No

If no, is the property leased?

Yes No

If yes, please attach a copy of any written lease or a statement as to the terms and conditions of the lease.

10. If leased, does the property use comply with local zoning laws, if any?

Yes No Unknown

11. Is there ground rent?

Yes No Unknown

If so, what is the ground rent payment (*i.e.*, annually, semi-annually, monthly)?

$ per

12. Do you pay any mandatory fees or special assessments to a homeowners' or condominium association?

Yes No

If so, what are the fees or assessments? $ per (*i.e.*, annually, semi-annually, monthly)

Payable to whom:

For what purpose:

13. Does the property ever have standing water in either the front, rear or side yard more than forty-eight hours after heavy rain?

Yes No Unknown

If yes, describe:

14. Are wetlands located upon any part of the property?

Yes No Unknown

If yes, describe:

15. Are any features of the property (*i.e.*, walls, fences, roads or driveways) shared in common with adjoining landowners whose use or responsibility for maintenance may have an effect on the property?

Yes No Unknown

If yes, describe:

II. STRUCTURAL INFORMATION

1. Have you experienced any water penetration problems in the walls, windows, doors, basement or crawl space?

Yes No Unknown

What water damage related repairs, if any, have been made? If so, when?

Is drain tile installed on the property?

Yes No Unknown

2. Are there any interior cracked walls or floors, or cracks or defects in exterior driveways, sidewalks, patios, or other hard surface areas?

Yes No Unknown

If yes, explain:

3. Have you ever experienced roof leakage?

Yes No Unknown

Type of roof covering: Age:
What roof repairs, if any, have been made?

If so, when?

Describe any existing unrepaired damage to the roof:

4. Have you received notice that fire retardant treated plywood may have been used during roof construction of the property?

Yes No

If yes, has an inspection of the roof been performed?

Yes No Unknown

If yes, explain results:

5. Is there insulation in:
   - the ceiling/attic?
     - Yes No Unknown
   - the walls?
     - Yes No Unknown
   - the floors?
     - Yes No Unknown

6. Are you aware of any termite or wood boring infestation or damage, either past or present?

Yes No Unknown

If yes, explain:

Has the property been treated for any termite or wood boring infestation or damage?

Yes No Unknown

If yes, who treated it and when?

7. Have you performed any work upon the property within the last five years that required a building, plumbing, electrical or any other permit?
Yes No

If yes, describe the work:

Was a permit obtained?

Yes No

Was the work approved by an inspector?

Yes No

Explain:

8. Has the property, structures, or improvements thereon ever been damaged? (*i.e.,* fire, smoke, wind, floods, hail or snow)

Yes No Unknown

If yes, has the damage been repaired?

Yes No

9. Are you aware of any problems with sewer blockage or backup, past or present?

Yes No Unknown

III. SYSTEMS/UTILITIES INFORMATION

NONE/NOT INCL. WORKING NOT WORKING UNKNOWN

A. ELECTRICAL SYSTEM
Burglar Alarm and/or Security System
Ceiling Fan
Garage Wiring
Garage Door/Opener Control(s)
Doorbell
Intercom
Light Fixtures
Sauna
Smoke and/or Fire Alarm
Switches and Outlets
Vent Fan
220 Volt Service

B. HEATING AND COOLING SYSTEM
Air Exchanger
Attic Fan
Air Purifier
Central Air—Electric
Central Air—Water Cooled
Fireplace
Fireplace Insert
Furnace/Heat—Electric or Gas
Humidifier
Propane Tank—Leased or Owned
Solar House—Heating
Woodburning Stove

C. WATER/SEWER SYSTEMS
Cistern
Hot Tub, Whirlpool, and Controls
Plumbing and Fixtures
Pool and Equipment
Septic/Leaching Field
Sump Pump
Underground Sprinkler and Heads
Water Heater—Electric or Gas
Water Purifier
Water Softener—Leased or Owned
Well and Pump
Sewer Systems/Drains

D. HAZARDOUS CONDITIONS
Are there any existing hazardous conditions of the property such as methane gas, lead paint, radon gas in the house or well, radioactive material, a landfill mineshaft, expansive soil, toxic materials, urea-formaldehyde foam insulation, asbestos insulation or buried fuel or chemical storage tanks?

Yes No Unknown

Have any tests been performed?

Yes No Unknown

Explain:

IV. MISCELLANEOUS INFORMATION

1. Is the street or road located at the end of the driveway to the property public or private?

Public Private Unknown

If private, is there a written road maintenance agreement?

Yes No Unknown

If yes, attach a copy of the maintenance agreement, if available.

2. Is this property located in or near a flood plain?

Yes No Unknown

3. When was the fireplace/wood stove/chimney flue last cleaned?
Date:

4. In the twelve months prior to the date of signing this document, did any of the following occur on the subject property?

	YES	NO	UNKNOWN
a. A human death by homicide			
b. Other felony committed against the property or a person on the property			

If yes to any of the above explain:

5. Are the improvements connected to a public or private water system?

6. If private, what is the date and result of the last water test?

7. Are the improvements connected to a public or private sewer system?

8. Are there broken window panes or seals?

Yes No Unknown

If so, specify:

9. Are there any items attached to the property that will not be left, such as: towel bars, mirrors, swag lamps and hooks, curtain rods, window coverings, light fixtures, clothes lines, swing sets, storage sheds, ceiling fans, basketball hoops, mail boxes, etc.

10. Are there any other problems that have not been disclosed above?

Yes No

If yes, explain:

CLOSING SECTION

The Seller hereby certifies that the information contained herein is true and correct to the best of the Seller's information, knowledge and belief as of the date of the Seller's signature below. If any of these conditions change before conveyance of title to this property, the change will be disclosed in a written amendment to this disclosure statement.

SELLER DATE

SELLER DATE

THE SELLER AND THE BUYER MAY WISH TO OBTAIN PROFESSIONAL ADVICE AND INSPECTIONS OF THE PROPERTY TO OBTAIN A TRUE REPORT AS TO THE CONDITION OF THE PROPERTY AND TO PROVIDE FOR APPROPRIATE PROVISIONS IN ANY CONTRACT OF SALE AS NEGOTIATED BETWEEN THE

SELLER AND THE BUYER WITH RESPECT TO SUCH PROFESSIONAL ADVICE AND INSPECTIONS.

I/We acknowledge receipt of a copy of this statement on the date appearing beside my/our signature(s) below. Any agent representing any party to this transaction makes no representations and is not responsible for any conditions existing in the property.

BUYER DATE

BUYER DATE

Form 7.6
Tennessee Disclosure form

Following is the form prescribed by the general assembly which is necessary to comply with the provisions of this part. The form used does not have to be the one included in this section, but it is the intent of the general assembly that any such form includes all items contained in the form bclow with all acknowlcdgmcnt provisions of such form:

Tennessee Residential Property Condition Disclosure

The Tennessee Residential Property Disclosure Act states that anyone transferring title to residential real property must provide information about the condition of the property. This completed form constitutes that disclosure by the Seller. This is not a warranty, or a substitute for any professional inspections or warranties that the purchasers may wish to obtain.

Instructions to the Seller:

Complete this form yourself and answer each question to the best of your knowledge. If an answer is an estimate, clearly label it as such. The Seller hereby authorizes any agent(s) representing any party in this transaction to provide a copy of this statement to any person or entity in connection with any actual or anticipated sale of the subject property.

Property Address

City

Seller's Name(s)

Property Age

Date Seller Acquired the Property

Do You Occupy the Property?

If Not Owner-Occupied, How Long Has It Been Since the Seller Occupied the Property?

A. The Subject Property Includes the Items Checked Below:

Range

Oven

Microwave

Dishwasher

Garbage Disposal

Trash Compactor

Water Softener

220 Volt Wiring

Washer/Dryer Hookups

Central Heating

Heat Pump

Central Air Conditioning

Wall/Window Air Conditioning

Window Screens

Rain Gutters

Fireplace(s) (Number)

Gas Starter for Fireplace

Smoke Detector/Fire Alarm

Burglar Alarm

Patio/Decking/Gazebo

Irrigation System

Sump Pump

Garage Door Opener(s)

Intercom

TV Antenna/Satellite Dish

Pool

Spa/Whirlpool Tub

Hot Tub

Sauna

Current Termite Contract

Access to Public Streets

Other

Other

Garage: Attached Not Attached Carport

Water Heater: Gas Solar Electric

Water Supply: City Well Private Utility Other

Waste Disposal: City Sewer Septic Tank Other

Gas Supply: Utility Bottled Other

Roof(s): Type Age (approx.)

Other Items:

To the best of your knowledge, are any of the above NOT in operating condition?

YES NO

If YES, then describe (Attach additional sheets if necessary):

B. Are You (Seller) Aware of Any Significant Defects/Malfunctions in Any of the Following?

Interior Walls
YES NO UNKNOWN

Ceilings
YES NO UNKNOWN

Floors
YES NO UNKNOWN

Windows
YES NO UNKNOWN

Doors
YES NO UNKNOWN

Insulation
YES NO UNKNOWN

Plumbing
YES NO UNKNOWN

Sewer/Septic
YES NO UNKNOWN

Electrical System
YES NO UNKNOWN

Exterior Walls
YES NO UNKNOWN

Roof
YES NO UNKNOWN

Basement
YES NO UNKNOWN

Foundation
YES NO UNKNOWN

Slab
YES NO UNKNOWN

Driveway
YES NO UNKNOWN

Sidewalks
YES NO UNKNOWN

If any of the above is/are marked YES, please explain:

C. Are You (Seller) Aware of Any of the Following?

1. Substances, materials or products which may be an environmental hazard such as, but not limited to: asbestos, radon gas, lead-based paint,

fuel or chemical storage tanks and/or contaminated soil or water on the subject property

YES NO UNKNOWN

2. Features shared in common with adjoining landowners, such as walls, fences and driveways, whose use or responsibility for maintenance may have an effect on the subject property

YES NO UNKNOWN

3. Any authorized changes in roads, drainage or utilities affecting the property, or contiguous to the property

YES NO UNKNOWN

4. Any changes since the most recent survey of this property was done

YES NO UNKNOWN

Most recent survey of the property: (check here if unknown)

5. Any encroachments, easements, or similar items that may affect your ownership interest in the subject property

YES NO UNKNOWN

6. Room additions, structural modifications or other alterations or repairs made without necessary permits

YES NO UNKNOWN

7. Room additions, structural modifications or other alterations or repairs not in compliance with building codes

YES NO UNKNOWN

8. Landfill (compacted or otherwise) on the property or any portion thereof

YES NO UNKNOWN

9. Any settling from any cause, or slippage, sliding or other soil problems

YES NO UNKNOWN

10. Flooding, drainage or grading problems

YES NO UNKNOWN

11. A flood insurance requirement

YES NO UNKNOWN

12. Major property or structural damage from fire, earthquake, floods or landslides

YES NO UNKNOWN

13. Any zoning violations, nonconforming uses and/or violations of "setback" requirements

YES NO UNKNOWN

14. Neighborhood noise problems or other nuisances

YES NO UNKNOWN

15. Subdivision and/or deed restrictions or obligations

YES NO UNKNOWN

16. A Homeowners Association (HOA) which has any authority over the subject property

YES NO UNKNOWN

Name of HOA:

HOA Address:

Monthly Dues: Special Assessments:

17. Any "common area" (facilities such as pools, tennis courts, walkways, or other areas co-owned in undivided interest with others)

YES NO UNKNOWN

18. Any notices of abatement or citations against the property

YES NO UNKNOWN

19. Any lawsuit(s) against the seller threatening or affecting this real property

YES NO UNKNOWN

If the answer to any of the above is YES, please explain:

D. Certification: I/we certify that the information herein, concerning the real property located at , is true and correct to the best of my/our knowledge as of the date signed. Should any of these conditions change prior to conveyance of title to this property, these changes will be disclosed in addenda to this document.

Transferor (Seller) Date

Transferor (Seller) Date

Parties may wish to obtain professional advice and/or inspections of the property and to negotiate appropriate provisions in the purchase agreement regarding advice, inspections or defects.

Transferee/Buyer's Acknowledgment:

I/we understand that this disclosure statement is not intended as a substitute for any inspection, and that I/we have a responsibility to pay diligent attention to and inquire about those material defects which are evident by careful observation.

I/we acknowledge receipt of a copy of this disclosure.

Transferee (Buyer) Date

Transferee (Buyer) Date

Form 7.7
Washington Form of Disclosure Statement

Effective January 1, 1995, in a transaction for the sale of residential real property, the seller must deliver to the buyer a completed real property transfer disclosure statement in the following form, unless:

✓ the buyer has expressly waived the right to receive the disclosure statement, or
✓ the transfer is exempt under RCW 64.06.010.

INSTRUCTIONS TO THE SELLER
Please complete the following form. Do not leave any spaces blank. If the question clearly does not apply to the property write "NA." If the answer is "yes" to any * items, please explain on attached sheets. Please refer to the line number(s) of the question(s) when you provide your explanation(s). For your protection you must date and sign each page of this disclosure statement and each attachment. Delivery of the disclosure statement must occur not later than . . . days (or five days if not filled in) of mutual acceptance of a written contract to purchase between a buyer and a seller.

NOTICE TO THE BUYER
THE FOLLOWING DISCLOSURES ARE MADE BY THE SELLER(S), CONCERNING THE CONDITION OF THE PROPERTY LOCATED AT...
.... ("THE PROPERTY"), LEGALLY DESCRIBED ON ATTACHED EXHIBIT A. DISCLOSURES CONTAINED IN THIS FORM ARE PROVIDED BY THE SELLER ON THE BASIS OF SELLER'S ACTUAL KNOWLEDGE OF THE PROPERTY AT THE TIME THIS DISCLOSURE FORM IS COMPLETED BY THE SELLER. YOU HAVE . . . BUSINESS DAYS, OR THREE BUSINESS DAYS IF NOT FILLED IN, FROM THE SELLER'S DELIVERY OF THIS SELLER'S DISCLOSURE STATEMENT TO REVOKE YOUR OFFER BY DELIVERING YOUR SEPARATE SIGNED WRITTEN STATEMENT OF REVOCATION TO THE SELLER, UNLESS YOU WAIVE THIS RIGHT AT OR PRIOR TO ENTERING INTO A SALE AGREEMENT. THE FOLLOWING ARE DISCLOSURES MADE BY THE SELLER AND ARE NOT THE

REPRESENTATIONS OF ANY REAL ESTATE LICENSEE OR OTHER PARTY. THIS INFORMATION IS FOR DISCLOSURE ONLY AND IS NOT INTENDED TO BE A PART OF ANY WRITTEN AGREEMENT BETWEEN THE BUYER AND THE SELLER.

FOR A MORE COMPREHENSIVE EXAMINATION OF THE SPECIFIC CONDITION OF THIS PROPERTY YOU ARE ADVISED TO OBTAIN AND PAY FOR THE SERVICES OF A QUALIFIED SPECIALIST TO INSPECT THE PROPERTY ON YOUR BEHALF, FOR EXAMPLE, ARCHITECTS, ENGINEERS, LAND SURVEYORS, PLUMBERS, ELECTRICIANS, ROOFERS, BUILDING INSPECTORS, OR PEST AND DRY ROT INSPECTORS. THE PROSPECTIVE BUYER AND THE OWNER MAY WISH TO OBTAIN PROFESSIONAL ADVICE OR INSPECTIONS OF THE PROPERTY AND TO PROVIDE FOR APPROPRIATE PROVISIONS IN A CONTRACT BETWEEN THEM WITH RESPECT TO ANY ADVICE, INSPECTION, DEFECTS OR WARRANTIES.

Seller .... is/ .... is not occupying the property.

I. SELLER'S DISCLOSURES:

*If "Yes" attach a copy or explain. If necessary use an attached sheet.

1. TITLE

[ ]Yes [ ]No [ ]Don't know
A. Do you have legal authority to sell the property?

[ ]Yes [ ]No [ ]Don't know
*B. Is title to the property subject to any of the following?

(1) First right of refusal
(2) Option
(3) Lease or rental agreement
(4) Life estate?

[ ]Yes [ ]No [ ]Don't know
*C. Are there any encroachments, boundary agreements, or boundary disputes?

[ ]Yes [ ]No [ ]Don't know
*D. Are there any rights of way, easements, or access limitations that may affect the owner's use of the property?

[ ]Yes [ ]No [ ]Don't know
*E. Are there any written agreements for joint maintenance of an easement or right of way?

[ ]Yes [ ]No [ ]Don't know
*F. Is there any study, survey project, or notice that would adversely affect the property?

[ ]Yes [ ]No [ ]Don't know
*G. Are there any pending or existing assessments against the property?

[ ]Yes [ ]No [ ]Don't know
*H. Are there any zoning violations, nonconforming uses, or any unusual restriction on the subject property that would affect future construction or remodeling?

[ ]Yes [ ]No [ ]Don't know
*I. Is there a boundary survey for the property?

[ ]Yes [ ]No [ ]Don't know
*J. Are there any covenants, conditions, or restrictions which affect the property?

2. WATER

A. Household Water

(1) The source of the water is
[ ]Public [ ]Community[ ]Private [ ]Shared

(2) Water source information:

[ ]Yes [ ]No [ ]Don't know
*a. Are there any written agreements for shared water source?

[ ]Yes [ ]No [ ]Don't know
*b. Is there an easement (recorded or unrecorded) for access to and/or maintenance of the water source?

[ ]Yes [ ]No [ ]Don't know
*c. Are any known problems or repairs needed?

[ ]Yes [ ]No [ ]Don't know
*d. Does the source provide an adequate year round supply of potable water?

[ ]Yes [ ]No [ ]Don't know
*(3) Are there any water treatment systems for the property?
[ ]Leased [ ]Owned

B. Irrigation

[ ]Yes [ ]No [ ]Don't know
(1) Are there any water rights for the property?

[ ]Yes [ ]No [ ]Don't know
*(2) If they exist, to your knowledge, have the water rights been used during the last five-year period?

[ ]Yes [ ]No [ ]Don't know
*(3) If so, is the certificate available?

C. Outdoor Sprinkler System

[ ]Yes [ ]No [ ]Don't know
(1) Is there an outdoor sprinkler system for the property?

[ ]Yes [ ]No [ ]Don't know
*(2) Are there any defects in the outdoor sprinkler system?

3. SEWER/SEPTIC SYSTEM

A. The property is served by: [ ]Public sewer main,
[ ]Septic tank system
[ ]Other disposal system (describe)
....

[ ]Yes [ ]No [ ]Don't know
B. If the property is served by a public or community sewer main, is the house connected to the main?

C. If the property is connected to a septic system:

[ ]Yes [ ]No [ ]Don't know
(1) Was a permit issued for its construction, and was it approved by the city or county following its construction?

(2) When was it last pumped:

.... , 19 ....

[ ]Yes [ ]No [ ]Don't know
*(3) Are there any defects in the operation of the septic system?

[ ]Don't know
(4) When was it last inspected?

.... , 19 ....

By Whom: ....

[ ]Don't know
(5) How many bedrooms was the system approved for?

.... bedrooms

[ ]Yes [ ]No [ ]Don't know
*D. Do all plumbing fixtures, including laundry drain, go to the septic/sewer system? If no, explain: ....

[ ]Yes [ ]No [ ]Don't know
*E. Are you aware of any changes or repairs to the septic system?

[ ]Yes [ ]No [ ]Don't know
F. Is the septic tank system, including the drainfield, located entirely within the boundaries of the property?

4. STRUCTURAL

[ ]Yes [ ]No [ ]Don't know
*A. Has the roof leaked?

[ ]Yes [ ]No [ ]Don't know
If yes, has it been repaired?

[ ]Yes [ ]No [ ]Don't know
*B. Have there been any conversions, additions, or remodeling?

[ ]Yes [ ]No [ ]Don't know
*1. If yes, were all building permits obtained?

[ ]Yes [ ]No [ ]Don't know
*2. If yes, were all final inspections obtained?

[ ]Yes [ ]No [ ]Don't know
C. Do you know the age of the house? If yes, year of original construction:

....

[ ]Yes [ ]No [ ]Don't know
*D. Do you know of any settling, slippage, or sliding of the house or other improvements? If yes, explain:

....

[ ]Yes [ ]No [ ]Don't know
*E. Do you know of any defects with the following: (Please check applicable items)

[] Foundations	[] Decks	[] Exterior Walls
[] Chimneys	[] Interior Walls	[] Fire Alarm
[] Doors	[] Windows	[] Patio
[] Ceilings	[] Slab Floors	[] Driveways
[] Pools	[] Hot Tub	[] Sauna
[] Sidewalks	[] Outbuildings	[] Fireplaces
[] Garage Floors	[] Walkways	[] Wood Stoves
[] Other		

[ ]Yes [ ]No [ ]Don't know
*F. Was a pest or dry rot, structural or "whole house" inspection done? When and by whom was the inspection completed? ....

[ ]Yes [ ]No [ ]Don't know
*G. Since assuming ownership, has your property had a problem with wood destroying organisms and/or have there been any problems with pest control, infestations, or vermin?

5. SYSTEMS AND FIXTURES

If the following systems or fixtures are included with the transfer, do they have any existing defects:

[ ]Yes [ ]No [ ]Don't know
*A. Electrical system, including wiring, switches, outlets, and service

[ ]Yes [ ]No [ ]Don't know
*B. Plumbing system, including pipes, faucets, fixtures, and toilets

[ ]Yes [ ]No [ ]Don't know
*C. Hot water tank

[ ]Yes [ ]No [ ]Don't know
*D. Garbage disposal

[ ]Yes [ ]No [ ]Don't know
*E. Appliances

[ ]Yes [ ]No [ ]Don't know
*F. Sump pump

[ ]Yes [ ]No [ ]Don't know
*G. Heating and cooling systems

[ ]Yes [ ]No [ ]Don't know
*H. Security system [ ] Owned [ ] Leased

*I. Other ....

6. COMMON INTEREST

[ ]Yes [ ]No [ ]Don't know
A. Is there a Home Owners' Association?
Name of Association ....

[ ]Yes [ ]No [ ]Don't know
B. Are there regular periodic assessments:
$ .... per [ ] Month [ ] Year

[ ] Other ....

[ ]Yes [ ]No [ ]Don't know
*C. Are there any pending special assessments?

[ ]Yes [ ]No [ ]Don't know
*D. Are there any shared "common areas" or any joint maintenance agreements (facilities such as walls, fences, landscaping, pools, tennis courts, walkways, or other areas co-owned in undivided interest with others)?

7. GENERAL

[ ]Yes [ ]No [ ]Don't know
*A. Is there any settling, soil, standing water, or drainage problems on the property?

[ ]Yes [ ]No [ ]Don't know
*B. Does the property contain fill material?

[ ]Yes [ ]No [ ]Don't know
*C. Is there any material damage to the property or any of the structure from fire, wind, floods, beach movements, earthquake, expansive soils, or landslides?

[ ]Yes [ ]No [ ]Don't know
D. Is the property in a designated flood plain?

[ ]Yes [ ]No [ ]Don't know
E. Is the property in a designated flood hazard zone?

[ ]Yes [ ]No [ ]Don't know
*F. Are there any substances, materials, or products that may be an environmental hazard such as, but not limited to, asbestos, formaldehyde, radon gas, lead-based paint, fuel or chemical storage tanks, and contaminated soil or water on the subject property?

[ ]Yes [ ]No [ ]Don't know
*G. Are there any tanks or underground storage tanks (*e.g.*, chemical, fuel, etc.) on the property?

[ ]Yes [ ]No [ ]Don't know
*H. Has the property ever been used as an illegal drug manufacturing site?

8. FULL DISCLOSURE BY SELLERS

A. Other conditions or defects:

[ ]Yes [ ]No [ ]Don't know
*Are there any other material defects affecting this property or its value that a prospective buyer should know about?

B. Verification:

The foregoing answers and attached explanations (if any) are complete and correct to the best of my/our knowledge and I/we have received a copy hereof. I/we authorize all of my/our real estate licensees, if any, to deliver a copy of this disclosure statement to other real estate licensees and all prospective buyers of the property.

DATE .... SELLER ....
SELLER ....

II. BUYER'S ACKNOWLEDGMENT

A. As buyer(s), I/we acknowledge the duty to pay diligent attention to any material defects which are known to me/us or can be known to me/us by utilizing diligent attention and observation.

B. Each buyer acknowledges and understands that the disclosures set forth in this statement and in any amendments to this statement are made only by the seller.

C. Buyer (which term includes all persons signing the "buyer's acceptance" portion of this disclosure statement below) hereby acknowledges receipt of a copy of this disclosure statement (including attachments, if any) bearing seller's signature.

DISCLOSURES CONTAINED IN THIS FORM ARE PROVIDED BY THE SELLER ON THE BASIS OF SELLER'S ACTUAL KNOWLEDGE OF THE PROPERTY AT THE TIME OF DISCLOSURE. YOU, THE BUYER, HAVE ... BUSINESS DAYS (OR THREE BUSINESS DAYS IF

NOT FILLED IN) FROM THE SELLER'S DELIVERY OF THIS SELLER'S DISCLOSURE STATEMENT TO REVOKE YOUR OFFER BY DELIVERING YOUR SEPARATE SIGNED WRITTEN STATEMENT OF REVOCATION TO THE SELLER UNLESS YOU WAIVE THIS RIGHT OF REVOCATION.

BUYER HEREBY ACKNOWLEDGES RECEIPT OF A COPY OF THIS REAL PROPERTY TRANSFER DISCLOSURE STATEMENT AND ACKNOWLEDGES THAT THE DISCLOSURES MADE HEREIN ARE THOSE OF THE SELLER ONLY, AND NOT OF ANY REAL ESTATE LICENSEE OR OTHER PARTY.

DATE .... BUYER ....

BUYER ....

The real property transfer disclosure statement shall be for disclosure only, and shall not be considered part of any written agreement between the buyer and seller of residential real property. The real property transfer disclosure statement shall be only a disclosure made by the seller, and not any real estate licensee involved in the transaction, and shall not be construed as a warranty of any kind by the seller or any real estate licensee involved in the transaction.

Form 7.8
EPA/HUD Sample Disclosure Format for Target Housing Sales

Disclosure of Information on Lead-Based Paint and/or Lead-Based Paint Hazards

Lead Warning Statement

Every purchaser of any interest in residential real property on which a residential dwelling was built prior to 1978 is notified that such property may present exposure to lead from lead-based paint that may place young children at risk of developing lead poisoning. Lead poisoning in young children may produce permanent neurological damage, including learning disabilities, reduced intelligence quotient, behavioral problems, and impaired memory. Lead poisoning also poses a particular risk to pregnant women. The seller of any interest in residential real property is required to provide the buyer with any information on lead-based paint hazards from risk assessments or inspections in the seller's possession and notify the buyer of any known lead-based paint hazards. A risk assessment or inspection for possible lead-paint hazards is recommended prior to purchase.

Seller's Disclosure

(a) Presence of lead-based paint and/or lead-based paint hazards (check (i) or (ii) below):

(i) ......Known lead-based paint and/or lead-based paint hazards are present in the housing (explain). ....

....

(ii) ......Seller has no knowledge of lead-based paint and/or lead-based paint hazards in the housing.

(b) Records and reports available to the seller (check (i) or (ii) below):

(i) ......Seller has provided the purchaser with all available records and reports pertaining to lead-based paint and/or lead-paint hazards in the housing (list documents below). ....

....

(ii) ...... Seller has no reports or records pertaining to lead-based paint and/or lead-based paint hazards in the housing.

Purchaser's Acknowledgment (initial)

(c) ...... Purchaser has received copies of all information listed above.

(d) ...... Purchaser has received the pamphlet Protect Your Family from Lead in Your Home.

(e) ...... Purchaser has (check (i) or (ii) below):

(i) ...... received a 10-day opportunity (or mutually agreed upon period) to conduct a risk assessment or inspection for the presence of lead-based paint and/or lead-based paint hazards in the housing (list document below). ....

....

(ii) ...... Seller has no reports pertaining to lead-based paint and/or lead-based paint hazards in the housing. Agent's Acknowledgment (initial)

(f) ...... Agent has informed the seller of the seller's obligations under 42 U.S.C. 4852d and is aware of his/her responsibility to ensure compliance.

Certification of Accuracy

The following parties have reviewed the information above and certify, to the best of their knowledge, that the information they have provided is true and accurate.

....	....	....	....
Seller	Date	Seller	Date
....	....	....	....
Purchaser	Date	Purchaser	Date
....	....	....	....
Agent	Date	Agent	Date

Form 7.9
EPA/HUD Sample Disclosure Format for Target Housing Rentals and Leases

Disclosure of Information on Lead-Base-Paint and/or Lead-Based Hazards

Lead Warning Statement

Housing built before 1978 may contain lead-based paint. Lead from paint, paint chips, and dust can pose health hazards if not managed properly. Lead exposure is especially harmful to young children and pregnant women. Before renting pre-1978 housing, lessors must disclose the presence of known lead-based paint and/or lead-based paint hazards in the dwelling. Lessees must also receive a federally approved pamphlet or poisoning prevention.

Lessor's Disclosure

(a) Presence of lead-based paint and/or lead-based paint hazards (Check (i) or (ii) below):

(i)...... Known lead-based paint and/or lead-based paint hazards are present in the housing (explain). ....

....

(ii) ...... Lessor has no knowledge of lead-based paint and/or lead-based paint hazards in the housing.

(b) Records and reports available to the lessor (Check (i) or (ii) below):

(i) ...... Lessor has provided the lessee with all available records and reports pertaining to lead-based paint and/or lead-based paint hazards in the housing (list documents below).

....

(ii) ...... Lessor has no reports or records pertaining to lead-based paint and/or lead-based paint hazards in the housing.

Lessee's Acknowledgment (initial)

(c) ...... Lessee has received copies of all information listed above.

(d) ...... Lessee has received the pamphlet Protect Your Family from Lead in Your Home.

Agent's Acknowledgment (initial)

(e) ...... Agent has informed the lessor of the lessor's obligations under 42 U.S.C. 4852d and is aware of his/her responsibility to ensure compliance.

Certification of Accuracy

The following parties have reviewed the information above and certify, to the best of their knowledge, that the information they have provided is true and accurate.

....	....	....	....
Lessor	Date	Lessor	Date
....	....	....	....
Lessee	Date	Lessee	Date
....	....	....	....
Agent	Date	Agent	Date

Form 7.10
EPA/HUD Sample Contract Contingency Language

This contract is contingent upon a risk assessment or inspection of the property for the presence of lead-based paint and/or lead-based paint hazards at the Purchaser's expense until 9 p.m. on the tenth calendar-day after ratification [Insert date 10 days after contract ratification or a date mutually agreed upon]. (Intact lead-based paint that is in good condition is not necessarily a hazard. See the EPA pamphlet Protect Your Family from Lead in Your Home for more information.) This contingency will terminate at the above predetermined deadline unless the Purchaser (or Purchaser's agent) delivers to the Seller (or Seller's agent) a written contract addendum listing the specific existing deficiencies and corrections needed, together with a copy of the inspection and/or risk assessment report. The Seller may, at the Seller's option, within ...... days after delivery of the addendum, elect in writing whether to correct the condition(s) prior to settlement. If the Seller will correct the condition, the Seller shall furnish the Purchaser with certification from a risk assessor or inspector demonstrating that the condition has been remedied before the date of the settlement. If the Seller does not elect to make the repairs, or if the Seller makes a counter-offer, the Purchaser shall have ...... days to respond to the counter-offer or remove this contingency and take the property in "as is" condition or this contract shall become void. The Purchaser may remove this contingency at any time without cause.

Chapter 8

Voluntary Cleanup Programs and the Duty to Audit

§ 8.1 Overview

Voluntary cleanup statutes do not directly regulate environmental auditing practices, but they have become a major force in defining "how clean is clean?" As such, voluntary cleanup statutes have become the most dramatic force in defining Phase II and Phase III environmental audits: those designed not to merely quantify liability but to design and complete the remediation efforts.

Approximately twenty-one states encourage the reuse and redevelopment of contaminated industrial property through enacting Brownfields restoration and voluntary cleanup legislation. Many additional states are debating Brownfields restoration or voluntary cleanup legislation. Depending upon the jurisdiction involved, volunteers satisfying program requirements may be shielded from private third-party actions and citizen suits under state law.

These state voluntary cleanup programs are new and have developed to complement or replace the old, enforcement-liability driven systems of the state Superfund and property transfer laws. Numerous states have announced Brownfield economic redevelopment programs during the last two years—many have enacted such legislation within the past year alone. Many of these programs apply not just to prospective purchasers but also to site owners and operators who report site contamination and agree to enter into a voluntary remediation program.

State voluntary cleanup laws are primarily aimed at protecting potential purchasers of used industrial property, developers, and commercial lenders. As such, any remedial activity conducted under these programs is designed to occur with minimal regulatory oversight.[1]

---

[1] For more detailed discussion of Brownfield cleanup and reuse *see* Office of Technology Assessment, State of the States on Brownfields Programs for Cleanup and Reuse of Contaminated Sites (June 1995).

Oversight

Year Begun	State Voluntary Cleanup Programs	State Oversight	No Oversight	Privatized Oversight
1988	Minnesota	X		
1989	Washington		X	
	Illinois	X		
1990	—			
1991	New Jersey	X		
	Oregon	X		
1992	Massachusetts			X
1993	Delaware	X		
	Indiana	X		
	Maine	X		
	Missouri	X		
1994	Nebraska	X		
	Wisconsin		X	
	Colorado	X		
	Tennessee	X		
	Connecticut		X	
	Ohio			X
	North Carolina			X
	California		X	

Year Begun	State Voluntary Cleanup Programs	State Oversight	No Oversight	Privatized Oversight
1995	Virginia	X		
	Arkansas	X		
	Michigan	X		
	Pennsylvania			X
	Texas	X		
	Arizona			
	Vermont			

Generally, state Brownfields restoration and voluntary cleanup programs:

✓ limit liability of nonresponsible property owners for the voluntary cleanup of hazardous substances and/or petroleum site contamination;[2]

✓ limit prospective purchasers' and developers' liability when purchasing, reusing, or redeveloping Brownfields;[3]

✓ protect lenders accepting industrial property as security for commercial financing from liability;[4]

✓ protect trustees from assuming liability when forced to administer estates suffering from site contamination.[5]

---

[2] *See* COLO. REV. STAT. ANN. § 25-16-310(1) (West 1994); IND. CODE ANN. § 13-7-8.9-18 (West 1994); MICH. COMP. LAWS ANN. § 299.614(a) (West 1994); MINN. STAT. ANN. § 115.175 [subdiv. 1] (West 1994); OHIO REV. CODE ANN. § 3746.12 (Baldwin 1994).

[3] *Id.*

[4] *See* IND. CODE ANN. § 13-7-8.9-18(2)(A) (West 1994) (pursuant to §18(2)(A), liability protection afforded to parties undertaking voluntary remedial activity may be legally transferred to third parties, such as lenders); MINN. STAT. ANN. § 115B.175 [subdiv. 6](2) (West 1994); OHIO REV. CODE ANN. § 3746.25(A)(1) (Baldwin 1994).

[5] *See* OHIO REV. STAT. ANN. § 3746.27(A) (Baldwin 1994); see also IND. CODE ANN. § 13-7-8.9-18(2)(A) (West 1994).

Typically, state agencies will provide a release and covenant not to sue to parties who are prepared to make some predefined investment in site remediation activities. Alternatively, some states have been willing to issue letters stating that no action will be required at a specific site. In some states, the programs have been implemented by statute. In other states, the programs have no legislative authority and are simply the result of policy initiatives. Below is a table describing the historic evolution of state voluntary cleanup programs, including the statutory or regulatory authority they are operated under.

Chronology of Voluntary Cleanup Programs and Authority

Year Begun	**State Voluntary Cleanup Programs**	**Authority**
1988	Minnesota	Minn. Stat. Sec. 115.B175
1989	Washington	Wash. Rev. Code Sec. 173-340-300
	Illinois	415 ILSC Sec. 5/1 22.2(m)
1990	—	
1991	New Jersey	N.J. Admin. Code tit. 7 Sec. 26
	Oregon	Ore. Admin. R.Sec. 340-122-010
1992	Massachusetts	Mass. Ann. Laws Ch. 21E
1993	Delaware	Del. Code Ann. tit. 91
	Indiana	Ind. Code Sec. 13-7-8.9
	Maine	Me. Rev. Stat. Ann. tit. 38 Sec. 342-343
	Missouri	Mo. Rev. Stat. Sec. 260.565.575

Year Begun	State Voluntary Cleanup Programs	Authority
1994	Nebraska	Sec. 81-15.181.188
	Wisconsin	Wis. Stat. Sec. 144-442
	Colorado	Colo. Rev. Stat. Sec. 25-16-301-310
	Tennessee	Tenn.Code Ann. Sec. 68-212-104
	Connecticut	Conn. Gen. Stat. Sec. 22a-471
	Ohio	1994 Ohio Laws Sec. 3746
	North Carolina	N.C. Gen. Stat. Sec. 130A-310
	California	1994 Cal. Stat. 453
1995	Virginia	H.B. 1845
	Arkansas	H.B. 1076
	Michigan	H.B. 4596
	Pennsylvania	S.B. 1
	Texas	H.B. 2296
	Arizona	
	Vermont	

Most programs provide that a state agency will issue a formal covenant not to sue that becomes effective once agreed-upon site remediation activities or payment obligations have been completed. Even without a formal program, some states have demonstrated a willingness to work with prospective purchasers and investors to provide comfort in the form of a written letter. Following is a table of the type of "comfort documents" given by the states upon completion of the cleanup:

Type of Relief of Liability Granted

Year Begun	State Voluntary Cleanup Programs	Certificate of Completion	Covenant Not to Sue	No Further Action (NFA) Letter[6]	None[7]
1988	Minnesota	X	X[8]		
1989	Washington				X
	Illinois			X	
1990	—				
1991	New Jersey		X		
	Oregon		X[9]	X	
1992	Massachusetts		X		
1993	Delaware	X			
	Indiana		X		
	Maine		X	X	
	Missouri			X	

6 Other terms for an NFA include:
* No action assurance letter (Maine)
* Close out letter (Wisconsin)
* Notification of a No-Action Determination (Colorado).

See, e.g., ME. REV. STAT. ANN. tit. 38 Sec. 342-343; WIS. STAT. § 144-442; COLO. REV. STAT. § 25-16-301-310.

7 No liability release is given by the state.

8 Covenants not to sue are entered into on a case-by-case basis.

9 Covenants not to sue are entered into on a case-by-case basis.

Year Begun	State Voluntary Cleanup Programs	Certificate of Completion	Covenant Not to Sue	No Further Action (NFA) Letter	None
1994	Nebraska			X	
	Wisconsin		X		
	Colorado		X		
	Tennessee			X	
	Connecticut			X	
	Ohio		X		
	North Carolina			X	
	California	X			
1995	Virginia	X			
	Arkansas	X			
	Michigan			X	
	Pennsylvania		X		
	Texas		X		
	Arizona				
	Vermont				

Through the use of specific numeric cleanup standards or site-specific risk assessment techniques, most state programs require assurances that reuse or redevelopment of Brownfields will safeguard the public health, safety, welfare, and environment. Other programs provide the volunteer with a choice among cleaning up to

✓ background levels;

✓ cleanup levels based upon numerical equations, with default values based upon contamination and land use; or

✓ site specific risk assessment.

The following table describes the varying cleanup standards used in the states with enacted voluntary cleanup legislation.

Cleanup Standards

Year Begun	**State Voluntary Cleanup Programs**	**Site-Specific Risk Assess-ments (RA)**	**Risk-Based Rules Provide Volunteer with a Choice**	**Cleanup to Back-ground or Ground-water Std.**	**No RA**
1988	Minnesota	X			
1989	Washington	X			
	Illinois		X		
1990	—				
1991	New Jersey	X			
	Oregon	X			
1992	Massachusetts		X		
1993	Delaware	X			
	Indiana		X		
	Maine	X			
	Missouri	X			
1994	Nebraska	X			
	Wisconsin			X	X
	Colorado		X		
	Tennessee	X			
	Connecticut			X	
	Ohio	X			
	North Carolina	X			
	California	X			

Year Begun	State Voluntary Cleanup Programs	Site-Specific Risk Assess-ments (RA)	Risk-Based Rules Provide Volunteer with a Choice	Cleanup to Back-ground or Ground-water Std.	No RA
1995	Virginia	X			
	Arkansas	X			
	Michigan			X	
	Pennsylvania			X	
	Texas			X	
	Arizona				
	Vermont				

Property owners satisfying program requirements generally qualify for Certificates of Completion or No Further Action (NFA) letters issued by the states. Certificates of Completion or NFA letters are submitted to state environmental regulators in lieu of the issuance of a Covenant Not to Sue to the "volunteer(s)."

Several programs provide economic incentives for property owners to voluntarily clean up, reuse, and redevelop Brownfields sites. The most obvious incentive applicable to all programs is increased land values. The remediation of hazardous substances can, however, be a very expensive proposition. For small and mid-sized businesses lacking the financial capacity to underwrite environmental assessments and voluntary cleanups, some states have created Voluntary Cleanup Funds or other economic incentives to assist qualified voluntary cleanup program participants. For example, Illinois provides "volunteers" with a 25 percent tax credit for the costs of remediation. In the State of Ohio, "volunteers" receive a 10 percent tax abatement for the increase in land values, with an additional ten-year tax abatement from local municipalities for qualifying properties.

Certain programs create an express duty on the part of the state to petition EPA on behalf of the volunteers to (1) withhold from listing a Brownfields site on either CERCLIS or the NPL and (2) petition EPA to waive pursuing federal enforcement proceedings against participating parties who have successfully completed the voluntary cleanup program. Similarly, under certain state cleanup programs, satisfactory completion of voluntary cleanup programs may protect property owners from actions brought by third parties and citizen groups for further site remediation. Other voluntary cleanup programs do not preclude participants from seeking cost recovery from the parties responsible for on-site contamination due to the presence of hazardous substances and/or petroleum.

Some Brownfields sites may be ineligible for participation in state voluntary cleanup programs. For instance, in Ohio, the existence of groundwater contamination will currently bar a property owner from participating in voluntary cleanups, although ultimately sites with groundwater contamination will be included in the program as ultimately promulgated. Involvement in ongoing state or federal environmental enforcement actions or site listing on the NPL generally prevents a property owner from engaging in a state voluntary cleanup program. Finally, most programs apply only to property owners or other volunteers who are not responsible for existing site contamination. Under most state Brownfields restoration and voluntary cleanup programs, parties owning or operating the property during the period of site contamination may not elect to remediate the site.

§ 8.2 State by State Review

Following is a brief description of each of the voluntary cleanup programs run by certain states.

§ 8.2.1 California

The California Voluntary Expedited Remedial Action Reform Act of 1994,[10] which was enacted on September 7, 1994, applies only to sites selected by the Site Designation Committee of the California Environmental Protection Agency (CEPA) for expedited remediation. Once selected, CEPA should notify all known potentially responsible parties for a selected site and enter into an enforceable agreement with a responsible party(ies) to undertake response actions at the site. The act specifies procedures for, and the consequences of, filing a Certificate of Completion for a response action. The act also requires the apportionment of liability for response costs and specifies criteria for such apportionment. An Expedited Site Remediation Trust Fund has been created to carry out CEPA's statutory mandate of expedited remediation of selected sites. Certain land use controls on sites subject to expedited remedial procedures must be recorded.[11] For example, to expedite the conversion of property into productive use and to provide funds for response activities, CEPA may approve a site owner's request to modify the boundaries of a selected site for response action only if the following criteria have been met:

- ✓ A remedial plan has been approved by CEPA;
- ✓ The holder of the first deed of trust, if any, approves;
- ✓ Portions of the site proposed to be removed by the boundary modification do not require any response action;

---

[10] For example, to expedite the conversion of property into productive use and to provide funds for response activities, CEPA may approve a site owner's request to modify the boundaries of a selected site for response action only if the following criteria have been met:

- A remedial plan has been approved by CEPA;
- The holder of the first deed of trust, if any, approves;
- Portions of the site proposed to be removed by the boundary modification do not require any response action;
- The boundary modification will not result in an unacceptable risk of human exposure to hazardous substances at the site;
- Hazardous substances have not migrated, and are not expected to migrate onto portions of the site proposed for boundary redesignation; and
- The modification of boundaries of the site will not significantly interfere with the response action at the site. CAL. HEALTH & SAFETY CODE § 25398.5(a)(1)(A)-(F) (West 1994)

[11] 1993 CA S.B. 923; CAL. HEALTH & SAFETY CODE § 25261 *et seq.* (West 1994).

✓ The boundary modification will not result in an unacceptable risk of human exposure to hazardous substances at the site;

✓ Hazardous substances have not migrated, and are not expected to migrate onto portions of the site proposed for boundary redesignation; and

✓ The modification of boundaries of the site will not significantly interfere with the response action at the site.[12]

The California program precludes state enforcement actions if program requirements are met.[13] Participants satisfying program requirements are protected from contribution actions brought by third-parties for cleanup costs.[14] Moreover, parties entering into a valid voluntary remediation agreement and satisfying program requisites are entitled to seek contribution from other responsible parties.[15] Volunteers may sue recalcitrant responsible parties failing or refusing to comply with any order or agreement relating to a site selected by CEPA for expedited remediation for treble damages.[16]

The Voluntary Expedited Remedial Action Reform Act establishes a program whereby volunteers enter the program,[17] become subject to state oversight through micromanagement of remedial activities,[18] and receive a Certificate of Completion[19] upon satisfactory completion of the program. Specifically, property designated for expedited cleanup must employ remedies that are

✓ Protective of human health and the environment;

✓ Provide long-term reliability at a reasonable cost;

---

[12] CAL. HEALTH & SAFETY CODE § 25398.5(a)(1)(A)-(F) (West 1994).

[13] *Id.* at § 25398.2(b)(1)(C). Responsible parties are deemed liable to CEPA for oversight response costs and for other costs incurred by the state, which are allocated to responsible parties in a final liability apportionment. *Id.* at § 25399(a).

[14] *Id.* at § 25398.6(d). A responsible person settling with CEPA and paid any response cost as a part of the settlement agreement is not liable for claims of contribution from any other party for costs of the response action.

[15] *Id.* at § 25398.17(a).

[16] *Id.* One-half of any recovery of treble damages shall be deposited in the California Expedited Site Remediation Trust Fund. *Id.* § 25398.17(b).

[17] *Id.* at § 25396.6.

[18] *Id.* at § 25398.6.

[19] *Id.* at §§ 25398.14 and 25398.15.

✓ Provide reasonable protection to the waters of the state; and

✓ Leave the site in a condition that allows for permanent use free of any significant risk to human health or any potential for future environmental damage.[20]

A response action may achieve applicable cleanup standards through the use of:

✓ Engineering and appropriate land use controls applicable to the sites future use

✓ Treatment reducing the toxicity, mobility, or volume of hazardous substances;

✓ Removal techniques for hazardous substances;

✓ Any combination of the above; or

✓ Other approved methods of protection.[21]

The California program was not specifically designed to encourage economic development of urban industrial areas. The California program merely resembles an expedited state Superfund program. For instance, the Expedited Remedial program specifies criteria for the apportionment of liability between potentially responsible parties for response costs.[22]

During the apportionment phase, CEPA will weigh relevant equitable factors considered appropriate under the circumstances for which the remedial action was initiated.[23] In addition, the Expedited Remedial program requires that certain land use controls implemented at a site be recorded in the site's deed and other chain of title documents to provide notice to subsequent purchasers and users.[24]

Other Voluntary Cleanup Programs in California. The Department of Toxic Substances Control (DTSC) created the Voluntary Cleanup Program (VCP) in 1993.[25] The VCP permits parties, known as project proponents, to investigate and remediate sites after paying DTSC's

---

[20] *Id.* at § 25398.6(a).

[21] *Id.* at § 25398.6(b)(1)-(5).

[22] *Id.* at § 25398.8(b). Under § 25398.7(c), CEPA will weigh equitable factors considered appropriate under the circumstances of the release for which the remedial action was initiated.

[23] *Id.*

[24] *Id.* at § 25398.7(a).

[25] Cal. Health & Safety Code, ch.6.8 §§25300 *et seq.*

oversight fees and entering into an agreement with DTSC to assess a site, prepare plans, and conduct specific remedial or removal actions.[26] Following the cleanup, when DTSC reviews and approves of the action, it may issue either a "No Further Action" letter or Remedial Action Certification.[27] Unlike the Expedited Remedial Action Program agreement, however, the VCP does not provide a covenant not to sue.

The following is a breakdown of the actual steps one must follow to participate in the VCP. First, a project proponent applies to DTSC to participate in the VCP. If the application is accepted, the applicant then negotiates the terms and services of the VCP with the DTSC. Based on the estimated cleanup costs, either a Remedial Action Workplan (RAW) or Remedial Action Plan (RAP) must be prepared. A RAW must be prepared for projects where the estimated total cleanup cost is less than $1 million.[28] A RAP is prepared for projects estimated to cost more than $1 million.[29]

The plan is put through notice and comment. It is then subject to final approval from DTSC. Finally, cleanup activities commence, under DTSC's supervision and per the requirements of the Hazardous Substance Account Act and the National Contingency Plan.[30]

Sites that are ineligible for the Voluntary Cleanup Program include the following:

✓ DTSC's Annual Workplan sites;
✓ NPL sites;
✓ Federal facilities;
✓ Sites containing petroleum products or waste or otherwise outside of DTSC's jurisdiction.[31]

---

[26] Jane Kroesche, California, at 3, citing DTSC, Policy and Procedure for Managing Voluntary Site Mitigation Projects, The Voluntary Cleanup Program, #E0-95-006-PP, at 1-2 (Sept. 20, 1995) (unpublished manuscript, on file with author).

[27] *Id.* at 4, citing Barbara Coler and Steve Koyasako, Redevelopment and Revitalization of Brownfields, Department of Toxic Substance Control Initiatives, at 2 (Oct. 1995).

[28] CAL. HEALTH & SAFETY CODE §§ 25356.1(h), 25323.1.

[29] *Id.*

[30] *Id.* § 25356.1(d).

[31] Kroesche, *supra* note 26, at 4.

The Unified Agency Review Law. The California Unified Agency Review Law[32] provides overview, final approval, and certification by an Administering Agency to responsible parties that successfully undertake investigation and cleanups.[33] In addition to providing qualified immunity to the responsible party, the certification binds state and local agencies.[34] The law is designed to enable participants to rely on the decisions of one lead agency within the California system.

The Private Site Manager Program. The Private Site Management Program,[35] which is expected to be in effect in 1997, allows private site managers to investigate sites and provide cleanup recommendations to DTSC.[36] If cleanup is recommended, either a RAW or RAP must be prepared, and remediation of the site occurs.[37] The private site manager will later request a certificate of completion that details the cleanup activities and confirms the site remediation.[38] DTSC's certification of the site must be made known to the public and may call for continued site maintenance.[39]

§ 8.2.2 Colorado

Colorado's Voluntary Cleanup Program provides private parties contemplating the development of existing industrial sites with a mechanism for determining what the cleanup responsibilities will be when planning the actual reuse of existing sites.[40] The Colorado program endeavors to eliminate impediments regarding sale or redevelopment of

---

[32] CAL. HEALTH & SAFETY CODE §25260 *et seq.*
[33] *Id.* §§ 25262, 25264(a)(b).
[34] *Id.* § 25264(b), (c)(1-6).
[35] CAL. HEALTH & SAFETY CODE §§ 25395.1 *et seq.*
[36] Kroesche, *supra* note 26, at 20; CAL. HEALTH & SAFETY CODE § 25395.2.
[37] CAL. HEALTH & SAFETY CODE §§ 25395.5, 25395.6.
[38] *Id.* §25395.8(a); Kroesche, *supra* note 26, at 21.
[39] CAL. HEALTH & SAFETY CODE, §25395.8(d); Kroesche, *supra* note 26, at 21.
[40] COLO. REV. STAT. ANN. § 25-16-301 *et seq.* (West 1994).

Brownfields, encourages prompt cleanup activities, and minimizes the administrative process.[41] Brownfield sites listed on the NPL, subject to state or federal corrective action or subject to a consent order issued by the Colorado Water Quality Control Division, are ineligible for voluntary cleanups.[42] RCRA treatment, storage, and disposal (TSD) facilities are also expressly excluded from voluntary cleanup.[43]

In Colorado, voluntary cleanup plans must provide for the cleanup and control of hazardous substances and/or petroleum consistent with state cleanup standards.[44] Under the Colorado program, voluntary cleanups must be conducted under a state-approved voluntary cleanup plan. Generally, the plan must include an environmental assessment of contaminated property, a proposal to remediate any contamination or condition that has led or could lead to a release that poses an unacceptable risk to human health and the environment, and a description of applicable state standards establishing acceptable concentrations of contaminants in soils, surface water, or groundwater.[45] If state standards do not exist for the contaminants located at the site, the Colorado program requires volunteers to undertake risk assessment procedures to determine proposed cleanup levels and any current risk to human health or the environment. The reduction of contaminants must meet concentration levels that do not present unacceptable risks to health and the environment based upon the current and proposed uses of the site.[46]

---

[41] *Id.* at § 25-16-302.
[42] *Id.* at § 25-16-303(3)(b)(I)-(II).
[43] *Id.* at § 25-16-303(3)(b)(III)-(IV).
[44] *Id.* at § 25-16-304(2)(c).
[45] *Id.* at § 25-16-304.
[46] *Id.*

Voluntary cleanup plans require the submission of an ESA,[47] a remediation plan proposal,[48] and remedial alternatives.[49] The ESA must be prepared by a qualified professional,[50] who is defined as a person with education, training, and experience in preparing environmental studies and assessments.[51]

A "No Action" determination may be submitted by a volunteer to the state under two circumstances:[52]

1. at the successful completion of voluntary remedial activity; or
2. upon determination that existing site contamination migrated onto the property from an off-site source.[53]

---

[47] *Id.* at 25-16-302(2)(a). An environmental assessment for voluntary response actions should include

- A legal description and map of the site;
- The physical characteristics of the site and contiguous areas, including any existing surface water bodies or groundwater aquifers;
- Location and description of the use of any wells on the site or on areas within an one-half mile radius of the site;
- Current and proposed use of on-site groundwater;
- Operational history of the site, including the current use of areas contiguous to the site;
- The present and proposed uses of the site;
- The nature and extent of any contamination and releases of hazardous substances and/or petroleum, including impacts on areas contiguous to the site;
- The results of any sampling or other data characterizing the condition of the soil, groundwater, or surface water on the site; and
- A description of human and environmental exposure to contamination at the site based on the property's current and future use. *Id.* at § 25-16-308(2).

[48] *Id.* at 25-16-304.

[49] *Id.* at § 25-16-305. Based on the actual risk to human health and the environment, remediation alternatives must consider the following factors:

- Present or proposed uses of the site;
- Potential pathways of migration for contaminants exceeding applicable state standards resulting in exposure to humans and the surrounding environment or, in the absence of state standards, representing an unacceptable risk to human health or the environment; and
- Potential risks associated with the proposed cleanup alternatives, the economic and technical feasibility, and reliability of such alternatives.

[50] *Id.* at § 25-16-308(1).

[51] *Id.*

[52] *Id.* at § 25-16-307.

[53] *Id.*

This determination is appropriate when a qualified professional indicates that the existence of contaminants does not exceed applicable state standards or the contamination does not pose an unacceptable risk to health or the environment.[54]

Upon satisfactory completion of voluntary remedial activities, the State of Colorado must actively pursue a determination by the U.S. EPA that the property subject to the voluntary plan should not be subject to further federal action.[55] It is unclear whether a private cause of action may lie in mandamus for failure to "actively" pursue a determination by the U.S. EPA to take no further federal action. The U.S. EPA is not, however, bound by legal authority to obey DPHE's petition. Any such action taken by a volunteer could, however, be asserted as a defense to or evidence of mitigating factors to federal environmental enforcement proceedings.

§ 8.2.3 Connecticut

The Connecticut General Assembly recently added voluntary cleanup measure to its already existing Hazardous Waste Establishment Transfer Act and Hazardous Waste Site Remediation Act.[56] The amendments provide for an owner of property identified on the inventory of hazardous waste disposal sites[57] who wishes to perform voluntary cleanup of the property to

[54] *Id.* at § 25-16-307(2)(a). Generally, a "No Action" determination states that no further action is required to ensure that the property, when used for the purposes identified in the no action petition, is protective of existing and proposed uses and does not pose an unacceptable risk to human health or the environment.*Id.* at § 25-16-307(2)(b).

[55] *Id.* at § 25-16-309(2). In pertinent part, the relevant provision states:

> If the United States [E]nvironmental [P]rotection [A]gency indicates that it is investigating a site which is the subject of an approved voluntary cleanup plan or no action petition, the department *shall actively pursue* a determination by the United States . . . that the property not be addressed under the federal act, that no further federal action be taken with respect to the property at least until the voluntary cleanup plan is completely implemented.

[56] P.A. No. 95-183, approved June 30, 1995; repealing CONN. GEN. STAT. § 22a-134, and substituting CONN. GEN. STAT. §§ 22a-134a to 22a-134d. For a further discussion of Connecticut cleanup programs, see Terry J. Tondro, *Reclaiming Brownfields to Save Greenfields: Shifting the Environmental Risks of Acquiring and Reusing Contaminated Land*, 27 CONN. L. REV. 789 (1995).

[57] This inventory is maintained pursuant to CONN. GEN. STAT. § 22a-133c.

submit to the Commissioner of Environmental Protection an environmental condition assessment form.[58] The commissioner shall then notify the owner within thirty days whether a review and approval of any such remedial action will be necessary.[59]

If the commissioner determines that prior approval is not necessary, the property owner is required to submit a statement of proposed actions for investigation and remediating his or her property, and a schedule of implementation for such activities.[60] This statement must be submitted within ninety days of the commissioner's notification.[61] However, the commissioner may require the owner to submit copies of technical plans and reports related to the cleanup activities.[62] In addition, the property owner has the duty to notify the commissioner of any modifications to the proposed schedule.[63]

The commissioner may also determine that formal review and approval of the site remediation is necessary. In such case, the property owner has the obligation to submit to the commissioner a proposed schedule for:

1. Investigating and remediating the property; and
2. All technical plans, technical reports, and progress reports related to the investigation and remediation.[64]

If approved, the property owner must submit further technical plans, technical reports, and progress reports to the commissioner in accordance with the approved schedule, and perform all actions identified in the approved technical plans, reports, and progress reports in accordance with the approved schedule.[65] If modifications are made to the original plans, they must be approved in writing by the commissioner.[66]

---

[58] S.B. No. 1189 § 3(a).
[59] *Id.*
[60] *Id.* §3(b).
[61] *Id.*
[62] *Id.*
[63] *Id.*
[64] *Id.* §3(c).
[65] *Id.*
[66] *Id.*

Connecticut utilizes a "Licensed Environmental Professional"[67] to verify that parcels that have petitioned for voluntary cleanup meet the standards that were proposed to, and approved by, the commissioner.[68] A copy of such verification must be submitted to the commissioner at the completion of the remediation work.[69]

At the completion of the environmental remediation, and verification of such work, the property owner may petition the commissioner for a Form II statement of remediation.[70] This document provides that any discharge, spill, uncontrolled loss, seepage, or filtration of hazardous waste that had occurred on the property has been properly remediated by the property owner, and that this remediation has been approved by the state, and verified by a certified environmental professional.[71]

The fee for this environmental remediation assessment is $2,000, and must be paid at the time that the initial assessment form is submitted to the state.[72]

§ 8.2.4 Illinois

Illinois currently has two voluntary cleanup programs. The Voluntary Cleanup Program, enacted in 1989, eventually will be phased out as the Site Investigation and Remedial Activities Program (Site Remediation Program) enacted in 1995 takes effect. Participants in the Voluntary Cleanup Program may opt into the Site Remediation Program.

---

[67] "Environmental Professional" is defined in the statute at § 4 as:

> "a person who is qualified by reason of his knowledge, as specified in subsection (e) of this section, to engage in activities associated with the investigation and remediation of pollution and sources of pollution including the rendering or offering to render to clients professional services in connection with the investigation and remediation of pollution and sources of pollution."

Id. § 4(a).

[68] *Id.* § 3(b), (c).

[69] *Id.*

[70] *Id.* § 3(d).

[71] *Id.* § 1(12), 3(d).

[72] *Id.* § 3(e).

The Site Remediation Program formally sets forth the requirements and procedures for voluntary site investigation and remediation. Participants in the Illinois program, known as Remediation Applicants,[73] may be owners or operators involved in the site.[74] The following sites are excluded from the program:

- ✓ sites on CERCLA's NPL list;
- ✓ sites subject to RCRA or state solid or hazardous waste laws;
- ✓ sites subject to federal or state underground storage tank laws;
- ✓ sites where investigation and remediation is ordered by federal court or the U.S. EPA.[75]

Successful participation in the Site Remediation Program yields a "No Further Remediation letter" (NFRL),[76] which releases the participant from further responsibility and is evidence that a threat no longer exists and no further action is required.[77] The NFRL becomes effective once the participant officially records it with the Office of the Recorder or the Registrar of Titles.[78] The NFRL may contain control or use restrictions and is voidable if the restrictions are not followed.[79] Failure to pay program fees and oversight costs can also result in the voiding of the NFRL.[80]

The 1989 Voluntary Cleanup Program provides for traditional review and evaluation of plans for site investigation and cleanup of hazardous substances and/or petroleum.[81] Once the nature and extent of contamination is determined, a remediation goal is pursued jointly by the Illinois EPA and

---

73 ILL. ANN. STAT. Ch. 415 para. 5/58.2.
74 *Id.*
75 *Id.* para. 5/58.1(a)(2).
76 *Id.* para. 5/58.7(d)(4).
77 *Id.* para. 5/58.10(d).
78 *Id.* para. 5/58.8.
79 *Id.* para. 5/58.10(b).
80 *Id.* para. 5/58.10(e).
81 ILL. ANN. STAT. ch. 415, para 22.2b (Smith-Hurd 1994).

the property owner.[82] Site-specific voluntary cleanup goals may be achieved through the use of engineering controls, institutional controls, future land use restrictions placed in the chain of title of the property, or numeric controls.[83] Following the satisfactory completion of voluntary remedial activities, the Illinois EPA would provide a release from liability to volunteers co-extensive with the remedial actions taken.[84]

Proposals have been submitted in the Illinois legislature to modify this program. H.B. 1153, entitled the "Brownfields Redevelopment Act," would (1) provide assistance in identifying and assessing properties potentially contaminated by commercial, industrial, or other uses; (2) promote appropriate cleanup of commercial, industrial, or other properties through the Illinois' Voluntary Cleanup Program; and (3) create and administer the Brownfields Redevelopment Fund.[85] H.B. 1153 is symptomatic of the growing landowners' rights movements in the various state legislatures.[86] As such, it addresses the issue of Brownfields redevelopment from an owner's or operator's perspective. The bill creates a rebuttable presumption against state claims and conclusive against private claims, that a defendant has made all appropriate inquiry at the time of acquisition of contaminated

---

[82] *Id.* at § 22b(a)(3). Under § 22(m)(1), the following provisions are deemed conditions precedent for Illinois EPA review and evaluation of proposed voluntary cleanup plans:

- Submission of remedial work plan for voluntary cleanup at the site;
- Proposed site visit or other site evaluation by the Illinois EPA;
- Agreement to perform the workplan as approved by the Illinois EPA;
- Payment of reasonable cost incurred and documented by the Illinois EPA in providing such services; and
- Advanced payment for anticipated services of the Illinois EPA in the amount of $5,000 or one-half of the total anticipated costs, whichever is less.

[83] *Id.* at § 22b(d).

[84] *Id.* at § 22b(a)(1).

[85] *Id.* at § 5(c).

[86] *See generally* Hertha L. Lund, *The Property Rights Movement and State Legislation.* Cf., Nancie G. Marzulla, *The Property Rights Movement: How It Began and Where It Is Headed,* in BRUCE YANDLE, LAND RIGHTS (1995).

industrial property.[87] H.B. 1153 provides very detailed procedures for parties conducting Phase I and Phase II ESAs. A Phase I ESA that does not disclose the presence of contaminants would be deemed to have met any legal presumption against public or private claims, thus releasing the volunteer from liability.[88] Similarly, a Phase II ESA demonstrating that the property meets or exceeds the applicable standards will be deemed to have met all legal presumptions relieving the volunteer of liability.[89]

A party choosing to enter into an agreement with the state to perform a response action plan may upon successful completion be issued a "No Further Remediation Letter."[90] The issuance of a "No Further Remediation Letter" will be considered a release from further responsibility under H.B. 1153 for further preventive or corrective action at the site.[91] The protections afforded by the "No Further Remediation Letter" will extend to, among others, lenders, trustees, and prospective and subsequent purchasers.[92]

§ 8.2.5 Indiana

The Indiana Voluntary Cleanup Program requires that program participants provide notice to the state and submit workplans for the review

---

[87] 1995 IL H.B.1153, §§ 22.2(j)(1)(C) & 22.2(j)(6)(E)(i). To determine whether a party took "all appropriate inquiry" into the previous ownership and operation of the contaminated site consistent with good commercial or customary practice, the following factors will be taken into account:

- Any specialized knowledge of the party;
- The relationship between the purchase price and the value of the property if uncontaminated;
- The obviousness of the presence or likely presence of contamination at the property; and
- The ability to detect such contamination. *Id.* at § 22.2(j)(6)(B).

[88] *Id.* at § 22.2(j)(6)(E)((i)(I).

[89] *Id.* at § 22.2(j)(6)(E)(i)(II).

[90] Similar to the current Voluntary Cleanup Program, a volunteer must submit a response action plan, arrange for a site visit or other site evaluation by the Illinois EPA, agree to perform a response action plan, and pay any reasonable costs incurred through Illinois EPA oversight. *Id.* at § 22.2(m)(1). Following the successful completion of an Illinois EPA-approved response action plan, the agency may issue a "No Further Action Letter" applicable to the site. *Id.* at § 22.2(m)(7)(A).

[91] *Id.* at 22.2(m)(7)(A).

[92] *Id.* at 22.2(m)(B).

and approval of the investigation and voluntary remediation of contaminated industrial property.[93] During the investigatory phase of voluntary remedial activities an environmental assessment (the contents of which are specified by law)[94] must be conducted.[95] Participants are required to reimburse the state for oversight costs associated with voluntary cleanups.[96] Upon the satisfactory completion of voluntary cleanup activities a Certificate of Completion would be issued.[97] Receipt of a Certificate of Completion is a condition precedent to the issuance of a Covenant Not to Sue by the governor of Indiana.[98] The covenant generally releases

---

[93] IND. CODE ANN. §13-25-5-2 (West 1994). Specifically, an eligible party seeking to remediate used industrial property under the Indiana voluntary cleanup program must provide the following information:

- General information on the parties, the site to be remediated, and other requested background information;
- An environmental assessment of the actual or threatened release of the hazardous substance or petroleum; and
- An application fee of one thousand dollars ($1,000).

Applicant eligibility should be determined within thirty days of receipt of an application and application fee. *Id.* at § 13-25-5-4. Property subject to pending state or federal enforcement actions, having conditions creating an imminent and substantial threat to human health or the environment, or where a federal grant requires an enforcement action are ineligible for voluntary cleanup. *Id.* at 13-25-5-5(a)(1)-(3).

[94] *Id.* at § 13-25-5-3. The elements of an environmental assessment under the Indiana Voluntary Cleanup Program include:

- A legal description of the site;
- A description of the physical characteristics of the site;
- The operational history of the site to the extent known by the applicant;
- Information regarding the nature and extent of any relevant contamination and relevant releases at the site and immediately contiguous to the site; and
- Relevant information concerning the potential for human exposure to contamination at the site.

Id. at § 13-25-5-3(1)-(5).

[95] *Id.* at 13-25-5-2(b)(2)(B).

[96] *Id.* at § 13-25-5-8(a)(1). Volunteers are required to reimburse the state for all reasonable costs exceeding the application fee submitted by participating parties. Specifically, the state may recover costs incurred in the review and oversight of the workplan and in association with the voluntary remediation agreement.

[97] *Id.* at § 13-25-5-16(a). The issuance of a Certificate of Completion certifies the successfully completion of an approved voluntary remediation workplan. The issuance of a Certificate of Completion will be deemed final agency action.

[98] *Id.* at § 13-25-5-18(a).

volunteer(s) from liability, including future liability, for a claim relating to contamination specifically redressed by an approved voluntary cleanup.[99] Further, Indiana Covenants Not to Sue are transferable and apply to both current and future owners of Brownfield sites.[100]

Prior to approving a proposed voluntary remediation workplan, the state and the volunteer(s) must enter into a voluntary remediation agreement.[101] This agreement sets the terms and conditions for the evaluation and implementation of the work plan.[102] Once the voluntary remediation agreement is entered, the department may evaluate the

---

[99] *Id.*

[100] *Id.* at § 13-25-5-18(b). A covenant not to sue issued under this section should bar all public or private claims arising in connection with the release or threatened release of hazardous substances and/or petroleum that was the subject of an approved voluntary remediation workplan. *Id.* Protection under the Covenant Not to Sue specifically applies to persons receiving such covenants through a legal transfer of a Certificate of Completion or by acquiring property to which the Certificate of Completion applies. *Id.*

[101] *Id.* at § 13-25-5-8.

> Indiana offers volunteers three tiers of cleanup standards. Tier one concerns background levels of contamination. Background contamination has been defined as normally occurring pollution in the ambient environment. Background standards are generally recognized as being very stringent standards. Tier two cleanup standards are health-based standards. Specifically, property cleaned up for residential purposes must meet a 1 x 10-6 health-based standard. Nonresidential uses of contaminated property must meet a lower standard of 1 x 10-5. Tier three concerns site-specific risk assessment procedures. These options allow volunteers to select a risk level that is comfortable and compatible with the intended use of a redeveloped facility. The more extensive the voluntary cleanup workplan is, the more extensive the protection for future liability received by program participants. Conversely, a more limited cleanup results in limited protection from future liability.

Conference on Brownfields Redevelopment and Environmental Risk Management in Business and Real Estate Transactions (March 22-23, 1995) (Statement of S. Richard Heymann, Esq.).

[102] *Id.* A mechanism to resolve disputes arising from the evaluation, analysis, and oversight of the implementation of the workplan, including arbitration, adjudication, or alternative dispute resolution procedures is required to be incorporated in the voluntary remediation agreement. *Id.* at § 13-25-5-8(a)(2). Also required to be incorporated in the voluntary remediation agreement are provisions concerning the indemnification of parties, retention of record, and timetables for the reasonable review and evaluation of the adequacy of the voluntary workplan.

proposed voluntary remediation workplan.[103] Voluntary remediation workplans submitted for approval are subject to notice, comment, and, if requested, public hearings.[104] If approved, volunteers are notified in writing that remedial activities may be commenced, as well as the date by which the work must be completed.[105] Oversight and progress reporting duties regarding the implementation of the voluntary remediation workplans lie with the state.[106] Upon successful completion of an approved voluntary remediation workplan, volunteers should qualify for the issuance of a Certificate of Completion and a Covenant Not to Sue.[107] The protection afforded by the Covenant Not to Sue does not extend to

[103] *Id.* at § 13-25-5-7. A proposed workplan must include the following criteria:

- Detailed documentation of the remedial investigation and a description of the work to be performed to determine the nature and extent of the actual or threatened release;
- A proposed statement of goals to be accomplished by the workplan;
- Plans concerning:
  - -Quality assurance measures for the implementation of the proposed remediation project;
  - -Descriptions of sampling and analysis;
  - -Health and safety considerations;
  - -Community relations;
  - -Data management and recordkeeping; and
  - -Proposed work schedule concerning the implementation-specific remedial activities.

Id. at § 13-25-5-7(b).

[104] *Id.* at § 13-25-5-11. Notice to local government units located in the county affected the by the proposed remediation workplan is required. *Id.* at § 13-25-5-11(a)(1). A copy of the voluntary workplan must be provided to each municipality and placed in, at least, one public library. *Id.* at § 13-25-5-11(a)(2). Notice requesting comments concerning the proposed voluntary workplan must be published, followed by a thirty-day comment period. *Id.* at § 13-25-5-11(a)(3), (b). Interested persons may submit written comments or request a public hearing. *Id.* at § 13-25-5-11(b)(1)(2). If at least one public comment is received, a public hearing may be held in the geographical area affected by the proposed voluntary remediation work plan on the issue of approval or denial of the plan itself. *Id.* at § 13-25-5-11(c).

[105] *Id.* at § 13-25-5-13.

[106] *Id.* at § 13-25-5-15.

[107] *Id.* at §§ 13-25-5-16, 13-25-5-18.

1. future liability for property previously remediated under a voluntary workplan;[108]
2. contamination not known at the time the Certificate of Completion was issued;[109]
3. liability to the federal government for claims based on federal law.[110]

Volunteers are, however, protected from public or private actions brought during the implementation of an approved voluntary remediation workplan.[111] Volunteers are protected from any cause of action relating to the release or threatened release of hazardous substances and/or petroleum.[112] Further, administrative actions brought by the state are barred during the pendency of the voluntary remediation workplan.[113] Similarly, private actions for legal or equitable relief brought by third parties or citizens groups are barred during voluntary cleanup activities.[114]

Nothing under the Indiana Voluntary Cleanup Program prevents volunteers bringing an action or raising a claim, including a claim for contribution, against third parties responsible for existing site contamination.[115] Volunteers successfully completing voluntary cleanups are afforded contribution protection for matters specifically addressed in a workplan or Certificate of Completion.[116]

§ 8.2.6 Kentucky

Kentucky's newly proposed bill centers around the application for a No Further Remediation Letter to be given by the state to entities who successfully and voluntarily clean up facilities to the appropriate level.[117] The bill allows for the application by any party to the state for a No Further Remediation Letter. The entity's application must include

[108] *Id.* at § 13-25-5-18(c)(1).
[109] *Id.* at § 13-25-5-18(c)(2).
[110] *Id.* at § 13-25-5-18(d).
[111] *Id.* at § 13-25-5-18(e).
[112] *Id.*
[113] *Id.*
[114] *Id.*
[115] *Id.* at § 13-25-5-20(a).
[116] *Id.* at § 13-25-5-20(b).
[117] Kentucky H.B. 725, introduced February 28, 1996.

1. A legal description of the property;
2. A copy of the deed to the property;
3. An environmental site assessment of the property sufficient to show the nature and extent of any hazardous waste at the site;
4. A proposed plan to remediate the environmental contamination at the site; and
5. The subsequent proposed use of the property after obtaining the No Further Remediation Letter.[118]

The state has the ability to approve the application, to deny the application after public comment, or to enter into negotiations with the party regarding the proposed remediation activities.[119] If approval of the application is made, the applicant has the affirmative duty to follow the provisions of the approved plan to afford remediation.[120] The state will issue the No Further Remediation letter only after the successful completion of the remediation activities.[121]

The bill further states that the No Further Remediation letter shall be construed as prima facie evidence that the property does not constitute a threat to human health and the environment and, additionally, the property does not require further remediation activities.[122] Such a letter is to be limited to and include

1. Acknowledgment that the requirement of the remediation plan was satisfied;
2. Legal description of the property location;
3. Any remediation objectives, including monitoring requirements or land use restrictions imposed;
4. Statement that the No Further Remediation Letter constitutes completion of the remediation activities with no further cleanup efforts required, secures a release from further responsibilities under the laws, and stands as prima facie evidence of no threat to public health or environment;

---

[118] H.B. 725 § 2(1).
[119] *Id.* § 2(2).
[120] *Id.* § 2(3).
[121] *Id.* § 2(4).
[122] *Id.* § 3(1).

5. A prohibition against uses that are inconsistent with the land use limitations imposed on the property, without additional remedial activities and conveyance of the parcel to a third party; and
6. A description of any preventive, engineering, and institutional controls required in the remediation plan and notification that failure to manage and maintain the controls with the terms of the remediation plan may result in voidance of the No Further Remediation letter.

The letter shall apply in favor of the applicant who was issued the letter; the mortgagee, the trustee, or the successor in interest of a deed of trust; any successor in interest to the property; any transferee of the property through bankruptcy, partition, settlement or adjudication of a civil action gift or bequest; and any financial institution gaining control over the property by and through foreclosure or other terms of security interest held on the property.[123] Finally, the bill states that the No Further Action Letter may be voidable if the property is managed contrary to the provisions of this statute.[124]

§ 8.2.7 Maine

Maine's Voluntary Response Action Program (VRAP) was established in 1993.[125] Depending on the level of cleanup undertaken, the VRAP may exempt individuals who initiate site investigation and cleanup from liability under Maine's environmental laws.[126] It also provides a "no-action" guarantee from the Maine Department of Environmental Protection (DEP).[127] Participants in the VRAP pay an initial oversight fee to the DEP, and may incur additional costs for the DEP's services.[128]

---

[123] *Id.* § 3(3).
[124] *Id.* § 3(4).
[125] ME. REV. STAT. ANN. tit. 38, § 343-E.
[126] *Id.* § 343-E (1)(2)(6)(8)(9); James T. Kilbreth and Juliet T. Browne, Maine (unpublished manuscript, on file with author).
[127] ME. REV. STAT. ANN, § 343-E(9).
[128] Kilbreth and Browne, *supra* note 126, at 4.

The Maine DEP has yet to create regulations for VRAP. Thus, to date, any site may be eligible for cleanup, so long as the cleanup is consistent with federal environmental laws.[129] The DEP currently decides program eligibility on a site-by-site basis.[130]

As with other voluntary cleanup programs, a proposed action plan for either partial or complete cleanup must be submitted to the DEP.[131] The plan must detail the investigation of the site, the results of the investigation, and an analysis of the findings.[132] Implementation of the complete plan begins once it is approved by the DEP. The DEP provides a "no-action" letter upon the approval of the plan, in which the DEP assures the parties that it will not take legal action if the plan and any additional site requirements are followed.[133] Once the site is cleaned per the plan's requirements and meets DEP satisfaction, the DEP issues a certificate of completion.[134]

§ 8.2.8 Massachusetts

Massachusetts' Clean Sites Initiative encourages redevelopment of contaminated properties by providing a covenant not to sue from the Commonwealth of Massachusetts upon successful completion of the cleanup. The project is limited for use in Economic Target Areas (ETA), which are identified by the Massachusetts Economic Assistance Coordinating Council. Exceptions may, however, be made if property not included in an ETA would provide an "exceptional economic development opportunity."[135]

As with other voluntary cleanup programs, a party's application is reviewed, and if approved, a covenant not to sue is issued. Cleanup commences, and when finished, the party responsible for the cleanup must prepare a Response Action Outcome (RAO) Statement for the DEP.

---

[129] *Id.* at 3-4; ME. REV. STAT. ANN. § 343-E(1).

[130] Kilbreth and Browne, *supra* note 126, at 4.

[131] ME. REV. STAT. ANN., § 343-E(1),(2),(4).

[132] *Id.* § 343-E(4).

[133] *Id.* § 343-E(9).

[134] *Id.* § 343-E(5).

[135] Ned Abelson et al., Massachusetts, at 19 (unpublished manuscript, on file with author). For a further discussion on Massachusetts cleanups, see Terry J. Tondro, *Reclaiming Brownfields to Save Greenfields: Shifting the Environmental Risks of Acquiring and Reusing Contaminated Land*, 27 CONN. L. REV. 789 (1995).

Upon receipt of the RAO Statement, the DEP issues a Certificate of Completion. The certificate merely reflects DEP's receipt of the RAO statement. DEP does not approve any action conducted. The agency does not evaluate the cleanup. Nor does DEP protect against future site audits.[136]

§ 8.2.9 Michigan

The Michigan State Legislature took affirmative steps to revise its former equivalent of "Superfund," or the Michigan Environmental Response Act (MERA), by enacting substantive amendments to the act.[137] The amendments change significantly the liability scheme in MERA to completely eliminate liability for owners and operators of property contaminated with hazardous waste, unless these persons caused the contamination.[138] In addition, persons who purchase or begin operations at property after June 5, 1995, will be exempt from liability for existing contamination if they conduct a baseline environmental assessment (BEA) of the property.[139] These new property owners/operators must also exercise "due care" with regard to any contamination that exists at the time of their purchase or occupancy.[140] The BEA must provide sufficient information about the site and the nature and extent of contamination to determine if, at some time in the future, the new owner has contributed to the contamination on the property. The party preparing the BEA may elect to limit the scope of the BEA if he or she limits the use of the hazardous substances at the facility in the future, or if no activity will occur at the property during his or her ownership/operation.[141]

The new amendments also establish cleanup standards that are intended to protect the public health and environment, yet also allow for a reduction in the cost of cleanups by allowing the cleanup standards to reflect the expected future use of the property.[142] The Michigan Department of Natural Resources (MDNR) is authorized to establish cleanup standards

---

[136] *Id.* at 21.

[137] 1995 P.A. 71, as amended. Part 201, effective June 5, 1995 (HB 4569); codified at MICH. COMP LAWS ANN. §§ 20101-20133.

[138] *Id.* § 20126.

[139] *Id.* § 20126(1)(c).

[140] *Id.* § 20107a.

[141] *Id.* § 20101(1)(d).

[142] *Id.* § 20120.

in land use-based categories.[143] The categories for cleanup criteria are residential, commercial, industrial, and recreational.[144] Voluntary cleanup of property will be determined in accordance with the correlating cleanup category and standards.[145]

In addition to the BEA, liability protection is offered to persons who clean up contaminated property in the form of Covenants Not to Sue[146] and contribution protection.[147] Finally, the amendments provide protection to lenders,[148] local governments,[149] and contractors.[150]

Finally, the new voluntary program is fee-based. The statutes establish a $750 fee for the nonliability letter, which is issued after the BEA has been accepted by the MDNR.[151]

§ 8.2.10 Minnesota

The Minnesota Environmental Response and Liability Act (MERLA), as amended by the 1992 Land Recycling Act (LRA),[152] offers incentives to private parties conducting voluntary investigations and cleanups of contaminated industrial sites. Through Minnesota's Voluntary Investigation and Cleanup (VIC) Program, the LRA attempts to facilitate the transfer and development of property by providing resource and technical assistance at the request of program participants.[153] Moreover, program participants

[143] *Id.* § 20120a.

[144] *Id.* § 20120a(1) (a-e).

[145] *Id.*

[146] *Id.* §§ 20132 (developers of property), 20133 (voluntary cleanup of property). These covenants are offered to property owners prior to cleanup of the property. *Id.*

[147] *Id.* §§ 20129(5), (6). The new amendments have also attempted to provide for such protection against the federal government. *Id.*

[148] *Id.* § 20101(a), (b).

[149] *Id.* § 20126(a), (b), (d).

[150] *Id.* § 20128.

[151] *Id.* § 20126(3).

[152] Minn. Stat. Ann. § 115B.175 *et seq.* (West 1994).

[153] Minn. Stat. Ann. § 115B.175 (West 1994). The LRA was specifically enacted to clarify the cleanup liability of specific parties and obtain statutory mechanisms to obtain liability protection. A number of these amendments offer incentives to promote the voluntary investigation and cleanup of hazardous substances and/or petroleum. These amendments also encourage and facilitate the transfer and development of property by providing additional resources and technical assistance regarding contaminated property.

undertaking and completing cleanups in accordance with an approved voluntary cleanup plan, and who are not otherwise responsible for a release under MERLA, are eligible for protection from liability under the LRA.[154]

As amended in 1993, the LRA allows *responsible persons* to conduct cleanups and gain liability protection for associated parties who are not otherwise liable for the release, including lenders, developers, prospective purchasers, and municipalities.[155] Specifically, the LRA allows volunteers to choose from five separate types of written assurances for protection from environmental liability for parties successfully meeting the VIC program requisites:[156]

- ✓ Technical Assistance Approval Letters;
- ✓ No Action Letters;
- ✓ Off-site Source Determination Letters;
- ✓ Certificates of Completion; and
- ✓ No Association Determination Letters.[157]

---

[154] MINN. STAT. ANN. § 115B.175[subdiv. 1] (West 1994). Interestingly, the Minnesota Program started modestly with two staff personnel and twenty proposed projects waiting for approval when the doors opened. In 1995, the program has twenty staff members and has recently just received its 500th application for assistance. On average, the Minnesota Pollution Control Agency (MPCA) receives three to six applications per week. The program was designed to assist parties interested in developing contaminated land, and determining the environmental status of land before purchasing the property. *Environmental Law Institute: Redeveloping Brownfields Workshop* (March 28, 1995) (Statement of Joseph Zachmann, Project Manager, Site Response Section, MPCA).

[155] MINN. STAT. ANN. § 115B.175[subdiv. 6a] (West 1994). The benefit to responsible parties under the LRA stems from the release from liability for future, unanticipated cleanup of contamination specifically addressed under the VIC program. Thus, through remediating contaminated property under the VIC program, responsible persons preclude subsequent claims based on retroactive liability for site contamination.

[156] *Id.* at § 115B.175[subdiv. 6]. These assurances can be issued in a variety of forms, such as letters, agreements, or certificates, and can apply to technical, administrative, and liability matters.

[157] *Minnesota Pollution Control Agency: Voluntary Investigation and Cleanup,* Guidance Document No. 4, § 3.0 (Revised January 1994). These five broad categories of written assurances may be issued by the MPCA Commissioner or designated MPCA staff following approval of an investigation and a Response Action Plan (RAP). *Id.* Except for No Action letters or agreements, these assurances are issued under discretionary authority expressly provided under Minnesota Statutes. *Id.* No Action letters are also discretionary, but are issued pursuant to the inherent authority of the MPCA to exercise discretion in its enforcement action. *Id.; See* MINN. STAT. ANN. § 115B.175[subdiv. 3] (West 1994).

Upon successful completion of the VIC Program, volunteers will receive a "No Action Letter" from the Minnesota Pollution Control Agency (MPCA).[158] No Action Letters provide qualified liability protection from future Superfund or administrative enforcement actions brought by the MPCA.[159] The assurances provided by No Action Letters or Agreements, however, apply only to parties expressly included within the VIC Program.[160] Unless explicitly stated, a No Action Letter or Agreement does not extend to other parties associated with properties, such as lenders and successors to the owner of the property.[161] Cases where No Action Letters may be issued include:

✓ No contamination is detected and an approved Phase I Investigation Report determines the past and current use of the property would not have contributed to a release to soils or groundwater.[162]

✓ Contamination is detected but the levels are not significant as determined by MPCA staff, and as a result no cleanup is required.[163]

[158] MINN. STAT. ANN. § 115B.175[subdiv. 6] (West 1992).

[159] *Minnesota Pollution Control Agency: Voluntary Investigation and Cleanup*, Guidance Document No. 4, § 3.2.

[160] *Id.*

[161] *Id.*; It should be noted that lenders receive express statutory liability protection under the VIC Program. *See* MINN. STAT. ANN. § 115B.175[subdiv. 6](2) (West 1994).

[162] *See Minnesota Pollution Control Agency: Voluntary Investigation and Cleanup*, Guidance Document No. 8 (Revised January 1994) (information on the procedures and criteria involving a Phase I Investigation Report).

[163] *Minnesota Pollution Control Agency: Voluntary Investigation and Cleanup Program*, Guidance Document No. 4, § 3.2. A No Action letter issued in this circumstance should state that a release of hazardous substances was detected, but the levels in groundwater were, for example, below the Minnesota Department of Health Recommended Allowable Limits. A Phase I Investigation Report must also be reviewed and approved by the MPCA staff to qualify for this No Action letter.

✓ The property has been remediated as approved by the MPCA and no additional cleanup is required.[164]

The No Action Letter will state that the MPCA will refrain from referring the contaminated industrial property to CERCLIS, U.S. EPA's Superfund site tracking system and preparing a Hazard Ranking System (HRS) score for the site.[165] Specific LRA requirements mandate the submission of a voluntary response plan and a report describing the results of an investigation of actual or threatened releases of hazardous substances and/or petroleum.[166] Program participants are required to meet cleanup standards protective of public health, public welfare, and the environment.[167]

Since the Minnesota VIC Program was to create incentives for redevelopment and reuse of inactive or abandoned industrial sites,[168] specific statutory protection is included for lenders.[169] This protection compensates for the MPCA's position that No Action Letters or

[164] *Id.* Within the above three categories, a No Action letter may be tailored according to site-specific conditions. *Id.* A "Limited No Action" letter may be issued depending upon the completeness of the investigation or cleanup performed, the specific boundaries of the investigation or cleanup, or an abbreviated list of contaminants analyzed. *Id.* Cases where "Limited No Action" letters have been issued include:

- Sites where only soil is investigated and remediated. n such cases, a Limited No Action letter will address only the soil investigation and response actions and will contain disclaimers stating that groundwater was not investigated or remediated.
- Sites where only a discrete area of the property is investigated and remediated. An example includes an area of a proposed building expansion investigated and cleaned up, where other areas of the property were not addressed. As such, a Limited No Action letter would only address the expansion area and would contain disclaimers regarding other areas of the property that were not investigated or cleaned up.
- Sites where only a limited number of contaminants were investigated or cleaned up. A Limited No Action letter would only address the contaminants investigated and would contain disclaimers regarding other contaminants that have been detected or may be present. *Id.*

[165] *Minnesota Pollution Control Agency: Voluntary Investigation and Cleanup,* Guidance Document No. 4, § 3.2.

[166] MINN. STAT. ANN. § 115B.175[subdiv. 3] (West 1994).

[167] *Id.* at § 115B.175[subdiv. 3](c).

[168] *Environmental Law Institute: Redeveloping Brownfields Workshop* (Statement Joseph Zachmann), supra note 159.

[169] MINN. STAT. ANN. § 115B.175[subdiv. 6](2) (West 1994).

Agreements do not extend to the parties who have specifically applied to participate in the VIC program.[170] Specifically, the LRA's lender liability provision is aimed at facilitating the lending of funds to third parties undertaking a voluntary cleanup or using the land in a manner that does not affect the contamination already present.[171]

Under the LRA, parties responsible for site contamination cannot obtain liability protection for themselves from state or federal environmental enforcement actions.[172] Responsible parties may, however, obtain a Covenant Not to Sue for future cleanup liability when they conduct and successfully complete a voluntary response action.[173] Protection from liability takes effect when the completion of a voluntary cleanup is certified by the MPCA and a Certificate of Completion is issued.[174] The issuance of a Certificate of Completion must coincide with the approval of a voluntary Response Action Report (RAP).[175] The LRA's liability protection assurances will only vest when a RAP is approved and implemented.[176] If a response action is not required to remedy site contamination, then the contaminated site is not eligible for assurances of liability protection under the LRA.[177]

§ 8.2.11 Missouri

The Missouri Voluntary Cleanup Program[178] affords state oversight of any and all voluntary cleanup measures commenced by private parties. Any

---

[170] *Minnesota Pollution Control Agency: Voluntary Investigation and Cleanup*, Guidance Document No. 4, § 3.2 (Revised January 1994).

[171] *Environmental Law Institute: Redeveloping Brownfields Workshop* (March 28, 1995) (Statement of Joseph Zachmann) *supra* note 159.

[172] MINN. STAT. ANN. § 115B.175[subdiv. 6a] (West 1994).

[173] *Id.* at § 115B.175[subdiv. 6a](b); *Minnesota Pollution Control Agency: Voluntary Investigation and Cleanup*, Guidance Document No 4, § 3.5 (Revised January 1994).

[174] MINN. STAT. ANN. § 115B.175[subdiv. 5] (West 1994).

[175] *Minnesota Pollution Control Agency: Voluntary Investigation and Cleanup*, Guidance Document No. 4, § 3.5.

[176] *Id.*A RAP is only required when the MPCA has determined that the property is the site of a release or threatened release that needs to be addressed to protect human health, welfare, or the environment.

[177] *Id.*

[178] MO. ANN. STAT. §§ 260.565-260.575 (Vernon Supp. 1994).

person, including persons acquiring, disposing of, or in current possession of property, may apply to the Missouri Department of Natural Resources (MDNR) to commence voluntary remediation of real property.[179] The MDNR will review the application of the applicant, and shall approve or deny the application within ninety days of submission.[180] If the application is approved, the applicant is to submit copies of all reports prepared in preparation of remediation activities.[181] The MDNR has the duty to review this material, and to comment on any additional site assessments that will be needed, within 180 days of the original submission of these assessment documents.[182] In addition, the applicant will be required at the time of the submission of these documents to place a deposit with the MDNR of not more than $5,000, which will be utilized to cover the costs of the department's oversight.[183]

After the approval of assessment documents, the applicant has the obligation to submit for approval his or her proposed remedial action plan.[184] This proposed plan is to include all work plans, safety plans, testing protocols, and appropriate monitoring plans for postremediation activities.[185] If the remediation plans are acceptable, the applicant may commence operations to restore the property[186] but must file quarterly progress reports of his or her activities with the department.[187]

At the completion of the remedial activities, the person(s) conducting the remediation of the property is eligible for a "No Further Action" letter to be promulgated by the state.[188] If the department determines that the

---

179 *Id.* § 260.567(1).

180 *Id.* § 260.567(2).

181 *Id.* § 260.567(3). These reports can and shall include any site assessment results, investigations, sample collections, and analyses, including Phase I audits. *Id.*

182 *Id.*

183 *Id.* In addition the department may seek reimbursement for any costs incurred in the administration and oversight of voluntary cleanup activities. *Id.* § 260.569(1). These costs shall be at the rate of the lesser of the actual costs to the department, or one hundred dollars per hour. *Id.*

184 *Id.* § 260.567(6).

185 *Id.*

186 *Id.* § 260.567(7).

187 *Id.* § 260.567(8).

188 *Id.* § 260.573.

remedial activity was performed according to the original plans, and the applicant has met with the appropriate requirements under the act, such a letter can be drafted.[189] However, a restriction may be placed in the letter requiring that the participant(s) perform future monitoring activities and report any additional contamination that is discovered.[190]

Finally, the act provides for the promulgation of specific guidance and regulations pertaining to cleanup protocol, standards, and cost administration to be applied by the department during voluntary cleanup activities.[191]

§ 8.2.12 Nebraska

Nebraska's Remedial Action Plan Monitoring Act,[192] which is administered by the Nebraska Department of Environmental Quality (DEQ), has been in effect since January 1995. Unlike other voluntary cleanup programs, the act places no restrictions on who can participate in the program.[193] Also, there is no restriction on what sites may be selected for cleanup activities.[194]

As with other voluntary cleanup programs, parties apply to DEQ to participate in the program. The application includes an action plan, non-refundable application and participation fees, and a payment schedule for additional oversight expenses.[195] Upon completion of the cleanup, the parties receive a "no further action" letter. As in other states, this letter contains numerous reopeners, including the possibility of requiring that additional action be taken should contamination recur.[196]

---

[189] *Id.*

[190] *Id.*

[191] *Id.* § 260.571.

[192] NEB. LAWS, § 81-15,181 to 81-15,188.

[193] James T. Price and Annette Kovar, Nebraska, at 1 (unpublished manuscript, on file with author).

[194] *Id.* at 4.

[195] NEB. LAWS, § 81-15,184.

[196] *Id.* § 81-15,186.

§ 8.2.13 New Jersey

Although the state of New Jersey historically approached the cleanup of contaminated property through the use of a mandatory program under the New Jersey Environmental Cleanup Responsibility Act (ECRA),[197] the state has recently passed legislation to encourage voluntary cleanup as well. This new legislation provides certain tax incentives for volunteers. This most recent Act is designed to complement the earlier business transfer statutes, which by themselves were insufficient to deal with New Jersey's Brownfield redevelopment problems.

New Jersey's Department of Environmental Protection (NJDEP) also has developed a voluntary cleanup program.[198] Parties that wish to participate in the program apply to NJDEP. Once an application is accepted, the New Jersey Voluntary Cleanup Program requires parties to enter into a Memorandum of Agreement (MOA) with the NJDEP. The MOA does not protect against liability, although it does expressly provide that performance under the MOA does not constitute any admission of fact or liability.[199]

§ 8.2.14 New York

The New York legislature first promulgated legislation to provide for the financing of Brownfield redevelopment studies with the state in 1994.[200] This legislation has been carried over to today.[201] The bill provides for

---

[197] N.J. STAT. ANN. § 13:1K-6 to 11; N.J. ADMIN. CODE, tit. 7, § 26C-E.

[198] Procedures for Department Oversight of Contaminated Sites, N.J. ADMIN. CODE 7:26C *et seq.*

[199] I. Leo Motiuk and Sean T. Monaghan, New Jersey, at 4 (unpublished manuscript, on file with author).

[200] *See* 1994 S.B. 7787, 215 Gen. Ass., 2d Sess. N.Y. Laws. *See also,* Robert S. Berger, *Brownfields: The New York Approach* 6 ENVTL. LAW IN NEW YORK 1 (January 1995) (a discussion of the proposal to study Brownfields legislation within the state of New York); Robert S. Berger et al., *Recycling Industrial Sites in Erie County: Meeting the Challenge of Brownfield Redevelopment,* 3 BUFF. ENVTL. L. J. 69 (1995).

[201] 1995 S.B. 5607, 219 Gen. Ass., 2d Sess., 1996.

financial assistance to municipalities within the state for the undertaking of Brownfields Economic Redevelopment Studies, in order to assess and develop strategies for the promotion of voluntary cleanups of properties contaminated with hazardous substances.[202]

New York's Department of Environmental Conservation (NYDEC) has initiated its Voluntary Cleanup Program under its enforcement discretion.[203] Under this informal program, parties may enter into agreements with the NYDEC to clean up limited sites while paying NYDEC's oversight fees. In return, the NYDEC provides a no further action letter.[204] Despite efforts, the New York agency-driven program has yet to be codified.[205]

§ 8.2.15 Ohio

Signed into law on June 29, 1994, the Ohio Voluntary Action Program[206] (S.B. 221) will be implemented in two stages: (1) an Interim Voluntary Action Program,[207] which will sunset upon (2) the adoption of the Final Program by the Ohio Environmental Protection Agency (Ohio EPA).[208] When enacting S.B. 221, the Ohio legislature intended to create a program that

√ allows private parties to voluntarily clean up property contaminated by hazardous substances and/or petroleum to standards meeting the actual use of the property;

✓ removes lender liability surrounding contaminated property;

✓ privatizes the cleanup of contaminated property;

✓ consolidates the permit process for remedial activities; and

✓ offers tax abatements and low interest loans to potential volunteers.[209]

---

[202] *Id.* § 24, tit. 17.

[203] Margaret Murphy, New York, at 1 (unpublished manuscript, on file with author).

[204] *Id.* at 2.

[205] *Id.* at 1.

[206] OHIO REV. CODE ANN. § 3746.01 *et seq.* (Baldwin 1994).

[207] *Id.* § 3746.04.

[208] *Id.*

[209] *Conference on Brownfields Redevelopment and Environmental Risk Management in Business and Real Estate Transactions* (March 22-23, 1995) (Statement of William Hinger, M.En., RELA, Env'tl. Spec., Huntington Technical Services, Huntington Bancshares Incorporated).

"Volunteers" satisfactorily completing the Voluntary Action Program will receive a Covenant Not to Sue from the Ohio EPA.[210] Program participants are required to remit a program fee of $2,000 to receive Covenants Not to Sue, plus additional fees if their cleanups involve a Consolidated Standards Permit or an "operation and maintenance agreement."[211] Conditioned upon continued compliance with applicable cleanup standards, the Covenants Not to Sue provide volunteers and subsequent purchasers with protection from civil liability for future, unanticipated cleanup work.[212]

Program participants must use state certified professionals and laboratories to verify the proper completion of cleanup work and determine that the property is remediated to acceptable standards.[213] All professionals must have at least five years of experience with investigating and remediating hazardous substances and three years experience directly supervising remediation projects.[214] Certified professionals include engineers,[215] professional geologists,[216] certified industrial hygienists,[217] and

---

[210] OHIO REV. CODE ANN. § 3746.12.

[211] *Id.* at § 3746.07(H).

[212] *Id.* at § 3746.12(B). Releases from liability under Covenants Not to Sue remain in effect only for as long as the property or portion thereof subject to the Covenant complies with the applicable standards issued as part of the Covenant. *Id.* at § 3746(B)(1). If a property falls out of compliance with the terms of a Covenant Not to Sue, a volunteer has thirty days, after being so notified, to bring the property into compliance and enter into a compliance schedule agreement with the Ohio EPA. *Id.* at § 3746.12(B)(3). Failure to bring the property into timely compliance with terms set forth in the Covenant will result in the revocation of Covenant Not to Sue.
Id. at § 3746.12(B)(4).

[213] OHIO REV. CODE ANN. §§ 3746.04(B)(5) and 3746.07(D) & (E) (Baldwin 1994). "Acceptable standards" are determined on a case-by-case basis. Specifically, a volunteer may choose to remediate a site consistent with numeric standards promulgated by Ohio EPA or use site-specific risk assessment procedures depending upon the future, proposed use of the facility.

[214] *See generally id.* at § 3746.07(D).

[215] *Id.* at § 3746.07(D)(1). Specifically, § 3746.7(D)(1) requires that professional engineers must registered under OHIO REV. CODE ANN. § 4733 (Baldwin 1994).

[216] *Id.* at § 3746.07(D)(2). Professional geologists participating in the Voluntary Action Program must be certified by the American Institute of Geologists. *Id.*

[217] *Id.* at § 3746.07(D)(3). Industrial hygienists participating in the Voluntary Action Program are required to be certified by the American Board of Industrial Hygiene.

toxicologists.[218] Certified professionals and laboratories are required pay a fee of $2,000 to participate in the Interim Program.[219]

Until Ohio EPA promulgates the Final Voluntary Action Program, the existence of groundwater contamination will render property ineligible for voluntary cleanup.[220] Moreover, properties listed on U.S. EPA's National Priorities List (NPL) list[221] and properties subject to closure requirements under solid or hazardous waste laws are not eligible for voluntary cleanup under both the Interim and Final Programs.[222] Properties subject to Ohio's underground storage tank (UST) law[223] and sites where the Director of the Ohio EPA has initiated an environmental enforcement action are also excluded.[224]

Carcinogenic hazardous substances must be cleaned up to a risk range of 1 x 10-4 to 1 x 10-6.[225] Petroleum must be cleaned up to standards set by the Ohio Bureau of Underground Storage Tank Regulations (BUSTR). Parties participating in the Voluntary Action Program must conduct Phase I Environmental Site Assessments (ESA) in accordance with ASTM standards (E 1527-94).[226] Primarily a document review process, a Phase I ESA includes an analysis of a property's history of environmental

[218] *Id.* at § 3746.07(D)(4). Toxicologists participating in the Voluntary Action Program must hold at least a Master's Degree in Toxicology or a related field from an accredited school. *Id.*

[219] *Id.* at § 3746.07(D)(4).

[220] *Id.* at § 3746.07(A)(3).

[221] *Id.* at § 3746.02(A)(1)(d).

[222] *Id.* at § 3746.02(A)(2).

[223] *See* the Ohio Bureau of Underground Storage Tanks Regulations (BUSTR), OHIO REV. CODE ANN. §§ 3737.88, 3737.882 and 3737.889 (Baldwin 1994).

[224] OHIO REV. CODE ANN. § 3746.02(A)(5) (Baldwin).

[225] *Id.* at § 3746.07(A)(2) (cancer-causing hazardous substances must be cleaned up to the point where one person in 10,000 to one person in 1,000,000 is at risk of cancer after a lifetime of exposure to hazardous contaminants). The cleanup standards relating to carcinogenic substances will be determined by the proposed future use of the facility, *i.e.*, whether it is for industrial, commercial, or residential purposes.

[226] OHIO REV. CODE ANN. § 3746.04(B)(3) (Baldwin 1994).

compliance, chain of title, previously conducted ESAs, and any aerial photographs taken of the site.[227] Employee interviews may also be conducted.[228]

If the Phase I ESA demonstrates that there is a reason to believe that contamination exists, a Phase II ESA must be performed following criteria set forth in the Ohio Voluntary Action Program.[229] Minimal procedural steps must be followed to demonstrate that any contamination present at a site does not exceeded applicable standards warranting remedial activity, or that remedial activities at the property have achieved compliance with applicable cleanup standards.[230] Phase II ESA procedures include a review of all documentation prepared in connection with the Phase I ESA 180 days prior to commencing the Phase II, quality assurance procedures for samples or measurements taken, sampling procedures ensuring representative sampling of potentially contaminated media, analytical and data assessment procedures, and data objectives ensuring that samples collected are biased toward potentially contaminated areas.[231] The criteria set forth under the Ohio Voluntary Action Program allow for the succinct quantification of the nature and extent of hazardous substance and/or petroleum contamination at Brownfield sites.

---

[227] *Id.* The standards for Phase I property assessments, at a minimum, require all of the above-listed requirements. Stated in greater detail, Phase I requirements specifically include:

- Review and analysis of deeds, mortgages, easements of record, and similar documentation related to the chain of title that are publicly available;
- Review and analysis of previous environmental assessments, property assessments, and environmental or geological studies of the property and any land within 2,000 feet of the boundaries of the contaminated site that are publicly available.
- Review of current and past environmental histories of *persons* who owned or operated the property [emphasis added];
- Review of aerial photographs of the property that indicate prior usage of the site;
- Interviews with managers of activities conducted at the property who have knowledge of environmental conditions of the property;
- Conducting a walkover inspection of the property;
- Identifying the current and past uses of the property, adjoining tracts of land, and the area surrounding the property including, without limitation, interviews with persons who reside or have resided, or who are or were employed, within the area surrounding the property regarding the current past uses of the site and adjacent tracts of land. *Id.* at § 3746.04(B)(3)(a)-(g).

[228] *Id.*

[229] *Id.* at § 3746.04(B)(3)(g).

[230] *Id.* at § 3746.04(B)(4).

[231] *Id.*

Upon adoption of the Voluntary Action Final Program by the Ohio EPA the following additional provisions will be implemented:

✓ creation of a "multidisciplinary board" comprised of public, private, environmental, scientific, and business advocates to advise the Director of the Ohio EPA regarding the adoption of rules for the final program;[232]

✓ numeric cleanup standards meeting the U.S. EPA Performance Standards;[233]

✓ criteria for certification of professionals[234] and laboratories,[235] as well as for revoking or suspending certification;

✓ information regarding the content of NFA letters in order to receive a Covenant Not to Sue;[236]

---

[232] OHIO REV. CODE ANN. § 3746.03 (Baldwin 1994).

[233] *Id.* at § 3746.04((B)(1). The generic numerical standards established under § 3746.04(B)(1) must be consistent and equivalent in scope, content, and coverage to any applicable standard established under federal environmental laws and regulations including without limitation:

- Federal Water Pollution Control Act, as amended, 33 U.S.C. § 1251;
- Resource Conservation and Recovery Act, as amended, 42 U.S.C. § 6921;
- Toxic Substances Control Act, as amended, 15 U.S.C. § 2601 *et seq.;*
- Comprehensive Environmental Response, Compensation and Liability Act, as amended, 42 U.S.C. § 9601 et seq.; and
- Safe Water Drinking Act, as amended, 42 U.S.C. § 300f *et seq.*

[234] OHIO REV. CODE ANN. at § 3746.04(B)(5) (Baldwin 1994). Factors considered for certification of environmental professionals include evaluating previous performance records regarding remedial investigations and activities and environmental compliance histories. *Id.* at § 3746.04(B)(5)(a)(i).

[235] *Id.* at § 3746.04(B)(6). The criteria and procedures for issuance, denial, suspension and revocation of laboratory certifications can be found at OHIO REV. CODE ANN. § 3745 (Baldwin 1992).

[236] *Id.* at § 3746.04(B)(7). Specifically the information included in the NFA letter should specify:

- A summary report of remedial activities prepared by the certified professional responsible for preparing the NFA letter;
- If performed in lieu of generic numeric cleanup standards, notification that a risk assessment was performed;
- The contaminants addressed, their source (if known), and contaminant levels prior to remediation;
- Identification of the person(s) who performed the remedial activity in support of the NFA letter;
- A list of data, information, records, and documents relied upon by the certified professional in preparing the NFA letter.

Id. at § 3746.04(B)(7)(a)-(e).

✓ fees to defray all direct and indirect costs of the program;[237]
✓ procedures for Ohio EPA to audit properties;[238]
✓ criteria for addressing groundwater contamination, including provisions for public participation on a case-by-case basis; [239]
✓ variance procedures for volunteers providing sufficient evidence for a variance decision by the director of Ohio EPA.[240]

Other provisions affecting both the Interim and Final Brownfields Programs provide incentives for parties to undertake voluntary cleanups for the redevelopment and reuse of Brownfields. These incentives include

✓ low interest loans from the Ohio Department of Development, Ohio Water Pollution Control Loan Fund, and the Ohio Water Development Authority;[241]
✓ increase in land value with provisions allowing local abatement on real and personal property taxes for development projects for up to ten years;[242]
✓ liability protection for the state, local governments, and contractors involved in voluntary cleanup projects;[243]
✓ lender/trustee/fiduciary liability protection for property held to protect a security interest;[244]

---

[237] *Id.* at § 3746.04(b)(8).

[238] *Id.* at § 3746.17. Ohio EPA should conduct audits in connection with the issuance of NFA letters to ascertain the following:

- The property subject to voluntary action meets applicable cleanup standards;
- Assess the qualifications of and work performed by certified professionals; and
- Review the quality of the work performed by certified laboratories in connection with the Voluntary Action Program. *Id.* at § 3746.17(A)(1)-(3).

Pursuant to § 3746.171, property subject to institutional controls will be audited by Ohio EPA to confirm continuing compliance with specified land use controls every five years.

[239] *Id.* at § 3746.04(B)(12).

[240] *Id.* at § 3746.09.

[241] *See* OHIO REV. CODE ANN. §§ 6111.036, 6123.032, 6123.041 (Baldwin 1994).

[242] *Id.* at §§ 5709.87(C)(2) and 5709.883.

[243] *Id.* at § 3746.24.

[244] *Id.* at § 3746.26. Paralleling the CERCLA § 101(20(A), 42 U.S.C. § 9601(20)(A), "secured creditor exemption," this provision states that any person who, without participating in the management of a property, holds indicia of ownership in a property primarily to protect a security interest is protected from liability for any civil action brought in connection with the property subject to the Voluntary Action Program, for investigative or

✓ Consolidated Standards Permit agreements;[245]

✓ NFA letters must be prepared and submitted to the state by certified professionals;[246]

✓ Covenants Not to Sue are conditioned upon the continued land use specified in the NFA and only protect volunteers from state civil actions, not federal or third party actions;[247]

---

remedial costs whether or not such costs were incurred in connection with property subject to the Voluntary Action Program, and for conducting or causing to be conducted all such nonremedial activity at the property in compliance with applicable law. *Id.* at § 3746.26(A)(1). The term "indicia of ownership" is defined at § 3746.26(B).

Borrowing much of the basic structure and definitions from the U.S. EPA's 1992 lender liability rules, this exemption from environmental liability applies only if, after taking title to the property, the holder of a security interest conducts all activities occurring at the property in compliance with the applicable requirements of the Ohio Revised Code, and the exemption will be lost if new violations occur after the holder acquires title to the property. Section 3746.23(G)(6)-(7) exempts holders of security interests, trustees, and fiduciaries for liability for costs of conducting a voluntary cleanup under Chapter 3746, provided the requirements of §§ 3746.26 & 3746.27 are met. *Conference on Brownfields Redevelopment and Environmental Risk Management* (March 22-23, 1995) (Statement of William Hinger).

[245] *Id.* at § 3746.15. If volunteers are required to obtain permits for remedial activities under the Federal Water Pollution Control Act, as amended, 33 U.S.C. § 1251, Resource Conservation and Recovery Act, as amended, 42 U.S.C. § 6921, or the Clean Air Act, as amended, 42 U.S.C. § 1857, Ohio EPA will issue a consolidated standards permit for the activities in connection with the voluntary action. *Id.* at § 3746.15(A).

[246] OHIO REV. CODE ANN. § 3746.10(A) (Baldwin 1994).

[247] *Id.* at § 3746.12(B)(2). Compliance with the terms of the covenant not to sue issued by Ohio EPA should allow a volunteer to raise a "permit shield" defense to subsequent state actions seeking additional cleanup of contaminants. However, this provision is silent with regard to protection from federal environmental enforcement and civil third-party actions. Although compliance with the terms of Covenant Not to Sue issued under the Voluntary Action Program may be asserted as a mitigating factor in federal or third-party enforcement proceeding, it will not meet the legal requisites of the permit shield defense.

Generally, a permit shield defense states that compliance with the terms of a permit or other enforceable agreement is deemed to be compliance with applicable law, even if the law is subsequently modified. Thus, until the permit renewal or the agreement must be renegotiated, a party is deemed to be in compliance with the law, so long as the party does not violate the express terms of the permit or agreement. Interestingly, the question arises whether federal sign-off on a Covenant Not to Sue issued by Ohio EPA under the Voluntary Action Program, via a Memorandum of Agreement (MOA) or Memorandum of Understanding (MOU), could be used to assert a permit shield defense from subsequent federal environmental enforcement actions.

✓ Covenants Not to Sue and NFAs must be recorded with county records and will run with the land to ensure the property is used in a manner consistent with the use specified in the NFA letter;[248]

✓ Ohio EPA has complete access to all cleanup documents and properties for audit purposes;[249]

✓ Information generated during the cleanup is privileged and is not admissible in court for subsequent enforcement actions;[250] and

✓ Parties participating in voluntary cleanups may recovery remedial costs from people who caused the pollution (except natural gas and petroleum cleanups).

Compliance with the Ohio Voluntary Action Program has been criticized as an overwhelming and burdensome task. Moreover, contrary to the intent of the drafters, the lending community has begun to insist that all properties eligible for participation in the program must participate and obtain a Covenant Not to Sue as a condition precedent to receiving a loan. The net effect of this innovative legislation is, thus, uncertain.

§ 8.2.16 Oregon

Oregon revised its Hazardous Waste and Material Removal or Remedial Action law last year to include provisions for the promotion of voluntary cleanup efforts.[251] The newly enacted provision states that the Department of Environmental Quality (DEQ) will issue a release from potential liability to the state to parties who assume the voluntary cleanup of their property.[252] The participants must meet the following criteria:

1. The participant cannot be currently liable for an existing release of hazardous waste at the facility;
2. The removal or remedial action must be necessary at the facility to protect human health or the environment;

[248] Ohio Rev. Code Ann. § 3746.14 (Baldwin 1994).

[249] *Id.* at § 3746.21.

[250] *Id.* at § 3746.28(C).

[251] Ch. 662, H.B. No. 3352, approved July 14, 1995. Codified at Ore. Rev. Stat. §§ 465.255, 465.325 (1995).

[252] H.B. 3352, § 4(1).

3. The proposed redevelopment or reuse of the facility will not contribute to or exacerbate existing contamination, increase health risks, or interfere with remedial measures necessary at the facility; and
4. There will be a substantial public benefit as a result of the agreed cleanup (*i.e.*, productive reuse of a vacant or abandoned industrial facility).[253]

The DEQ is to consult with local planning districts to determine the future land use plans of the area when considering a proposal for voluntary cleanup.[254] If the DEQ finds that such voluntary actions are feasible, an agreement between the parties may be promulgated. The agreement for voluntary cleanup must include:

1. A commitment by the participants to perform the remedial measures;
2. DEQ oversight of the remedial cleanup operations;
3. A waiver of the right to bring claims against the state that arise over contamination at the site that exist at the date of acquisition, ownership, or operation of the facility;
4. An irrevocable right of entry given to the DEQ;
5. Reservation of rights to any party not a part of the agreement between the participants and the state; and
6. A legal description of the party.[255]

At the successful completion of the remedial activities, the state shall absolve the participants from further liability for release of hazardous waste at the site as of the date of acquisition.[256] The participants have the burden of proving that the voluntary cleanup was completed successfully and that any further releases resulted from activities prior to acquisition of the facility.[257] The agreement between the state and the participants is to be made a permanent part of the deed, and is to be recorded in the county records office.[258] This agreement will run with the land.[259]

---

[253] *Id.*
[254] *Id.* § 4(2).
[255] *Id.*
[256] *Id.* § 4(4).
[257] *Id.*
[258] *Id.* § 4(5).
[259] *Id.*

Finally, the DEQ has the right to recovery of any and all costs and expenditures that resulted from such actions, including the costs of program development, rulemaking, and other administrative actions required. These costs may be recovered from any person who is undertaking remedial actions under the DEQ's oversight.[260]

§ 8.2.17 Pennsylvania

Pennsylvania's progressive voluntary cleanup program is comprised of three different statutes, which were signed into law as a package on May 19, 1995. These laws are

1. Land Recycling and Environmental Remediation Standards Act (Act 2);
2. The Economic Development Agency, Fiduciary and Lender Environmental Liability Protection Act (Act 3); and
3. Industrial Sites Environmental Assessment Act (Act 4).

The first statute, entitled the Land Recycling and Environmental Remediation Standards Act (Act 2), creates incentives to encourage responsible persons to voluntarily develop and implement cleanup plans without the use of taxpayer funds or the need for adversarial enforcement actions.[261] Arguably as comprehensive as the Ohio Voluntary Action Program, the act provides for the recycling of existing industrial and commercial sites, further defines cleanup liability of new industries and tenants, and establishes a framework for setting environmental remediation standards.[262] The Pennsylvania program also creates a Industrial Land Recycling Fund[263] and Voluntary Cleanup Loan Fund[264] to aid industrial site remediation.

---

[260] *Id.* § 8.

[261] 1995 PA S.B. 1.

[262] *See* Preamble, 1995 PA S.B. 1.

[263] 1995 PA S.B. 1, § 701.

[264] *Id.* at § 702. The purpose of the Voluntary Loan Fund is to assist volunteers through providing funding to persons undertaking voluntary remediation of a property. *Id.* at § 702(b) The following parties are eligible for low interest loans and grants for up to 75% of the costs incurred for conducting an environmental study and for implementing a cleanup plan:

- Local economic agencies;
- Municipalities;
- Public agencies; and
- Persons not responsible for site contamination.

Under the Land Recycling and Remediation Standards Act, volunteers must select and attain compliance with one or more specified cleanup standards.[265] These standards include background standards, statewide health-based standards, site-specific standards, or a combination thereof.[266] For contaminated sites remediated to a level consistent with statewide cleanup standards, volunteers may meet either residential or nonresidential cleanup standards.[267] Compliance with background standards and state health-based standards are not subject to deed notice

[265] *Id.* at § 302. The remediation standards established under this Act will be deemed to Applicable, Relevant and Appropriate Requirements (ARARs) for the Commonwealth of Pennsylvania under federal CERCLA, as amended, 42 U.S.C. § 9601 et seq. and the Pennsylvania Hazardous Sites Cleanup Act (P.L. 169, No. 32). *Id.* at § 106.

[266] *Id.* at § 301. Attainment of background standards must be demonstrated by the collection and analysis of representative samples of contaminated media located on site, including soils and groundwater in aquifers, through the use of statistical tests or other methods of generally recognized attainment procedures. *Id.* at § 302(a)(1). Statewide health standards, adopted by the Pennsylvania Environmental Quality Board, are designed to remove the threat of any substantial present or probable future risk to human health and the environment. *Id.* at § 301(a)(2). Site-specific standards achieve remediation levels based upon risk assessment procedures designed to eliminate or reduce any actual or potential threat to human health and the environment. *Id.* at § 301(a)(3).

[267] *Id.* at § 303(b). For statewide residential cleanup standards, the concentration of regulated substance in the soil cannot exceed either the direct contact medium-specific concentration based on residential exposure factors with a depth of fifteen feet from the existing ground surface, or soil-to-groundwater pathway numeric value throughout the soil column. *Id.* at § 303(b)(4). Residential soil-to-groundwater pathway numeric values may be determined by one of the following methods:

- A value 100 times the medium specific concentration;
- Soil concentrations not producing leachate in excess of medium specific concentrations for groundwater in an aquifer under the Synthetic Precipitation Leaching Procedures developed by U.S. EPA; or
- A generic value determined not to produce a concentration in groundwater in a aquifer in excess of medium specific concentrations based upon valid, peer-reviewed scientific methods. *Id.* at § 303(b)(4)(i)-(iii).

requirements.[268] Eliminating deed restriction requirements rewards the volunteers. Subsequent transfer of remediated property is not subjected to the stigma of being a formerly contaminated site. Therefore, a formerly impaired site can be sold and redeveloped free and clear of any indicia of past or present contamination.

Site-specific cleanup standards will require the development and implementation of specified procedures, including remedial investigation,[269] (if required) risk assessment procedures,[270] cleanup plans,[271] and a final report demonstrating completion of the approved remedial activities.[272] For contaminants known or suspected to be carcinogens, soil and groundwater cleanup standards must meet specific exposure factors.[273] Specifically, these

---

[268] *See id.* at §§ 302(d) & 303(g). Persons attaining or demonstrating compliance with background or residential statewide standards for all regulated substances subject to remediation are not subject to deed acknowledgment requirements set forth in the Pennsylvania Solid Waste Management Act (P.L. 380, No. 97) and the Hazardous Sites Cleanup Act (P.L. 756, No. 108). For parties meeting residential statewide cleanup standards, an existing acknowledgment in a deed prior to completion of voluntary remedial activities may be removed from the chain of title. *Id.* Parties remediating sites to nonresidential statewide cleanup standards are required to meet the deed notice requirements in both the Solid Waste Management Act and Hazardous Sites Cleanup Act. *Id.*

[269] *Id.* at § 304(l)(1). Specifically, a remedial investigation report should include:

- A description of the procedures and conclusions used to characterize the nature, extent, direction, rate of movement, volume, and composition of regulated substances;
- Concentrations of regulated substances in contaminated media, including summaries of sampling methodology, analytical results, and any information obtained from attempts to comply with either background or statewide standards;
- A description of existing or potential public benefits of use or reuse of the property;
- A fate and transport analysis demonstrating no present or future exposure pathways exist; and
- If no exposure pathways exist, a risk assessment report and cleanup plan are required and no remedy is required. *Id.* at § 304(l)(1)(i)-(v).

[270] *Id.* at § 204(l)(2). If required, a risk assessment report should demonstrate the potential adverse effects under both current and planned future conditions caused by the presence of a regulated substance in the absence of any further control, remediation, or mitigation measures. *Id.*

[271] *Id.* at § 304(l)(3). The cleanup plan should select a remedy that achieves the requirements necessary to meet the applicable cleanup standards consistent with intended future use of the facility.

[272] *Id.* at § 304(l)(4).

[273] *Id.* at § 304(b).

standards should represent an excess upper-bound lifetime risk of between 1 x 10-4 and 1 x 10-6.[274] The cumulative excess risk to populations exposed to Brownfields, including sensitive subgroups, cannot exceed 1 x 10-4.[275] For systemic toxicants, soil and groundwater cleanup standards must represent levels to which daily human exposure could occur without appreciable risk of deleterious effects.[276]

Volunteers satisfying program requirements are shielded from state environmental enforcement actions for thc unanticipated future cleanup of contaminants.[277] Furthermore, volunteers should also receive protection from lawsuits brought by citizen groups and contribution actions brought by third parties responsible for site contamination.[278] The act's liability protection provision extends to the current and future owners of contaminated industrial property, developers, successors, and assigns of any person to whom liability may apply, including public utilities performing activities on contaminated sites.[279]

§ 8.2.18 Rhode Island

The Rhode Islands Brownfields Program[280] is overseen by the Department of Environmental Management (RIDEM). There is no application fee for the program, although applicant is responsible to pay RIDEM's oversight costs.[281] Successful investigation and remediation of a site results in a settlement agreement with the state. This agreement includes a covenant not to sue and a limited contribution protection provision.[282]

---

[274] *Id.*

[275] *Id.*

[276] *Id.* at § 304(c).

[277] *Id.* at § 501(a).

[278] *Id.*

[279] 1995 PA S.B. 1, at § 501(a)(1)-(4).

[280] Rhode Island Industrial Property Remediation and Reuse Act, R.I. GEN. LAWS § 23-19.14-1 *et seq.* (19..)

[281] Ned Abelson et al., Rhode Island, at 9, citing Article III of DEM's Model Settlement Agreement (unpublished manuscript, on file with author).

[282] *Id.* at 5-7.

§ 8.2.19 Tennessee

In 1994, the Tennessee legislature amended the Hazardous Waste Management Act of 1983[283] to create the Voluntary Cleanup and Assistance Program.[284] The Tennessee Voluntary Cleanup Program allows willing and able parties to conduct investigations and voluntary cleanups of inactive contaminated industrial sites.[285] Certain activities conducted under the Voluntary Cleanup Program must occur with state oversight.[286] Specifically, investigatory and remedial measures, including monitoring, are among the activities requiring state oversight. State oversight costs incurred during the voluntary remediation of a contaminated industrial site must be reimbursed by volunteers.[287] Additionally, a $5,000 participation fee must be paid by volunteers.[288] This fee is used to establish a Voluntary Cleanup and Oversight Assistance Fund.[289]

The criteria for selecting containment and cleanup actions, including monitoring and maintenance options to be followed under the Voluntary Cleanup and Oversight Assistance Program, are specified by law.[290] Determined on a site-specific basis, the following relevant factors must be considered by program participants for state approval of voluntary remedial activities:

- ✓ Technological feasibility of each alternative;
- ✓ Cost-effectiveness of each alternative;
- ✓ The nature of the danger to the public health, safety, and the environment posed by existing site contamination; and
- ✓ The extent to which each alternative would achieve the goal of the Voluntary Cleanup and Assistance Program.[291]

Volunteers are required to evaluate reasonable alternatives and select those actions that adequately protect the public health, safety, and the environment from risks posed by hazardous substances.[292]

---

[283] TENN. CODE ANN. § 68-212-202 *et seq.* (West 1994).
[284] *Id.* § 68-212-224.
[285] *Id.* at § 68-212-224(a).
[286] *Id.*
[287] *Id.*
[288] *Id.* at § 68-212-224(b).
[289] *Id.* at § 68-212-224(c)(1).
[290] *Id.* at § 68-212-224(e).
[291] *Id.* at § 68-212-206(d)(1)-(4).
[292] *Id.* at § 68-212-206(d).

§ 8.2.20 Texas

Texas has enacted its own voluntary cleanup program.[293] The program requires that volunteers remit a $1,000 application fee[294] and enter into a enforceable agreement with the state regarding the remediation of contaminated industrial property.[295] Moreover, the agreement between the state and a volunteer imposes a work schedule and precludes environmental enforcement activities while voluntary remedial activities are conducted.[296]

Other features include reimbursement for state oversight costs, and expedited administrative processes for reviews and approvals of voluntary cleanups.[297] Upon satisfactory completion of remedial activities, a Certificate of Completion will be issued by the state to volunteers.[298]

---

[293] 1995 TX H.B. 2296; TEX. HEALTH & SAFETY CODE ANN. §§ 361.601-361.612 (West 1995).

[294] *Id.* at § 361.604(3) (West 1995); *see also* Heymann, Brownfields Redevelopment and Environmental Conference Remarks.

[295] TEX. HEALTH & SAFETY CODE § 361.606 (West 1995). Under this program, specific requirements comprising the elements of voluntary cleanup work plans and reports submitted under § 361.608 are not articulated. In fact, the Texas program neither explicitly sets forth any cleanup standards nor does it reference other state or federal provisions.

[296] *Id.* at § 361.606(C). Specifically, the voluntary cleanup agreement must:

- Identify all applicable federal and state statutes and rules;
- Describe any work plan submitted, including a final report providing all information necessary to verify that all work contemplated by the voluntary cleanup agreement has been completed;
- Schedule for the submission of required information;
- Describe technical standards applied in evaluating work plans and reports referring to the proposed future use of the site.

Id. at § 361.606(C)(1)-(4).

[297] *Id.* at § 361.603(B)(1).

[298] *Id.* at § 361.609. The Certificate of Completion must contain the following information:

- Acknowledgement of the release from liability as provided in § 361.610;
- Statement of the proposed future use of the site;
- A legal description of site.

Id. at § 361.609(B)(1)-(3).

Additionally, Texas provides releases from liability for volunteers, new owners, and lenders for the proper and timely completion of voluntary cleanup activities.[299]

In Texas voluntary cleanups may not be undertaken if a federal grant requires an enforcement action at the site.[300] Also, the Texas program fails to provide specific statutory protection for lenders securing commercial loans with impaired property. Although Certificates of Completion[301] and releases from liability[302] address lenders, the program makes no effort to address under what circumstances a lender may have triggered liability as a responsible party under the Texas law.[303]

§ 8.2.21 Utah

Under Utah's Voluntary Agreement Program, the Utah Department of Environmental Quality may enter into agreements with responsible parties to investigate or remediate sites contaminated with hazardous materials.[304] Before entering an agreement, however, potentially responsible parties must be notified of the cleanup proposal and must have an opportunity to

---

[299] *Id.* at § 361.610. The release from liability under this section applies to the person who is not "responsible party," as defined under TEX. HEALTH & SAFETY CODE ANN. §§ 361.271 and 361.275(G) (West 1995) if the person applies to perform a voluntary cleanup. *Id.* Moreover, the protection afforded extends to subsequent purchasers and lenders who make a loan secured by property subject to voluntary remediation. However, a release for liability under the Texas program will be void if the change in the land use increases the risk to human health and the environment. *Id.* at § 3631.610(C).

[300] *Id.* at § 361.605(A)(2). In contrast to COLO. REV. STAT. ANN. § 25-16-309(2), which creates a nondiscretionary duty on the part of the state to petition U.S. EPA to withhold enforcement action, § 361.605(A)(2) explicitly subrogates the Texas program if a federal grant would require enforcement. Thus, volunteers under the Texas program are at the beckoned mercy of federal legislators and regulators.

[301] TEX. HEALTH & SAFETY CODE ANN. § 361.609 (West 1995).

[302] *Id.* at § 361.610.

[303] Compare and contrast: OHIO REV. CODE ANN. § 3746.26(A)(1) (Anderson 1994); MICH. COMP. LAWS ANN. §§ 20101a and 20101b (West 1995); MINN. STAT. ANN. § 115B.175[subdiv. 6](2) (West 1992 and Supp. 1994); MONT. CODE. ANN. § 75-10-701(8)(b) (1995); 1995 NY S.B. § 19; VT. STAT. ANN. tit. 10, § 6615(C)(6)(A) (1995); and VA. CODE ANN. § 10.1-1429.1(A) (Michie 1995).

[304] UTAH CODE ANN. § 19-6-325.

comment.[305] Once an agreement is finalized, action according to the plan begins. The party undertaking the cleanup may seek contribution from other responsible parties for costs incurred that exceed its liability.[306] Upon completion of activities required by the agreement, the party requests a no further action letter from the Utah DEQ, which verifies that the party fulfilled the agreement.

§ 8.2.22 Vermont

Vermont's Redevelopment of Contaminated Properties Program (RCPP) is overseen by the Vermont Agency of Natural Resources. To be eligible for the program, the applicant may not be liable or potentially liable for the contamination at the site, nor should the applicant have had previous contact with the property.[307] For the site to be eligible for RCPP, the contamination must be caused by a release or threatened release of hazardous materials[308] and the property must be "vacant, abandoned, or substantially underutilized" or "be acquired by a municipality."[309] The program excludes sites that are on CERCLA's National Priority List (NPL), are subject to RCRA corrective action, or fall under Vermont's underground storage tank enforcement provisions.[310]

Once both the party and the site are deemed eligible for the program, the application process is similar to programs in other states: the application consists of oversight fees and a work plan.[311] After considering the work plan, the applicant must provide a site investigation report, and later provide a corrective action plan.[312] Clean-up actions begin upon approval of the corrective action plan.[313] When cleanup is completed, ANR provides the party with a certificate of completion. This certifies that (1) the cleanup was

[305] H. Michael Keller, Utah, at 2-4 (unpublished manuscript, on file with author).
[306] *Id.* at 2, 6-7, 10-11.
[307] VT. STAT. ANN. tit.10 § 6615a(f)(1)(B).
[308] *Id.* § 6615a(f)(1)(A).
[309] *Id.* § 6615a(f)(1)(c).
[310] *Id.* § 6615a(f)(2)(A-C).
[311] *Id.* § 6615a(g)(1)-(3).
[312] *Id.* § 6615a(g)(4)-(6), (h).
[313] *Id.* § 6615a(i).

conducted and completed per the corrective action plan, (2) no further activity is required to remediate the site, and (e) the party is shielded from liability arising under the Hazardous Waste Management Act.[314]

§ 8.2.23 Virginia

Recently, The Commonwealth of Virginia passed a bill to amend the Code of Virginia relating to the voluntary remediation of hazardous substances.[315] The Virginia Voluntary Cleanup Program allows persons who own, operate, or have a security interest in or enter into a contract for the purchase of contaminated property to voluntarily remediate releases of hazardous substances, hazardous wastes, solid wastes, or petroleum.[316] This program encourages the use of site-specific, risk-based standards "no more stringent than applicable federal standards" for the remediation of soil, groundwater, and sediments.[317] Virginia's program provides volunteers with flexibility regarding the selection of cleanup standards.[318] Each standard is based upon the proposed future use of the property and requires that the public health and environment be assured.[319]

---

[314] Id. § 6615a(k).

[315] 1995 VA H.B. 1847, VA. CODE ANN. § 10.1-1429.1 to 10.1-1429.3 (Michie 1995). A bill to amend the Code of Virginia by adding Chapter 14 to Title 10.1 and Article 4.1, consisting of §§ 10.1-1429.1 to 10.1-1429.3, relating to voluntary remediation of hazardous substances.

[316] *Id.* at § 10.1-1428.1(A).

[317] *Id.* at § 10.1-1429.1(A)(1). The establishment of site-specific risk-based remediation standards must consider the following relevant factors:

- Protection of the public health and the environment;
- Future industrial, commercial, residential, or other uses of the property remediated and surrounding property;
- Reasonably available and effective remediation technology and analytical quantitation technology;
- Availability of institutional or engineering controls protective of human health and the environment; and
- Natural background levels.

Id.

[318] *Id.* at § 10.1-1429.1(A)(1)(I)-(II).

[319] *Id.*

The Voluntary Cleanup Program permits the use of reasonably available and effective technology, as well as analytical quantitation technology to achieve applicable cleanup standards.[320] Furthermore, institutional and engineering controls are also permitted remedial measures.[321] Procedures have been established for expedited administrative processes,[322] the issuance of Certificates of Completion,[323] and the waiver or expedited issuance of any permits necessary to complete voluntary cleanups.[324] Finally, the Virginia program requires the remittance of a $5,000 application fee or 1 percent of the total cost of the remediation.[325] Satisfactory completion of voluntary cleanups and issuance of a Certificate of Completion will trigger statutory immunity from state environmental enforcement actions.[326]

§ 8.2.24 Washington

The State of Washington, similar to the State of New York, has promulgated legislation to promote the encouragement of community-based strategies for the voluntary cleanup of property.[327] The bill provides for state funding of local Brownfield revitalization efforts, appropriating $1 million from the budget of the Department of Ecology to be given as grants to local governments for the encouragement of developing and implementing area-wide strategies for the cleanup and reuse of industrial lands.[328]

---

[320] *Id.* at § 10.1-1429.1(A)(1)(III).

[321] *Id.* at § 10.1-1429.1(A)(1)(IV).

[322] *Id.* at § 10.1-1429.1(A)(2).

[323] *Id.* at § 10.1-1429.1(A)(3). Certificates of Completion should be granted at the completion of remedial activities. This decision should be based on then present site conditions and available information satisfying applicable cleanup standards, or where the state has determined that no further action is warranted. *Id.*

[324] *Id.* at § 10.1-1429.1(A)(4).

[325] *Id.* at § 10.1-1429.1(A)(5).

[326] *Id.* at § 10.1-1429.2.

[327] 1995 Wash. S.B. No. 6261, Washington 1996 Regular Session of the 54th Legislature.

[328] *Id.* § 2.

§ 8.3 U.S. EPA Brownfield Pilot Projects

In August 1994, U.S. Environmental Protection Agency (EPA) Administrator Carol Browner announced the federal Brownfields initiative. The EPA Brownfields Economic Redevelopment Project conducts studies on how to return old industrial sites to productive use and appropriates grants to qualified communities to help achieve this goal. The federal Brownfields initiative serves numerous coextensive purposes:

✓ to revitalize urban economies in the Northeast and Midwest through manufacturing and other industries;

✓ to prevent the unnecessary development of virgin land and resources; and

✓ to relieve the increased demand for new manufacturing and industrial infrastructure by reusing and redeveloping Brownfields.

EPA explained in a press release:

The current Superfund law extends liability to both past and prospective owners of contaminated sites. As a result, the market value of older industrial sites, or "Brownfields," has declined. To address the effects of these declining property values found in inner-city industrial areas, EPA has announced its Brownfields Action Agenda, which commits funding for 50 pilot projects. Under this initiative, EPA will work with stakeholders to address liability barriers and encourage economic redevelopment of these industrial facilities.

EPA also announced the availability of limited federal funding (up to $200,000 in funding to support two-year demonstration projects) for approximately fifty pilot projects in 1995 and 1996. The goal of the pilot program is to develop a national policy on Brownfields development that will serve as guidance to states and localities struggling with this issue. As such, the pilot programs are designed to help

✓ test redevelopment models by removing regulatory barriers without sacrificing environmental protection;

✓ encourage community groups, investors, lenders, developers, and other affected parties to work together to clean up contaminated sites and return them to productive use;

✓ create models for state and localities struggling with this issue; and,

✓ establish guidance to cities cleaning up and returning contaminated, abandoned property back to productive use.

§ 8.4 Comprehensive Environmental Response, Compensation, and Liability Information System (CERCLIS) Site Deletion

In a demonstration of its commitment to this initiative, EPA regional offices in partnership with the states have assessed and determined that about 25,000 to 38,000 potential hazardous waste sites do not require federal involvement and should be removed from the Superfund Inventory. Some sites may have received a "no further action" designation because state or local authorities had assumed enforcement responsibility. On Feb. 15, 1995, EPA announced that it had removed approximately 25,000 sites from the Site Tracking System known as CERCLIS. These sites were screened out of active investigation and designated as "No Further Remedial Action Planned" or "NFRAP." This is the first time EPA has pruned the list of sites determined not to warrant Superfund cleanup.

Since all lenders check the CERCLIS as part of a Phase I environmental assessment, even these "no further action" sites were stigmatized, severely limiting financing opportunities and blocking most redevelopment activities. Pruning the CERCLIS may be only partially effective in reducing Brownfields anxiety, because in many cases EPA has delayed final determinations of "no further action" while focusing its attention on remedial action sites. Although in the future EPA may be willing to clear more sites in response to specific inquiries, deletion of these sites from CERCLIS is not a guarantee that such sites are uncontaminated. Nor does it protect lenders or current property owners from third-party suits for cleanup of any contamination on site. For example, Cornell University successfully negotiated with EPA to be deleted from the CERCLIS, but was later sued—and held liable for contribution costs—by Alcan, a private third-party plaintiff.[329]

§ 8.5 Community Reinvestment Act (CRA)

The Administration recently promulgated agency guidance for EPA dealing with lenders and prospective purchasers. On May 4, 1995, new regulations were amended to provide an incentive for bankers and

---

[329] *See* United States v. Alcan Aluminum Corp., 964 F.2d 252 (3d Cir. 1992).

developers to help rescue cities with polluted industrial properties. "For the first time, lenders subject to the federal Community Reinvestment Act (CRA)—aimed at directing capital into poor neighborhoods—can claim CRA credit for loans made to help clean up and redevelop urban, industrial property." The new rule, orchestrated by EPA Administrator Carol Browner, was designed to complement the EPA's new "Brownfields Action Agenda."

Under this program, EPA is one of many federal agencies committing resources to economic revitalization of urban areas. EPA coordinates its efforts with the U.S. Department of Housing and Urban Development and the Department of Agriculture on designed "empowerment zone programs." The programs are designed to demonstrate that sustainable development is possible by linking redevelopment and job creation to environmental improvement and responsible stewardship.

EPA has not only worked to advise the two awarding departments about the environmental considerations in the applications, but will also work with those communities that win "Empowerment Zone" or "Enterprise Community" (EZ/EC) status to help ensure the environmental sustainability of their activities. EPA explained:

> We are confident that the Brownfields Initiative and Superfund reforms will lower many of the barriers to reclamation and redevelopment of urban lands by eliminating disincentives to buying and selling well-remediated properties; by facilitating voluntary cleanups to put land back into use; and by introducing procedures that will lead more quickly to cleanups more appropriate for the site and its future use.
>
> EPA is charged with protecting human health and the environment. That does not mean we need to be the spoiler of the redevelopment of abandoned industrial property. After all, land redevelopment is one of the most important forms of recycling around. Brownfield redevelopment should not only lead to cleaned up inner city properties and preservation of our remaining green spaces, but should also stimulate local economic growth and thereby help reverse the cycle of decay which endangers our urban cores.[330]

---

[330] Testimony of Timothy Fields, Jr., Deputy Assistant Administrator, U.S. Environmental Protection Agency, Before the U.S. House of Representatives, Committee on Commerce, Subcommittee on Commerce, Trade, and Hazardous Material (June 15, 1995).

§ 8.6 Revised Prospective Purchaser Agreement Guidance

On May 25, 1995, EPA released new criteria that will govern consideration of candidates for a Prospective Purchaser Agreement:

Elements Dictated by the May 25, 1995, Guidance	**Detailed Description**
Enforcement Action Is Anticipated by EPA	EPA will enter into a settlement agreement at sites where any form of federal involvement has occurred or is expected to occur and where there is a realistic probability of incurring Superfund expense. EPA intends to substantially expand the universe of sites qualifying for prospective purchaser agreements to include sites that have not yet been the subject of formal site investigation and remediation.
Direct Cleanup Benefit or Indirect Economic Benefit Will Occur	EPA must now consider prospective purchaser agreements both where substantial direct benefit will occur to EPA in the form of a cleanup commitment or funds for cleanup and, alternatively, where a lesser indirect benefit may be conferred on the affected community in terms of cleanup, creation of jobs, and development of otherwise unusable property.
Risk Posed to Persons Likely to Be Present at the Site	EPA must now consider potential risks of exposure and adverse health effects for employees and the public.
The Prospective Purchaser Is Financially Viable	EPA must consider whether it can recoup its costs in the event that the prospective purchaser breaches its obligations under the agreement.

The new guidance supersedes the agency's 1989 guidance, which stipulated that EPA was to enter into agreements only where enforcement action was anticipated and where the new owner provides the government with direct, substantial benefits. The new guidance is intended

- ✓ to provide EPA greater flexibility in offering Covenants Not to Sue to parties interested in buying contaminated land for cleanup or redevelopment;

✓ to ease these earlier restrictions, alleviate financial pressures for the individual parties, and enhance incentives to clean up contaminated sites;

✓ to allow more frequent application of the prospective purchaser agreements;

✓ to allow EPA into the agreements where any agency activity is anticipated or under way.

✓ to allow EPA to accept less direct benefit from the parties, provided that there is otherwise substantial benefit (which can include cleanup or response costs, or other activities considered to be justifiable benefits to the community, such as improving the quality of life or expanding employment opportunities).

EPA included a model Prospective Purchasers Agreement with the May 25 Guidance. This model agreement provides for a Covenant Not to Sue, release, and contribution protection with respect to subsequent third-party claims (such as those from prior landowners or adjoining property owners). The agreement would serve to release not only claims by EPA but related claims by any other federal authority (including claims under other environmental laws such as those pertaining to clean water, clean air and disposition of hazardous waste). The agreement would not, however, seek to provide a release from site conditions "unknown" at the time the agreement was signed. Nor would the protection of the agreement be assignable to other parties.

§ 8.7 Final Policy Toward Owners of Property Containing Contaminated Aquifers

On May 25, 1995, EPA released Guidance entitled "Final Policy Toward Owners of Property Containing Contaminated Aquifers." This guidance was designed for owners of uncontaminated properties situated above groundwater system that have been contaminated by sources outside the property.[331] The guidance was based, in part, on the success of cooperative Brownfields cleanup projects in Wichita, Kansas.

---

[331] Neither the prospective purchasers nor aquifer owner Guidances had Office of Solid Waste and Emergency Response (OSWER) directive numbers printed on them.

Where uncontaminated property may have been contaminated by subsurface groundwater migration in a non-drinking water source, the EPA guidance directs the enforcement staff not to take action against owners, provided that the owner has not caused contamination or made the contamination worse. Such contribution could be through the handling and disposition of the same or similar substances at the property. This policy is based on EPA's enforcement discretion and is, of course, in no way binding on third parties that may want to sue for cleanup of subsurface contamination.

In order to be eligible for this program, EPA officials must be satisfied that the person who caused the release of the hazardous substance that contaminated the groundwater system is not an agent, employee, or party connected by contractual relationship or otherwise related to the owner whose liability is to be resolved. The Guidance permits EPA to consider affording contribution protection to qualified owners by entering into a "de minimis settlement agreement." It is, however, discretionary on the part of the agency and beyond the control of the party seeking to take advantage of the guidance. According to EPA Administrator Carol Browner, even with the administrative guidance in place, "we still need legislative reform of the Superfund law to ensure that it makes economic and environmental sense for our cities, small businesses and communities."[332]

§ 8.8 State Deferral Guidance

EPA released a state deferral guidance designed to speed site cleanups while increasing the involvement of states, American Indian tribes, and local communities.[333] This guidance allows EPA to defer from adding sites to Superfund's NPL if states (1) oversee cleanup activities or (2) cleanups will be supervised by the state and conducted and paid for by PRPs. The deferral program is designed to speed cleanups by encouraging PRPs to settle early by allowing them to avoid listing on the NPL.

---

[332] State Role, Liability, Cleanup Guidance Among Major Reforms Announced by EPA, BNA State Environment Daily (May 30, 1995).

[333] OSWER Directive No. 9375.6-11 (May 3, 1995).

To be eligible for state or tribal deferral, sites must be on the CERCLIS, but not yet listed on the NPL. The sites must have PRPs who agree to pay for and conduct cleanups. Cleanups must be as protective as other Superfund cleanups and comply with all applicable federal and state requirements. In addition, deferral of the site cleanup to the state must have the support of the affected community. Federal facilities are not eligible for deferral.

The state deferral policy differs from the "state empowerment" proposal also being considered by EPA under the administration's "reinventing government initiatives." The empowerment proposal would require legislative changes before implementation is possible. It would allow states to run the Superfund program within their jurisdiction. The deferral policy needs no legislative change for implementation. It can be implemented under current law under EPA's enforcement discretion.

§ 8.9 Land Use Directive

Another of the reform initiatives issued on May 25, 1995, is EPA's land use directive,[334] which takes EPA's policy regarding consideration of future site usage into account when developing a cleanup plan. Under the guidance document, discussions regarding future land use should take place as early as possible in the investigation and study phases of cleanup. Residents as well as local planning authorities and officials should be consulted, as disagreement over future land use often exist at sites.

The land use policy would allow for less stringent cleanups in cases where a site has been and would probably remain industrial. According to the directive, "EPA has been criticized for too often assuming future land use will be residential."[335] Some of the factors listed in the document to take into account when deciding future land use include:

- ✓ current land use,
- ✓ zoning decisions,
- ✓ community master plans,

[334] OSWER Directive No. 9355.7-04 (May 25, 1995).
[335] *Id.*

- ✓ accessibility to existing infrastructure,
- ✓ site location in relation to other areas,
- ✓ development patterns,
- ✓ cultural factors,
- ✓ environmental justice issues, and
- ✓ proximity to wetlands, critical habitats, and other sensitive environmental areas.

The guidance allows certain portions of a site to be cleaned up to meet certain standards to the future land use, but other parts may require deed restrictions and containment of contamination. In addition, because of cost or practicality issues, some anticipated land uses may not be possible, and a more restricted use must be implemented.

If the cleanup implemented is based on less than unrestricted land use (a residential or eatable dirt standard), the guidance calls for institutional controls to be included in the site cleanup and maintenance plan. Accordingly, in cases where waste is left on-site at levels that would require limited use and restricted exposure, EPA will conduct reviews at least every five years to monitor the site for changes. If landowners or others decide in the future to change the land use in a way that would require additional cleanup, EPA would retain the authority to take further response action to ensure protectiveness. EPA would not expect, however, to become actively involved in those cleanups.

Forms to Be Used in Brownfields Redevelopments

Form 8.1
Model Prospective Purchaser Agreement

Region .....

In the matter of: [name] [Docket Number] under the authority of the Comprehensive Environmental Response, Compensation, and Liability Act of 1980, 42 U.S.C. 9601, *et seq.*, as amended. [state law, if appropriate]

Agreement and Covenant Not to Sue [Insert Settling Respondent's Name]

I. Introduction

This Agreement and Covenant Not to Sue ("Agreement") is made and entered into by and between the United States Environmental Protection Agency (EPA), [state of .....] and .... [insert name of Settling Respondent] (collectively the "Parties").

EPA enters into this Agreement pursuant to the Comprehensive Environmental Response, Compensation, and Liability Act of 1980, as amended (CERCLA), 42 U.S.C. § 9601, *et seq.* [If the state is a party, insert "The State of ....., enters into this Agreement pursuant to (cite relevant state authority.)" and make appropriate reference to state with respect to affected provisions, including payment or work to be performed].

[Provide introductory information, consistent with Definitions and Statement of Facts, about the party purchasing the contaminated property, including name ["Settling Respondent"], address, corporate status if applicable, and include proposed use of the property by prospective purchaser. Provide name, location and description of Site.]

The Parties agree to undertake all actions required by the terms and conditions of this Agreement. The purpose of this Agreement is to settle and resolve, subject to reservations and limitations contained in Sections VII, VIII, IX, and X [if this Agreement contains a separate section for Settling Respondent's reservations, add section number], the potential liability of the Settling Respondent for the Existing Contamination at the Property which would otherwise result from Settling Respondent becoming the owner of the property.

The Parties agree that the Settling Respondent's entry into this Agreement, and the actions undertaken by the Settling Respondent in accordance with the Agreement, do not constitute an admission of any liability by the Settling Respondent.

The resolution of this potential liability, in exchange for provision by the Settling Respondent to EPA [and the state] of a substantial benefit, is in the public interest.

II. Definitions

Unless otherwise expressly provided herein, terms used in this Agreement which are defined in CERCLA or in regulations promulgated under CERCLA shall have the meaning assigned to them in CERCLA or in such regulations, including any amendments thereto.

1. "EPA" shall mean the United States Environmental Protection Agency and any successor departments or agencies of the United States.

2. "Existing Contamination" shall mean any hazardous substances, pollutants or contaminants, present or existing on or under the Site as of the effective date of this Agreement.

3. "Parties" shall mean EPA, [State of .....], and the Settling Respondent.

4. "Property" shall mean that portion of the Site which is described in Exhibit 1 of this Agreement. [*34796]

5. "Settling Respondent" shall mean ..... .

6. "Site" shall mean the [Superfund] Site, encompassing approximately ..... acres, located at [address or description of location] in [name of city, county, and state], and depicted generally on the map attached as Exhibit 2. The Site shall include the Property, and all areas to which hazardous substances and/or pollutants or contaminants have come to be located [provide a more specific definition of the Site where possible; may also wish to include within Site description structures, USTs, etc.].

7. "United States" shall mean the United States of America, its departments, agencies, and instrumentalities.

III. Statement of Facts

8. [Include only those facts relating to the Site that are relevant to the covenant being provided the prospective purchaser. Avoid adding information that relates only to actions or parties that are outside of this Agreement.]

9. The Settling Respondent represents, and for the purposes of this Agreement EPA [and the state] relies on those representations, that Settling Respondent's involvement with the Property and the Site has been limited to the following: [Provide facts of any involvement by Settling Respondent with the Site, for example performing an environmental audit, or if Settling Respondent has had no involvement with the Site so state.].

IV. Payment

10. In consideration of and in exchange for the United States' Covenant Not to Sue in Section VIII herein [and Removal of Lien in Section XXI herein if that is part of the consideration for the agreement], Settling Respondent agrees to pay to EPA the sum of $ ....., within ..... days of the effective date of this Agreement. [A separate section should be added if the consideration is work to be performed.] The Settling Respondent shall

make all payments required by this Agreement in the form of a certified check or checks made payable to "EPA Hazardous Substance Superfund," referencing the EPA Region, EPA Docket number, and Site/Spill ID #..... [Insert 4-digit no.; first 2 numbers represent Region, second 2 numbers are Region's Site/Spill ID no.], [DOJ case number ....., if applicable] and name and address of Settling Respondent. [Insert Regional Superfund Lockbox address where payment should be sent.] Notice of payment shall be sent to those persons listed in Section XV (Notices and Submissions) and to EPA Region ..... Financial Management Officer [insert address].

11. Amounts due and owing pursuant to the terms of this Agreement but not paid in accordance with the terms of this Agreement shall accrue interest at the rate established pursuant to Section 107(a) of CERCLA, 42 U.S.C. 9607(a), compounded on an annual basis.

[..... .] [Work to Be Performed]

[Include this section and other appropriate provisions relating to performance of the work, such as financial assurance, agency approvals, reporting, etc., where work to be performed is the consideration for the Agreement.]

[..... . Statement of Work attached as Exhibit 3.]

V. Access/Notice to Successors in Interest

12. Commencing upon the date that it acquires title to the Property, Settling Respondent agrees to provide to EPA [and the state], its authorized officers, employees, representatives, and all other persons performing response actions under EPA [or state] oversight, an irrevocable right of access at all reasonable times to the Property and to any other property to which access is required for the implementation of response actions at the Site, to the extent access to such other property is controlled by the Settling Respondent, for the purposes of performing and overseeing response actions at the Site under federal [and state] law. EPA agrees to provide reasonable notice to the Settling Respondent of the timing of response actions to be undertaken at the Property. Notwithstanding any provision of this Agreement, EPA retains all of its authorities and rights, including

enforcement authorities related thereto, under CERCLA, the Solid Waste Disposal Act, as amended by the Resource Conservation and Recovery Act, 42 U.S.C. 6901, (RCRA) *et seq.,* and any other applicable statute or regulation, including any amendments thereto.

13. Within 30 days after the effective date of this Agreement, the Settling Respondent shall record a certified copy of this Agreement with the Recorder's Office [or Registry of Deeds or other appropriate office], ... County, State of .... Thereafter, each deed, title, or other instrument conveying an interest in the Property shall contain a notice stating that the Property is subject to this Agreement. A copy of these documents should be sent to the persons listed in Section XV (Notices and Submissions).

14. The Settling Respondent shall ensure that assignees, successors in interest, lessees, and sublessees of the Property shall provide the same access and cooperation. The Settling Respondent shall ensure that a copy of this Agreement is provided to any current lessee or sublessee on the Property as of the effective date of this Agreement and shall ensure that any subsequent leases, subleases, assignments or transfers of the Property or an interest in the Property are consistent with this Section, and Section XI (Parties Bound/Transfer of Covenant), of the Agreement [and where appropriate, Section ..... (Work to Be Performed)].

VI. Due Care/Cooperation

15. The Settling Respondent shall exercise due care at the Site with respect to the Existing Contamination and shall comply with all applicable local, state, and federal laws and regulations. The Settling Respondent recognizes that the implementation of response actions at the Site may interfere with the Settling Respondent's use of the Property, and may require closure of its operations or a part thereof. The Settling Respondent agrees to cooperate fully with EPA in the implementation of response actions at the Site and further agrees not to interfere with such response actions. EPA agrees, consistent with its responsibilities under applicable law, to use reasonable efforts to minimize any interference with the Settling Respondent's operations by such entry and response. In the event the Settling Respondent becomes aware of any action or occurrence which causes or threatens a release of hazardous substances, pollutants or

contaminants at or from the Site that constitutes an emergency situation or may present an immediate threat to public health or welfare or the environment, Settling Respondent shall immediately take all appropriate action to prevent, abate, or minimize such release or threat of release, and shall, in addition to complying with any applicable notification requirements under Section 103 of CERCLA, 42 U.S.C. 9603, or any other law, immediately notify EPA of such release or threatened release.

VII. Certification

16. By entering into this agreement, the Settling Respondent certifies that to the best of its knowledge and belief it has fully and accurately disclosed to EPA [and the state] all information known to Settling Respondent and all information in the possession or control of its officers, directors, employees, [*34797] contractors and agents which relates in any way to any Existing Contamination or any past or potential future release of hazardous substances, pollutants or contaminants at or from the Site and to its qualification for this Agreement. The Settling Respondent also certifies that to the best of its knowledge and belief it has not caused or contributed to a release or threat of release of hazardous substances or pollutants or contaminants at the Site. If the United States [and the state] determines that information provided by Settling Respondent is not materially accurate and complete, the Agreement, within the sole discretion of the United States, shall be null and void and the United States [and the state] reserves all rights it [they] may have.

VIII. United States' Covenant Not to Sue

17. Subject to the Reservation of Rights in Section IX of this Agreement, upon payment of the amount specified in Section IV (Payment) of this Agreement [if consideration for Agreement is work to be performed, insert, as appropriate, "and upon completion of the work specified in Section ..... (Work to Be Performed) to the satisfaction of EPA"], the United States [and the state] covenants not to sue or take any other civil or administrative action against Settling Respondent for any and all civil liability for injunctive relief or reimbursement of response costs pursuant to Sections 106 or 107(a) of CERCLA, 42 U.S.C. 9606 or 9607(a) [and state law cite] with respect to the Existing Contamination.

Since the Covenant Not to Sue is from the United States, Regions negotiating these Agreements should advise the Department of Justice of any other federal agency involved with the Site, or which may have a claim under CERCLA with respect to the Site and use best efforts to advise such federal agency of the proposed settlement.

IX. Reservation of Rights

18. The covenant not to sue set forth in Section VIII above does not pertain to any matters other than those expressly specified in Section VIII (United States' Covenant Not to Sue). The United States [and the state] reserves and the Agreement is without prejudice to all rights against Settling Respondent with respect to all other matters, including but not limited to, the following:

(a) claims based on a failure by Settling Respondent to meet a requirement of this Agreement, including but not limited to Section IV (Payment), Section V (Access/Notice to Successors in Interest), Section VI (Due Care/Cooperation), Section XIV (Payment of Costs), [and, if appropriate, Section ..... (Work to Be Performed)];

(b) any liability resulting from past or future releases of hazardous substances, pollutants or contaminants, at or from the Site caused or contributed to by Settling Respondent, its successors, assignees, lessees or sublessees;

(c) any liability resulting from exacerbation by Settling Respondent, its successors, assignees, lessees or sublessees, of Existing Contamination;

(d) any liability resulting from the release or threat of release of hazardous substances, pollutants or contaminants, at the Site after the effective date of this Agreement, not within the definition of Existing Contamination;

(e) criminal liability;

(f) liability for damages for injury to, destruction of, or loss of natural resources, and for the costs of any natural resource damage assessment incurred by federal agencies other than EPA; and

(g) liability for violations of local, state or federal law or regulations.

19. With respect to any claim or cause of action asserted by the United States [or the state], the Settling Respondent shall bear the burden of proving that the claim or cause of action, or any part thereof, is attributable solely to Existing Contamination.

20. Nothing in this Agreement is intended as a release or Covenant Not to Sue for any claim or cause of action, administrative or judicial, civil or criminal, past or future, in law or in equity, which the United States [or the state] may have against any person, firm, corporation or other entity not a party to this Agreement.

21. Nothing in this Agreement is intended to limit the right of EPA [or the state] to undertake future response actions at the Site or to seek to compel parties other than the Settling Respondent to perform or pay for response actions at the Site. Nothing in this Agreement shall in any way restrict or limit the nature or scope of response actions which may be taken or be required by EPA [or the state] in exercising its authority under federal [or state] law. Settling Respondent acknowledges that it is purchasing property where response actions may be required.

X. Settling Respondent's Covenant Not to Sue

22. In consideration of the United States' Covenant Not to Sue in Section VIII of this Agreement, the Settling Respondent hereby covenants not to sue and not to assert any claims or causes of action against the United States [or the state], its authorized officers, employees, or representatives with respect to the Site or this Agreement, including but not limited to, any direct or indirect claims for reimbursement from the Hazardous Substance Superfund established pursuant to the Internal Revenue Code, 26 U.S.C. § 9507, through CERCLA Sections 106(b)(2), 111, 112, 113, or any other provision of law, any claim against the United States, including, any department, agency or instrumentality of the United States under CERCLA

Sections 107 or 113 related to the Site, or any claims arising out of response activities at the Site, including claims based on EPA's oversight of such activities or approval of plans for such activities.

23. The Settling Respondent reserves, and this Agreement is without prejudice to, actions against the United States based on negligent actions taken directly by the United States, not including oversight or approval of the Settling Respondent's plans or activities, that are brought pursuant to any statute other than CERCLA or RCRA and for which the waiver of sovereign immunity is found in a statute other than CERCLA or RCRA. Nothing herein shall be deemed to constitute preauthorization of a claim within the meaning of Section 111 of CERCLA, 42 U.S.C. 9611, or 40 CFR 300.700(d).

XI. Parties Bound/Transfer of Covenant

24. This Agreement shall apply to and be binding upon the United States, [and the state], and shall apply to and be binding on the Settling Respondent, its officers, directors, employees, and agents. Each signatory of a Party to this Agreement represents that he or she is fully authorized to enter into the terms and conditions of this Agreement and to legally bind such Party.

25. Notwithstanding any other provisions of this Agreement, all of the rights, benefits and obligations conferred upon Settling Respondent under this Agreement may be assigned or transferred to any person with the prior written consent of EPA [and the state] in its sole discretion.

26. The Settling Respondent agrees to pay the reasonable costs incurred by EPA [and the state] to review any subsequent requests for consent to assign or transfer the Property.

27. In the event of an assignment or transfer of the Property or an assignment or transfer of an interest in the Property, the assignor or transferor shall continue to be bound by all the terms and conditions, and subject to all the benefits, of this Agreement except as EPA [the state] and the assignor or transferor agree otherwise and modify [*34798] this Agreement, in writing, accordingly. Moreover, prior to or simultaneous

with any assignment or transfer of the Property, the assignee or transferee must consent in writing to be bound by the terms of this Agreement including but not limited to the certification requirement in Section VII of this Agreement in order for the Covenant Not to Sue in Section VIII to be available to that party. The Covenant Not to Sue in Section VIII shall not be effective with respect to any assignees or transferees who fail to provide such written consent to EPA [and the state].

XII. Disclaimer

28. This Agreement in no way constitutes a finding by EPA [or the state] as to the risks to human health and the environment which may be posed by contamination at the Property or the Site nor constitutes any representation by EPA [or the state] that the Property or the Site is fit for any particular purpose.

XIII. Document Retention

29. The Settling Respondent agrees to retain and make available to EPA [and the state] all business and operating records, contracts, site studies and investigations, and documents relating to operations at the Property, for at least ten years, following the effective date of this Agreement unless otherwise agreed to in writing by the Parties. At the end of ten years, the Settling Respondent shall notify EPA [and the state] of the location of such documents and shall provide EPA [and the state] with an opportunity to copy any documents at the expense of EPA [or the state]. [Where work is to be performed, consider providing for document retention for ten years or until completion of work to the satisfaction of EPA, whichever is longer.]

XIV. Payment of Costs

30. If the Settling Respondent fails to comply with the terms of this Agreement, including, but not limited to, the provisions of Section IV (Payment), [or Section ..... (Work to Be Performed)] of this Agreement, it shall be liable for all litigation and other enforcement costs incurred by the United States [and the state] to enforce this Agreement or otherwise obtain compliance.

XV. Notices and Submissions

31. [Insert names, titles, and addresses of those to whom notices and submissions are due, specifying which submissions are required.]

XVI. Effective Date

32. The effective date of this Agreement shall be the date upon which EPA issues written notice to the Settling Respondent that EPA [and the state] has fully executed the Agreement after review of and response to any public comments received.

XVII. Attorney General Approval

33. The Attorney General of the United States or her designee has issued prior written approval of the settlement embodied in this Agreement.

XVIII. Termination

34. If any Party believes that any or all of the obligations under Section V (Access/Notice to Successors in Interest) are no longer necessary to ensure compliance with the requirements of the Agreement, that Party may request in writing that the other Party agree to terminate the provision(s) establishing such obligations; provided, however, that the provision(s) in question shall continue in force unless and until the party requesting such termination receives written agreement from the other party to terminate such provision(s).

XIX. Contribution Protection

35. With regard to claims for contribution against Settling Respondent, the Parties hereto agree that the Settling Respondent is entitled to protection from contribution actions or claims as provided by CERCLA Section 113(f)(2), 42 U.S.C. 9613(f)(2) for matters addressed in this Agreement. The matters addressed in this Agreement are [all response actions taken or to be taken and response costs incurred or to be incurred by the United States or any other person for the Site with respect to the Existing Contamination].

36. The Settling Respondent agrees that with respect to any suit or claim for contribution brought by it for matters related to this Agreement it will notify the United States [and the state] in writing no later than 60 days prior to the initiation of such suit or claim.

37. The Settling Respondent also agrees that with respect to any suit or claim for contribution brought against it for matters related to this Agreement it will notify in writing the United States [and the state] within 10 days of service of the complaint on them.

XX. Exhibits

38. Exhibit 1 shall mean the description of the Property which is the subject of this Agreement.

39. Exhibit 2 shall mean the map depicting the Site.

[...... Exhibit 3 shall mean the Statement of Work.]

XXI. Removal of Lien

40. [Use this provision only when appropriate.] Subject to the Reservation of Rights in Section IX of this Agreement, upon payment of the amount specified in Section IV (Payment) [or upon satisfactory completion of work to be performed specified in Section ..... (Work to Be Performed)], EPA agrees to remove any lien it may have on the Property under Section 107(l) of CERCLA, 42 U.S.C. 9607(l), as a result of response action conducted by EPA at the Property.

XXII. Public Comment

41. This Agreement shall be subject to a thirty-day public comment period, after which EPA may modify or withdraw its consent to this Agreement if comments received disclose facts or considerations which indicate that this Agreement is inappropriate, improper or inadequate.

It Is So Agreed:

United States Environmental Protection Agency

By:

....

Regional Administrator, Region — Date

It Is So Agreed:

By:

....

Name Date

Form 8.2
Administrative Order on Consent

In the Matter of: [Insert Site Name and Location]

Proceeding under Section 122(g)(4) of the Comprehensive Environmental Response, Compensation, and Liability Act of 1980, as amended, 42 U.S.C. 9622(g)(4)

I. Jurisdiction

This Administrative Order on Consent ("Consent Order") is issued pursuant to the authority vested in the President of the United States by Section 122(g)(4) of the Comprehensive Environmental Response, Compensation, and Liability Act of 1980, as amended by the Superfund Amendments and Reauthorization Act of 1986 (CERCLA), Pub. L. No. 99-499, 42 U.S.C. 9622(g)(4), to reach settlements in actions under Section 106 or 107(a) of CERCLA, 42 U.S.C. 9606 or 9607(a). The authority vested in the President has been delegated to the Administrator of the United States Environmental Protection Agency (EPA) by Executive Order 12580, 52 FR 2923 (Jan. 29, 1987), and further delegated to the Regional Administrators of the EPA by EPA Delegation No. 14-14-E (Sept. 13, 1987).

This Administrative Order on Consent is issued to [insert name] ("Respondent"). Respondent agrees to undertake all actions required by the terms and conditions of this Consent Order. Respondent further consents to and will not contest EPA's jurisdiction to issue this Consent Order or to implement or enforce its terms.

II. Definitions

"Site" shall mean that parcel of property located at [insert address and general description], more particularly described as [insert legal description of the property owned by Respondent]. [Note: Additional definitions may be required.]

III. Statement of Facts

1. [In one or more paragraphs, describe the NPL status of the site and briefly describe the historical hazardous substance activity at the site, including the date on which the hazardous substance activities were terminated.]

2. Hazardous substances within the definition of Section 101(14) of CERCLA, 42 U.S.C 9601(14), have been or are threatened to be released into the environment at or from the Site. [Note: Additional information about specific hazardous substances present on- or off-site may be included.]

3. As a result of the release or threatened release of hazardous substances into the environment, EPA has undertaken response action at the Site under Section 104 of CERCLA, 42 U.S.C. 9604, and will undertake response action in the future. [Note: A brief recitation of the specific response action undertaken or planned for the site, *e.g.*, whether an RI/FS and ROD have been completed, should be included.]

4. In performing this response action, EPA has incurred and will continue to incur response costs at or in connection with the Site. [Note: The dollar amount and costs incurred as of a specific date should be included.]

5. [Identify the Respondent, the nature of his (her, its) ownership interest in the site, the manner in which he (she, it) acquired the site, *e.g.*, by purchase, bequest, eminent domain proceedings, etc., and the date of acquisition. Add any other facts relevant to the requirements of Section 122(g).]

6. Respondent represents, and for the purposes of this order EPA accepts, that respondent's involvement with the Site is limited to the following: [State each fact. Make sure to address the elements of Section 122(g)(1)(B0, and if no cash consideration is involved, Sections 107(b) and 101(35).]

7. Payments required to be made by Respondent pursuant to this Consent Order are a minor portion of the total response costs at the Site which EPA, based upon currently available information, estimates to be between $ .... and $ .... [Note: This statement need not be included if EPA is settling only for access and due care assurances. The dollar figure inserted should include the total response costs incurred to date as well as EPA's projection of the total response costs to be incurred during completion of the remedial action at the site.]

IV. Determinations

Based upon the Findings of Fact set forth above and on the administrative record for this Site, EPA has determined that:

1. The Site as described in Section II of this Consent Order is a "facility" as that term is defined in Section 101(9) of CERCLA, 42 U.S.C. 9601(9).

2. Respondent is a "person" as that term is defined in Section 101(21) of CERCLA, 42 U.S.C. 9601(21).

3. Respondent is an "owner" of a facility within the meaning of Section 107(a)(1) of CERCLA, 42 U.S.C. 9607(a)(1), and a "potentially responsible party" within the meaning of Section 122(g)(1) of CERCLA, 42 U.S.C. 9622(g)(1).

4. The past, present, or future migration of hazardous substances from the Site constitutes an actual or threatened "release" as that term is defined in Section 101(22) of CERCLA, 42 U.S.C. 9601(22).

5. Prompt settlement with the Respondent is practicable and in the public interest within the meaning of Section 122(g)(1) of CERCLA, 42 U.S.C. 9622(g)(1).

6. This Consent Order involves at most only a minor portion of the response costs at the Site pursuant to Section 122(g)(1) of CERCLA, 42 U.S.C. 9622(g)(1). [Note: This statement need not be included if the Agency is settling only for access and due care assurances.]

7. Respondent is eligible for a de minimis settlement pursuant to section 122(g)(1)(B) of CERCLA, 42 U.S.C. 9622(g)(1)(B).

V. Order

Based upon the administrative record for this Site and the Findings of Fact and Determinations set forth above, and in consideration of the promises and covenants set forth herein, it is hereby AGREED TO AND ORDERED:

VI. Access and Notice

1. Respondent hereby grants to EPA, its representatives, contractors, agents, and all other persons performing response actions under EPA's oversight, an irrevocable right of access to the Site for the purposes of monitoring the terms of this Consent Order and performing response actions at the Site. Respondent shall file in the land records of .... County a notice, approved by EPA, to subsequent purchasers of the land, that hazardous substances were disposed of on the Site and that EPA makes no representations as to the appropriate use of the property. Nothing herein shall limit EPA's right of access under applicable law.

2. Nothing in this Consent Order shall in any manner restrict or limit the nature or scope of response actions which may be taken by EPA in fulfilling its responsibilities under federal law. Respondent recognizes that the implementation of response actions at the Site may interfere with the use of his [her, its] property. Respondent agrees to cooperate with EPA in the implementation of response actions at the Site and further agrees not to interfere with such response actions.

VII. Due Care

3. Nothing in this Consent Order shall be construed to relieve Respondent of his [her, its] duty to exercise due care with respect to the hazardous substances at the Site or his duty to comply with all applicable laws and regulations.

VIII. Payment

4. Respondent shall pay the sum of $ .... to the Hazardous Substance Superfund within ..... days [insert short time period, *e.g.*, 10, 30 or 45 days] of the effective date of this Consent Order. [Note: If EPA is settling only for access, notice and due care assurances, then this section may be omitted. If EPA is settling for an agreement by the owner to perform response activities (removal—since a consent decree is required for remedial activities) rather than a cash payment, then the following section should be substituted: "WORK TO BE PERFORMED: Respondent agrees to perform (insert general description of activities to be performed), as more fully described in the Scope of Work and schedules attached hereto as Exhibit A and incorporated herein, and in accordance with the schedules and standards set forth therein. Based on information provided by Respondent, EPA estimates the present value of this work to be approximately $ ...."]

5. The payment specified in Paragraph 4 shall be made by certified or cashier's check payable to "EPA Hazardous Substance Superfund." Each check shall reference the site name, the name and address of the Respondent, and the EPA docket number for this action, and shall be sent to:

[Insert address for Regional lock box]

6. Respondent shall simultaneously send a copy of the check to:

[Insert name and address of Regional Attorney or Remedial Project Manager]

IX. Civil Penalties

7. In addition to any other remedies or sanctions available to EPA, the Respondent shall be subject to a civil penalty of up to $25,000 per day for each failure or refusal to comply with any term or condition of this Consent Order pursuant to section 122(1) of CERCLA, 42 U.S.C. 9622(1). [Note: If the Respondent is to perform the removal action under the Consent Order, stipulated penalties should be considered.]

X. Certification of Respondent

8. The Respondent certifies that to the best of his [her, its] knowledge and belief he [she, it] has fully and accurately disclosed to EPA and stated in Paragraph 6, section III, all information currently in his [her, its] possession and in the possession of his [her] agents, [or in the possession of its officers, directors, employees, contractors or agents] which relates in any way to his [her, its] qualifications for a de minimis settlement under section 122(g)(1)(B) of CERCLA. [Note: In very limited circumstances this language may be omitted if EPA determines that the risk of discovering information which would disqualify the Respondent from a de minimis settlement is negligible.]

XI. Covenant Not to Sue

9. Subject to the reservation of rights in Paragraphs 11 and 12, section XII, of this Consent Order, upon payment of the amounts specified in Paragraph 4, section VIII, of this Consent Order [Note: If work is to be performed instead of a cash payment, this sentence should read: "upon satisfactory completion of the work specified in the Scope of Work." If EPA is settling only for access and due care assurances, this sentence should read: "upon the effective date of this Consent Order."], EPA covenants not to sue or take any other civil or administrative action against the Respondent for any and all civil liability for injunctive relief or reimbursement of response costs pursuant to sections 106 or 107(a) of CERCLA, 42 U.S.C. 9606 or 9607(a), or section 7003 of the Resource Conservation and Recovery Act, as amended, 42 U.S.C. 6973, with regard to the Site.

10. In consideration of EPA's Covenant Not to Sue in Paragraph 9, section XI, of this Consent Order, the Respondent agrees not to assert any claims or causes of action against the United States or its contractors or its employees or the Hazardous Substance Superfund arising out of expenses incurred or payments made [or work performed] pursuant to this Consent Order, or to seek any other costs, damages, or attorney's fees from the United States or its contractors or employees arising out of response activities at the Site.

XII. Reservation of Rights

11. Nothing in this Consent Order is intended to be nor shall it be construed as a release or Covenant Not to Sue for any claim or cause of action, administrative or judicial, at law or in equity, which the United States, including EPA, may have against Respondent for:

(a) Any liability as a result of failure to provide access, notice, or otherwise comply with Paragraphs 1 and 2, section VI, of this Consent Order;

(b) Any liability as a result of failure to exercise due care with respect to hazardous substances at the Site;

(c) Any liability as a result of failure to make the payments [or perform the work] required by Paragraph 4, section VIII, of this Consent Order;

(d) Any liability resulting from exacerbation by Respondent of the release or threat of release of hazardous substances from the Site;

(e) Any and all criminal liability; or

(f) Any matters not expressly included in the Covenant Not to Sue set forth in Paragraph 9, section XI, of this Consent Order, including, without limitation, any liability for damages to natural resources. [Note: This natural resource damage reservation must be included unless the Federal natural resource trustee has agreed to a Covenant Not to Sue pursuant to section 122(j)(2) of CERCLA. In accordance with section 122(j)(1) of CERCLA, where the release or threatened release of any hazardous substances at the site may have resulted in damages to natural resources under the trusteeship of the United States, the Region should notify the Federal natural resource trustee of the negotiations and encourage the trustee to participate in the negotiations.]

12. Nothing in this Consent Order constitutes a Covenant Not to Sue or to take action or otherwise limits the ability of the United States, including EPA, to seek or obtain further relief from the Respondent, and the

Covenant Not to Sue in Paragraph 9, section XI, of this Consent Order is null and void, if information different from that specified in Paragraph 6, section III, is discovered which indicates that Respondent fails to meet any of the criteria specified in section 122(g)(1)(B) of CERCLA.

13. Nothing in this Consent Order is intended as a release or Covenant Not to Sue for any claim or cause of action, administrative or judicial, civil or criminal, past or future, in law or in equity, which the United States, including EPA, may have against any person, firm, corporation or other entity not a signatory to this Consent Order.

14. EPA and Respondent agree that the actions undertaken by the Respondent in accordance with this Consent Order do not constitute an admission of any liability by the Respondent. The Respondent does not admit and retains the right to controvert in any subsequent proceedings, other than proceedings to implement or enforce this Consent Order, the validity of the Findings of Fact or Determinations contained in this Consent Order.

XIII. Contribution Protection

15. Subject to the reservation of rights in Paragraphs 11 and 12, section XII, of this Consent Order, EPA agrees that by entering into and upon carrying out the terms of this Consent Order, Respondent will have resolved his [her, its] liability to the United States for those matters set forth in the Covenant Not to Sue, Paragraph 9, section XI, as provided by section 122(g)(5) of CERCLA, 42 U.S.C. 9622(g)(5), and shall have satisfied his liability for those matters within the meaning of section 107(a) of CERCLA, 42 U.S.C. 9607(a).

XIV. Parties Bound

16. This Consent Order shall apply to and be binding upon the Respondent and his [her] heirs, agents, and assigns [its officers, directors, employees, agents, successors and assigns]. The signatory represents that he [she] is fully authorized to enter into the terms and conditions of this Consent Order and to legally bind the Respondent. [Note: The preceding sentence should use the bracketed phrase in the first sentence if the

respondent is a corporation or entity other than a natural person.] In the event that the Respondent transfers title or possession of the Site, he shall notify the United States EPA (at the address included in Paragraph 6, Section VIII) prior to any such transfer and shall continue to be bound by all of the terms and conditions of this Consent Order unless EPA agrees otherwise and modifies this Consent Order accordingly.

XV. Public Comment

17. This Consent Order shall be subject to a thirty-day public comment period pursuant to Section 122(i) of CERCLA, 42 U.S.C. 9622(i). In accordance with Section 122(i)(3) of CERCLA, 42 U.S.C. 9622(i)(3), EPA may withdraw or modify consent to this Consent Order if comments received disclose facts or considerations which indicate that this Consent Order is inappropriate, improper, or inadequate.

XVI. Attorney General Approval

18. The Attorney General or his [her] designee has issued prior written approval of the settlement embodied in this Consent Order in accordance with Section 122(g)(4) of CERCLA. [Note: Attorney General approval usually will be required for de minimis consent orders because the total past and projected response costs at the site will exceed $500,000, excluding interest. In the event that Attorney General approval is not required, the order shall not include this Paragraph 18, but should include the following as a separate numbered paragraph in the Determinations section (Section IV) above: "The Regional Administrator of EPA, Region ..... , has determined that the total response costs incurred to date at or in connection with the Site do not exceed $500,000, excluding interest, and that, based upon information currently known to EPA, total response costs at or in connection with the Site are not anticipated to exceed $500,000, excluding interest, in the future." Use of this determination requires changes to the model Statement of Facts in Section III above; specifically, Paragraph 3 of the Facts should delete "and will undertake response actions in the future." Paragraph 4 of the Facts should delete "and will continue to incur response costs at or in connection with the site."]

XVII. Effective Date

19. The effective date of this Consent Order shall be the date upon which EPA issues written notice to the Respondent that the public comment period pursuant to Paragraph 17, Section XV, of this Consent Order has closed and that comments received, if any, do not require modification of or EPA withdrawal from this Consent Order.

IT IS SO AGREED AND ORDERED:

[Respondent(s)]

By: [Name] ....

[Date] ....

U.S. Environmental Protection Agency

By: [Name] ....

[Date] ....

Form 8.3
CERCLA Section 122(g)(4) Consent Decree for Settlements with Landowners under Section 122(g)(1)(B)

United States of America, Plaintiff v. [Insert Name(s) of Defendant(s)] Defendant(s)

Civil Action No. XXXX

Judge XXXXXXXXXX

Consent Decree

[Note: If the complaint concerns causes of action which are not resolved by this document or names defendants who are not signatories to this document, the title should be "Partial Consent Decree."]

WHEREAS, the United States of America, on behalf of the Administrator of the United States Environmental Protection Agency ("Plaintiff" or "United States") filed a complaint on [insert date] against [insert defendant's name] ("Defendant") pursuant to [insert causes of action and relief sought, *e.g.*, Sections 106 and 107(a) of the Comprehensive Environmental Response, Compensation, and Liability Act of 1980, as amended by the Superfund Amendments and Reauthorization Act of 1986 (CERCLA), Pub. L. No. 99-499, 42 U.S.C. 9606 and 9607(a), and Section 7003 of the Resource Conservation and Recovery Act, as amended (RCRA), 42 U.S.C. 6973, seeking injunctive relief regarding the cleanup of the [insert site name] ("Site") and recovery of costs incurred and to be incurred in responding to the release or threat of release of hazardous substances at or in connection with the Site];

WHEREAS, the United States has incurred and continues to incur response costs in responding to the release or threat of release of hazardous substances at or in connection with the Site;

WHEREAS, the Regional Administrator of the United States Environmental Protection Agency, Region ..... ("Regional Administrator"), has determined that prompt settlement of this case is practicable and in the public interest;

WHEREAS, this settlement does not involve the payment of response costs [delete this clause if cash consideration is included pursuant to Section V];

WHEREAS, based on information currently available to the Environmental Protection Agency (EPA), the Regional Administrator has determined that Defendant qualifies for a de minimis settlement pursuant to Section 122(g)(1)(B) of CERCLA;

WHEREAS, the United States and the Defendant agree that settlement of this case without further litigation and without the admission or adjudication of any issue of fact or law is the most appropriate means of resolving this action;

NOW, THEREFORE, it is ORDERED, ADJUDGED and DECREED as follows:

I. Jurisdiction

This Court has jurisdiction over the subject matter and the parties to this action. The parties agree to be bound by the terms of this Consent Decree and not to contest its validity in any subsequent proceeding to implement or enforce its terms.

II. Parties Bound

This Consent Decree shall apply to and be binding upon the United States and the Defendant, his [her] heirs, agents, and assigns [its officers, directors, employees, agents, successors and assigns]. The signatory represents that he [she] is fully authorized to enter into the terms and conditions of this Consent Decree and to legally bind the Defendant. [Note: The preceding bracketed language should be used if the Defendant is a corporation or entity other than a natural person.]

III. Definitions

"Site" shall mean that parcel of property located at [insert address and general description], more particularly described as [insert legal description of the property owned by Defendant]. [Note: It may be necessary to include additional definitions.]

IV. Access and Notice

1. Defendant hereby grants to EPA, its representatives, contractors, agents, and all other persons performing response actions under EPA's oversight, an irrevocable right of access to the Site for the purposes of monitoring the terms of this Consent Decree and performing or monitoring performance of response actions at the Site. Defendant shall file in the land records of .... County a notice, approved by EPA, to subsequent purchasers of the land that hazardous substances were disposed of on the site and that EPA makes no representation as to the appropriate use of the property. Nothing herein shall limit EPA's right of access under applicable law. In the event that Defendant transfers title or possession of the Site, he [she, it] shall continue to be bound by all of the terms and conditions of this Consent Decree and shall notify the United States EPA prior to any such transfer.

2. Nothing in this Consent Decree shall in any manner restrict or limit the nature or scope of response actions which may be taken by EPA in exercising its authority under federal law. Defendant recognizes that the implementation of response actions at the Site may interfere with the use of his property. Defendant agrees to cooperate with EPA in the implementation of response actions at the Site and further agrees not to interfere with such response actions.

V. Payment

1. Respondent shall pay the sum of $ .... to the Hazardous Substance Superfund within ..... days [insert short time period, *e.g.*, 10, 30 or 45 days] of the effective date of this Consent Order. [Note: If EPA is settling only for access, notice and due care assurances, then this section may be omitted. If EPA is settling for an agreement by the owner to perform

response activities, rather than a cash payment, then the following section should be substituted: "WORK TO BE PERFORMED: Respondent agrees to perform [insert general description of activities to be performed], as more fully described in the Scope of Work and schedules attached hereto as Exhibit A and incorporated herein, and in accordance with the schedules and standards set forth therein. Based on information provided by Respondent, EPA estimates the present value of this work to be approximately $ ....

2. The payment specified in Paragraph 1 of this Section, shall be made by certified or cashier's check payable to "EPA Hazardous Substance Superfund." Each check shall reference the site name, the name and address of the Respondent, and the EPA docket number for this action, and shall be sent to:

[Insert address for Regional lock box]

3. Defendant shall simultaneously send a copy of the check to:

[Insert name and address of Regional Attorney or Remedial Project Manager]

VI. Due Care

Nothing in this Consent Decree shall be construed to relieve Defendant of his [her, its] duty to exercise due care with respect to hazardous substances at the Site or his [her, its] duty to comply with all applicable laws and regulations.

VII. Civil Penalties

In addition to any other remedies or sanctions available to the United States, Defendant shall be subject to a civil penalty of up to $25,000 per day for each failure or refusal to comply with any term or condition of this Consent Decree pursuant to Section 122(1) of CERCLA, 42 U.S.C. 9622(1). [Note: If the defendant is to perform remedial action under the Consent Decree, stipulated penalties, pursuant to Section 121(e)(2), must be included.]

VIII. Certification of Dependant

The Defendant certifies that, to the best of his [her, its] knowledge and belief, he [she, it] has fully and accurately disclosed to EPA all information currently in his [her, its] possession and in the possession of his [her] agents [and in the possession of its officers, directors, employees, contractors or agents] which relates in any way to his [her, its] qualifications for a de minimis settlement under Section 122(g)(1)(B) of CERCLA. [Note: In very limited circumstances this language may be omitted if EPA determines that the risk of discovering information which would disqualify the Defendant from a de minimis settlement is negligible. The bracketed language in this paragraph should be used if the Defendant is a corporation or entity other than a natural person.]

IX. Covenant Not to Sue

1. Subject to the reservation of rights in Section X, Paragraphs 1 and 2, of this Consent Decree, upon entry of this Consent Decree, the United States covenants not to sue or take any other civil or administrative action against the Defendant for any and all civil liability for reimbursement of response costs or for injunctive relief pursuant to Sections 106 or 107(a) or CERCLA, 42 U.S.C. 9606 or 9607(a), or Section 7003 of RCRA, 42 U.S.C. 6973, arising from conditions existing at the Site as of the date of entry of this Consent Decree.

2. In consideration of the United States' Covenant Not to Sue in Paragraph 1 of this Section, the Defendant agrees not to assert any claims or causes of action against the United States or its contractors or its employees or the Hazardous Substance Superfund arising out of expenses incurred or Superfund payments arising out of expenses incurred or payments made [or work performed] pursuant to this Consent Decree, or to seek any other costs, damages, or attorney's fees from the United States arising out of response activities at the Site.

X. Reservation of Rights

1. Nothing in this Consent Decree is intended to be nor shall it be construed as a release or Covenant Not to Sue for any claim or cause of

action, administrative or judicial, at law or in equity, which the United States, including EPA, may have against Defendant for:

a) Failure to provide access, notice or otherwise comply with Section IV, Paragraphs 1 and 2, of this Consent Decree;

b) Failure to exercise due care with respect to hazardous substances at the Site;

c) Exacerbation of the release or threat of release of hazardous substances from the Site;

d) Any liability resulting from the introduction of any hazardous substance, pollutant, or contaminant by any person at the Site after the entry of this Consent Decree;

e) Any and all criminal liability; or

f) Any matters not expressly included in the Covenant Not to Sue set forth in section IX, paragraph 1, of this Consent Decree, including, without limitation, any liability for damages to natural resources. [Note: This natural resource damage reservation must be included unless the Federal natural resource trustee has agreed to a Covenant Not to Sue pursuant to section 122(j)(2) of CERCLA. In accordance with section 122(j)(1) of CERCLA, where the release or threatened release of any hazardous substances at the site may have resulted in damages to natural resources under the trusteeship of the United States, the Region should notify the Federal natural resource trustee of the negotiations and encourage the trustee to participate in the negotiations.]

2. In the event that the United States asserts any claim or cause of action against the Defendant pursuant to Section X, Paragraph 1, of this Consent Decree, the Defendant shall bear the burden of proving that any release or threat of release which is the subject of the claim or cause of action is attributable solely to conditions existing at the Site as of the date of entry of this Consent Decree.

3. Nothing in this Consent Decree constitutes a Covenant Not to Sue or to take action or otherwise limits the ability of the United States, including EPA, to seek or obtain further relief from the Defendant, and the Covenant Not to Sue in Section IX, Paragraph 1, of this Consent Decree is null and void, if information not currently known to the United States is discovered which indicates that Defendant fails to meet any of the criteria specified in Section 122(g)(1)(B) of CERCLA.

4. Nothing in this Consent Decree is intended as a release from or Covenant Not to Sue for any claim or cause of action, administrative or judicial, civil or criminal, past or future, in law or in equity, which the United States, including EPA, may have against any person, firm, corporation or other entity not a signatory to this Consent Decree.

5. United States and Defendant agree that the actions undertaken by the Defendant in accordance with this Consent Decree do not constitute an admission of any liability by Defendant.

XI. Contribution Protection and Liens

Subject to the reservation of rights in section X, Paragraphs 1 and 3, of this Consent Decree, the United States agrees that by entering into and carrying out the terms of this Consent Decree, Defendant will have resolved his [her, its] liability to the United States for those matters set forth in the Covenant Not to Sue, section IX, Paragraph 1, as provided in section 122(g)(5) of CERCLA, 42 U.S.C. 9622(g)(5), and shall have satisfied his [her, its] liability for those matters within the meaning of Section 107(a) of CERCLA, 42 U.S.C. 9607(a).

XII. Public Comment

This Consent Decree shall be subject to a thirty-day public comment period. The United States may withdraw consent to this Consent Decree if comments received disclose facts or considerations which indicate that this Consent Decree is inappropriate, improper, or inadequate.

XIII. Effective Date

The effective date of this Consent Decree shall be the date of entry by this Court, following public comment pursuant to Section XII of this Consent Decree.

The United States of America

[Defendant]

By: ....

By: ....

SO ORDERED this .... day of .... 19..... .

[Name]
[Date]